Intellectual Curiosity and the Scientific Revolution
A Global Perspective

Seventeenth-century Europe witnessed an extraordinary flowering of discoveries and innovations. This study, beginning with the Dutch-invented telescope of 1608, casts Galileo's discoveries into a global framework. Although the telescope was soon transmitted to China, Mughal India, and the Ottoman Empire, those civilizations did not respond as Europeans did to the new instrument. In Europe, there was an extraordinary burst of innovation in microscopy, human anatomy, optics, pneumatics, electrical studies, and the science of mechanics. Nearly all of those aided the emergence of Newton's revolutionary grand synthesis, which unified terrestrial and celestial physics under the law of universal gravitation. That achievement had immense implications for all aspects of modern science, technology, and economic development. The economic implications are discussed in the concluding epilogue. All these unique developments suggest why, for at least four centuries, the West experienced a singular scientific and economic ascendancy.

Toby E. Huff is a research associate at Harvard University in the Department of Astronomy and Chancellor Professor Emeritus in Policy Studies at the University of Massachusetts, Dartmouth. He has lectured in Europe, Asia, and the Middle East and has lived in Malaysia. He is the author of *The Rise of Early Modern Science: Islam, China, and the West*, second edition (Cambridge University Press, 2003); coeditor with Wolfgang Schluchter of *Max Weber and Islam* (1999), and author of *An Age of Science and Revolutions, 1600–1800* (2005).

Intellectual Curiosity and the Scientific Revolution

A Global Perspective

TOBY E. HUFF

University of Massachusetts, Dartmouth

CAMBRIDGE
UNIVERSITY PRESS

CAMBRIDGE UNIVERSITY PRESS
Cambridge, New York, Melbourne, Madrid, Cape Town, Singapore,
São Paulo, Delhi, Dubai, Tokyo, Mexico City

Cambridge University Press
32 Avenue of the Americas, New York, NY 10013-2473, USA

www.cambridge.org
Information on this title: www.cambridge.org/9780521170529

First published 2011

Printed in the United States of America

A catalog record for this publication is available from the British Library.

Library of Congress Cataloging in Publication data
Huff, Toby E.
Intellectual curiosity and the scientific revolution : a global perspective / Toby E. Huff.
p. cm.
Includes bibliographical references and index.
ISBN 978-1-107-00082-7 (hardback) – ISBN 978-0-521-17052-9 (pbk.)
1. Science – Europe – History. 2. Science – Experiments – History. 3. Discoveries in
science – Europe – History – 17th century. 4. Science – Europe – History – 17th century.
5. Science – History. I. Title.
Q127.E8H84 2010
509.4–dc22   2010021876

ISBN 978-1-107-00082-7 Hardback
ISBN 978-0-521-17052-9 Paperback

Contents

Illustrations

Preface

Those who think about the long cycles of science and civilizations and the question of why the Western world succeeded as it did may need to anchor their speculations in several mundane facts. When the scientific revolution occurred in the seventeenth century, the United States of America did not yet exist. In 1609, when Galileo made his revolutionary telescopic discoveries, a hardy band of English settlers attempted to establish the Popham Colony on the forbidding coast of Maine. Owing to the harsh winters of New England, the ill-fated colony was gone a year later.

In 1776, when the thirteen colonies banded together to form the United States, the inhabitants of those often wilderness regions numbered perhaps six million. China and India at the time counted more than 100 million subjects each, dwarfing the population of the struggling American colonies. No one would have predicted that the educational, political, and economic institutions being fashioned in those embryonic United States would propel it to become the dominant power in the twentieth century.

Similarly, a population comparison of Western Europe with China and India in the seventeenth century would find a huge excess of nearly 50 percent more people in the Asian regions. Some would say that India and China were then richer in material goods than Europe.

Third, as the present narrative will show, whatever glories ancient China, India, or the Islamic Middle East may have enjoyed in the past, their contributions to the making of modern science were minor. This conclusion will seem shocking to many readers, largely because of the romantic views of China that can be found in histories of it. Likewise, as I suggest in Chapter 10, the Arab-Islamic achievements in mathematics and astronomy have often been discussed, but their direct influence on Copernicus, Tycho Brahe, Galileo, Kepler, and Newton, among others, has yet to be shown.

Nevertheless, there is little doubt but that the seventeenth-century scientific revolution of Europe gave that part of the world a huge bundle of intellectual capital that was not to be found outside the West for more than 350 years. All the great revolutionary advances in science that occurred from that time to the present were largely, if not wholly, fashioned in the ambience of the West. Given the resistance to the efforts to disseminate the telescope and other scientific advances to other parts of the world in the seventeenth century, described in this study, more searching reviews of the cultural heritages of China, India, and the Islamic Middle East may be needed. At the same time, those who think that we have entered a "Pacific century," with Asian powers greatly outstripping the Western world, will want to ask themselves just how this might be accomplished. The question is how those Asian societies and civilizations can so rapidly remake themselves as leaders in science, education, and political development against a background of stagnation for centuries between the sixteenth and the present centuries.

Can a resurgent Confucianism now emerging in China give it the necessary twenty-first-century grounding essential for a modern, democratic, borderless economy? Can the growing Hindu nationalism and ultranationalism (Hindutva) of India give it the foundation for the same modern postindustrial, global economy now emerging? And can the new Islamist orientation that has swept the Muslim world in the twentieth century provide the transformative intellectual foundations required for full participation in the increasingly secular, high-tech, knowledge-based economy?

Anyone who ponders the existence of the World Wide Web and its origins in the United States and Europe will doubtless come to the conclusion that many aspects of the extraordinary economic and technological growth of the early twenty-first century were made possible by scientific and technological advances designed in the West. Their globalization has brought seemingly infinite possibilities to all parts of the world. Great economic powers have come and gone, which makes one think that there may be far more gold in properly designed educational institutions and deep commitments to scientific inquiry than there appears to be in the ubiquitous marketplace.

On the roads to modernity, we are accustomed to identifying the Industrial Revolution of the eighteenth century as a great landmark. The present inquiry will lead us to consider whether that great transformation could have taken place without the scientific revolution and, above all, Newton's *Principia Mathematica* and the related developments in astronomy and the science of mechanics that occurred uniquely in Western Europe. It may be more than coincidence that the absence of those developments in other regions of the world had something to do with the economic and political stagnation that persisted outside Europe (and Europe overseas) all the way to the mid-twentieth century. Such are some of the questions that need to be examined in an age of apparent instant thought and communication that has everyone wired.

Acknowledgments

While writing this book, I benefited from the knowledge of a very broad range of scholars around the world. They generously answered my questions and supplied me with new materials, invaluable insights, important contacts, and translations of obscure documents. For all their assistance, I want to thank S. M. Razaullah Ansari, David Arnold, Cemil Aydin, Zaheer Baber, Zhang Baichun, Marvin Bolt, Rainer Brömer, Christopher Cullen, Eric Dusteler, Anne Goldgar, S. Irfan Habib, Gottfried Hagen, Al van Helden, Alan Hirshfeld, Minghui Hu, Ekmeleddin Ihsanoglu, Colin Imber, Cemal Kafadar, Mustafa Kajar, S. R. Karma, Rajesh Kochhar, Chai Choon Lee, Henrique Leitao, Rudi Lindner, Peter Louwman, Debin Ma, John Moffett, Ali Paya, Kapil Raj, Eugene Rudd, Seth Schulman, and Henik Zoomer. Numerous unnamed others aided my enterprise, for which I express my gratitude. To Elfie Raymond, I express appreciation for keen editorial insights. Thanks also go to Jill Rubalcaba for her editorial encouragement during the early phases of writing this book.

I owe a very special debt of gratitude to Owen Gingerich, who read (and reread) all my chapters while supplying me with a wealth of insights, corrected translations of critical passages in Galileo and Copernicus, and an abundance of detail about the history of astronomy. Our almost monthly luncheons during the last four years while I was a research associate in the Department of Astronomy at Harvard were a seminarian's delight, sometimes taking the form of Introduction to Ptolemy 101. Any errant or wayward statements remaining in the book are entirely mine.

I should also express my deep appreciation of the extraordinary resources of Widener Library and its staff, who helped me procure any book or journal article that aided my cause.

PART I

SOMETHING NEW UNDER THE SUN

I

Introduction

The seventeenth century was one of the most dynamic and eventful centuries in the history of the modern world. It can be called the great divide that separated Western Europe developmentally from the rest of the world for the next three and a half centuries. During the 100 years of the seventeenth century, the scientific revolution in Europe produced an enormous flow of discoveries that transformed scientific thought. These discoveries occurred in astronomy, optics, the science of motion, mathematics, and the newly created field of physics. The Newtonian synthesis brought forth for the first time an integrated celestial and terrestrial physics within the framework of universal gravitation. Advances were also made in hydraulics and pneumatics, medicine, microscopy, and the study of human and animal anatomy. Not least of all, big steps were taken toward the discovery of electricity.

Given this extraordinary pattern of discovery, it is easy to ask why all this did not happen elsewhere. Simply put, why the West? Why did the Western world take off and become the dominant scientific, economic, and political power on this planet? Why did the great civilizations of China, India, and the Muslim Middle East, with their long records of growth and accomplishment, fall behind? Today, the prevailing view is that whatever happened culturally and developmentally in the West must have taken place elsewhere because people are basically the same in all places. The sociologist and medieval historian Benjamin Nelson called this idea *uniformitarianism*.[1]

---

[1] Benjamin Nelson coined this term back in the 1970s; see Nelson, *On the Roads to Modernity: Conscience, Science and Civilizations: Selected Writings by Benjamin Nelson*, ed. Toby E. Huff (Totowa, NJ: Roman and Littlefield, 1981), pp. 241–42.

The idea that people are everywhere and at all times the same over-looks the fact that human creatures are born into cultural settings or symbolic universes that often have contrasting worldviews, epistemological assumptions, and moral underpinnings. Not all those cultural universes are equally encouraging of scientific inquiry, neither are they equally supportive of original ideas.

In this study, I take the view that developmentally, Western Europe took off in the seventeenth century, charting new directions in many areas but especially in science and technology. These advances resulted in the accumulation of an enormous amount of intellectual capital absent outside Europe.

This development built on the reconstruction of European civilization that had taken place in the twelfth and thirteenth centuries in philosophy, law, institution building, and education. The reconstitution of Western civilization gave a new impetus to intellectual and scientific development that, a little more than three and a half centuries later, flowered in the scientific revolution and then in the Enlightenment of the eighteenth century.

In this book, I lay out the comparative tracks of scientific development and educational practice in Europe and in the three other great civilizations of the world: China, Mughal India, and the Ottoman Empire. To make the comparison as concrete as possible, I trace the events in Europe centered on Galileo's astronomical advances and then consider what happened in astronomy and the science of motion outside Europe during that same period. By focusing on the unique invention of the telescope in Europe in 1608, the narrative of Europe's scientific ascendancy becomes palpably visible.

Moreover, because the telescope was quickly transported around the world in the early seventeenth century by European traders, missionaries, and ambassadors, we get to see non-European reactions to this world-altering scientific instrument. This was the era when Europeans were making their early forays into China, India, and Southeast Asia. In coming into contact with the telescope, aspiring scientists in China, the Ottoman Empire, and Mughal India could have joined Europe in its ecumenical, global pursuit of modern science that culminated in the Newtonian synthesis. This path to modern science was indispensable: it laid the foundations of the modern world order – in mechanics, the science of motion, pneumatics, and ultimately electricity and the electronic society.

But the intercivilizational encounters of the seventeenth century did not result in such a new world science. The three civilizational encounters, between Europe and China, Europe and the Mughals, and Europe and the Ottomans, did not bear much fruit. The discovery machine – that is, the telescope that set Europeans on fire with enthusiasm and curiosity – failed to ignite the same spark elsewhere. That led to a great divergence that was to last all the way to the end of the twentieth century. But it was not just the telescope's promise that was passed by: the same thing occurred with the microscope and the study of human and animal microscopy as well as electrical energy and pneumatics.

We should notice that when these encounters took place in the early seventeenth century, it was a precolonial world, a world in which Europeans were perceived and treated as supplicants. The rulers of China, Mughal India, and the Ottoman Empire all had the means of holding Europeans at bay and did so. There is little meaning to the term *European imperialism* applied in this era. The age of imperialism had yet to find a toehold outside North America.

Put differently, the worldview that Europeans brought with them (even before the completion of the scientific revolution) stood at odds with the metaphysical foundations of the other civilizations. By the end of the seventeenth century, the cultural and civilizational gap between Europe and the others was large, while the scientific and technological gap was greater than ever before. This also meant that Europeans would be on a very different intellectual track, for they no longer lived in a pre-Newtonian and prebacterial world.

At the same time, it is important to recognize that European societies were held together by a very different conception of *law and legal structure*, one that sometimes clashed with the perceptions of the other civilizations, for the idea of positive law enacted by an elected body of citizens had not emerged outside Europe. This was true even though the European legal system was itself in a constant state of reform and renewal. That idea of deliberately planned legal reform, with due consideration of the *rights* of many participants, citizens, professionals, and nobles, was quite different from the legal views prevailing in the Ottoman Empire, China, and Mughal India, as we shall see in Chapter 6.

While carrying out the analysis of this study, I do not intend to underestimate the accomplishments of Mughal India, Ming China, or the Ottomans. The Mughals did amass great wealth: the Taj Mahal is an extraordinary monument, and Mughal miniatures are exquisite objects of beauty and precision. The poetic tradition of Persia that migrated with the Mughals to India is also the product of finely tuned sensibilities that engaged probing intellects for centuries.

Likewise, Ming pottery and ceramics are things of remarkable quality and beauty. One may also acknowledge that China's hydraulic networks over the centuries were epic accomplishments. The great Confucian classics are extraordinary achievements of the intellect, and the Confucian scholars who certifiably mastered them in the Examinations were surely the best and brightest intellects of China.

In the greater Middle East, the Ottoman Turks' long reign as a self-sufficient empire (nearly 600 years), with its remarkably resilient governing structure, deserves notice. Its many architectural monuments, especially those designed by Sinan, the architect who died in 1588, stand among the most beautiful of such human constructions. Even Sinan's own story – that of a young boy "collected" from a Christian home by the Ottoman slave system, converted to Islam, and educated in the palace school – reflects an extraordinary system of education and recruitment, although it is foreign to modern sensibilities.

The issue for this study is not whether those great civilizations once achieved remarkable things; it is the question of whether their intellectual and scientific development opened the way for future progress, for themselves and for others. Whether we wish so, humankind tends to press forward, undertaking new projects, building on what is sustainable, and discarding what is not in the light of new discoveries and new ambitions. Economic systems, along with their ups and downs, evolve over time, leaving behind what were once viable enterprises. The seventeenth and eighteenth centuries were a time when the new capitalistic system was coming into being, creating a multitude of new opportunities but also making traditional patterns obsolete. In the famous words of the economist Joseph Schumpeter, capitalism is a system that thrives on creative destruction: technological innovations and new kinds of enterprises create unique business opportunities, but those novel ways of doing business close off the viability of the ones they replace.

For present purposes, and with this background in mind, focusing on the scientific revolution of the seventeenth century in close comparison with developments in all four civilizations provides a unique vantage point for understanding comparative civilizational trajectories. At the same time, it gives us a panoramic perspective that may help us come closer to understanding the contrasting developmental patterns, the attempts at alternative modernities, of the civilizations considered. Given the results presented in this study, a reader might conclude that a more probing reassessment of Middle Eastern and Asian intellectual traditions of the past is in order.

The question for this study, however, is just how Western civilization developed its inner dynamic, especially in the realms of science, technology, and education that gave it a surplus of intellectual capital, especially in the seventeenth century, before Western hegemony set in, and which surpassed that of other civilizations. From the standpoint of this study, that surplus of *human capital* was singular and did set the West on a unique developmental trajectory. As we shall see in the epilogue, the unique Western system of education and the abundant fruit spawned by the scientific revolution created a level of human capital unmatched anywhere in the world until the end of the twentieth century. Consequently, that advantage in the realm of human capacity and scientific insight enabled Europeans to achieve a level of economic prosperity from the seventeenth century onward that was to elude the rest of the world until the end of the twentieth century. The rise in European literacy in the sixteenth and seventeenth centuries was a major part of that success, as also is explored later.

There are, of course, legal and institutional dimensions to this unfolding new dynamic, and when they are considered along with the literacy revolution and the scientific revolution, we gather still further insight into the unique emergence of a public sphere that conferred still other advantages on the Western world. This is seen in the newspaper revolution of the 1640s in England, paralleled in most respects in other countries across Europe but absent in China and the Muslim world until the nineteenth century. Even then, those derivative vehicles remained pale reflections of the freedom of expression seen in the Western world.

Viewed from the angle of all these extraordinary new developments, it is apparent that Western Europe from the seventeenth century onward accrued multiple advantages on the path to intellectual, economic, and political modernization. Grasping those advantages helps us understand why the West succeeded in this singular manner.

Outline of a New Perspective

The scientific revolution of the seventeenth century stands at the center of the great transformation that we now recognize as the modern world order. For without it, the Industrial Revolution of the eighteenth century would have been impossible. As noted, during those 100 years of the seventeenth century, a large number of revolutionary *scientific* discoveries flowed out of Europe, transforming our understanding of the natural world. Those discoveries occurred in astronomy, optics, the new science of motion, and microscopy. The outstanding advances made in hydraulics and pneumatics, medicine, and electrical studies pushed Europe far beyond the intellectual frontiers known elsewhere in the world.

It would be pleasant to think that all the peoples of the world shared equally in the extraordinary advance of thought signified by the scientific revolution, yet the European contribution far exceeded that of all the other peoples and civilizations of the globe. In the context of today's multiculturalism, this statement will sound like a Eurocentric sentiment. But as this study shows, in vast areas of scientific inquiry, such as optics, the science of motion, human anatomy, microscopy, pneumatics, and electrical studies, there were no parallels to Europe's discoveries outside the West in the seventeenth century.

Of course, there were earlier scientific developments, especially in the Middle East, that built on the Greek legacy of a still earlier age. But when we come to the advances of the seventeenth century, there is little evidence that scientific developments in China, Mughal India, or the Ottoman Empire gave any impetus to the European scientific revolution. Moreover, we must remember that science is, as the philosopher Karl Popper put it, the "unended quest." In athletic terms, it is a race of unending hurdles. In that context, it is not particularly useful to say that in 1000 B.C., China was the first to cross one or two of the early hurdles. Neither is it insightful to say that China has the longest written history of any peoples. Equally, it is not germane to point out that there was something very close to a revolution in optics among opticians in the eleventh century in the Middle East. For in science, it is always, "What have you done for me lately?"

Nevertheless, if we want to understand why Europe took the ascendancy leading to the modern world condition, then we have to explore crucial details of a number of episodes in the history of science that set Europe off from other parts of the world, for those advances laid down some of the technical foundations of the modern economic order. Neither the Industrial Revolution of the eighteenth century nor the Internet age would have been possible without the scientific advances of the seventeenth century.

Beyond Rhetoric

To grasp the all-encompassing nature of scientific curiosity that then gripped Europe, we must focus carefully on what scholars like Galileo, Kepler, and Gilbert, among others, were doing while at the same time reviewing the activities of scholars in the three other civilizations. Only in this way can we overcome the great propensity to cast the activities of Europeans, or others, into caricatures that cloud understanding and greatly misrepresent what was actually taking place. Even the most dispassionate students of the histories of India, China, or elsewhere can hardly refrain from saying things like, when Europeans were busy burning Giordano Bruno "for heresy at the stake in the Campo dei Fiori" in 1600, the Mughals or Ottomans, or whoever, were showing acts of tolerance, amassing great wealth, and living productive lives.[2] No mention would be made of the fact that William Gilbert published a landmark study in that same year that launched the whole field of electrical studies, that the telescope would be invented eight years later, and that two years after that, Galileo would discover the craters on the moon, the satellites of Jupiter, and the phases of Venus. Meanwhile, no parallel discoveries or inventions occurred in those other places that would catapult them into the center of the scientific revolution.

To get beyond such caricatures and limited perspectives, I begin with the simple story of the invention of the telescope in Holland in 1608 and Galileo's quick use of it to make revolutionary discoveries. Then I focus on its initial transmission around the world in the following decades. For the telescope as a discovery machine is a powerful symbol of the scientific revolution. Reactions to it can serve as a sort of acid test of the levels of scientific curiosity in other parts of the world.

Eventually, we shall have to consider the nature of the educational experiences that characterized Europe and the non-European world in the century or so leading up to the time of Galileo. Only in that way can we fathom the quite different educational practices and goals that prevailed in the major civilizational areas of the world in the sixteenth and seventeenth centuries.

The New Geography

It is useful to begin our inquiry by remembering the changed global circumstances that brought Europeans and others into a new kind of proximity. The sixteenth century was a great age of discovery. It followed the voyages of Columbus, Vasco de Gama, and Amerigo Vespucci, and finally Magellan's circumnavigation of the globe. The world was fresh and new, many parts unexplored and unknown to Europeans, who were

[2] William Dalrymple, *The Last Mughal: The Fall of Dynasty: Delhi, 1857* (New York: Vintage Books, 2006), p. 5.

experiencing an unparalleled awakening. In the process, Europeans encountered hundreds of unknown peoples, many living exotic lives scattered around the world. Some of the inhabitants of the New World, on the islands of Cuba and Santo Domingo, were said to live completely naked, enjoying an Eden-like existence of warm sunshine, abundant fish, fruit, wild game, and flowers.

Less reliable reports told of warlike Amazonian women on the coast of South America as well as oddly shaped people with their head where their stomach ought to be. For Europeans sailing along the coasts of South America, West Africa, around the Cape of Good Hope to India, or through the Straits of Magellan to the wide Pacific, there were unending marvels of geography and peoples, of flora and fauna never before seen. Naturalists of all persuasions faced an overwhelming abundance of specimens, while the range of human variation gave birth to the first European inklings of the new science of man: anthropology.

While the Spanish conquistadors wreaked havoc on the native populations of Mesoamerica, the populations of Africa, India, and China and other parts of Asia were too numerous, widely dispersed, or organized for Europeans to overwhelm or subdue. Throughout the sixteenth and seventeenth centuries, apart from the Spanish military campaigns, Europeans were petitioners seeking admittance into places like Mughal India, China, and the Ottoman Empire. The last of these stretched around the Mediterranean, from Morocco to Istanbul, and then north through the Balkans to Hungary and the gates of Vienna. Only occasionally did the Ottoman Turks lose sea battles to Europeans, such as their brief setback at the battle of Lepanto off the coast of Greece in 1571. It was an extraordinary encounter involving a combined force of nearly 100,000 seaborne sailors and soldiers. The Europeans had mustered a multinational fleet of Italians, Spaniards, Germans, and others, led by Venetians. Spanish nationals, among others, had equipped merchant ships with cannons that caught the Turks off guard, destroying their fleet. But the Turks' loss at this battle had little impact on the Ottoman Empire, which was to survive another 350 years. The age of European imperialism would not start until the nineteenth century.

Prelude to the Industrial Revolution

In the interim, Europeans in the seventeenth century witnessed the scientific revolution, a harbinger of the Industrial Revolution of the eighteenth century. Both these transformations – the scientific revolution and the Industrial Revolution – contributed significantly to the Western ascendancy that was to last into the twenty-first century. That coalescence of economic development, political power, and scientific creativity has long been seen as a puzzle. It seems inexplicable that so many human and cultural factors could come together in the domain of the Europeans but bypass the other, much larger populations of the world, especially in India and China.

How shall we think about the singularity of the rise and dominance of the Western world? Was there something more than just imperialist hubris that propelled the West into economic, scientific, and technological dominance in the eighteenth century? Did Westerners know something? Had they discovered or invented something that was not discovered in other parts of the world? Were there some intellectual breakthroughs that paved the way for the scientific revolution and the later economic salience of the European powers? One wonders if it was the mix of so many local groups and cultures that led Europeans to craft inclusive cultural and legal structures that set them apart from other parts of the world. Or were Europeans simply more inclined to think imaginatively about the political and legal structures needed to build a viable social and political world?

Max Weber's Legacy

A century ago, when it was not thought to be insensitive to ask big questions about how the world had gotten to be the way it is, the German sociologist Max Weber laid out his thoughts about these profound questions. He did so at the end of his short life of only fifty-six years, after he had searchingly probed the religious and intellectual history of China and India, Islam and ancient Judaism. He had also closely studied medieval and early modern European law and commerce. At the end of that long journey of discovery, he concluded that there were a number of striking intellectual features that arose only in the West and yet had a *universal* significance, a global impact as we would say today. As a cosmopolitan European of the early twentieth century, Weber felt compelled to ask "to what combination of circumstances the fact should be attributed that in Western civilization, and in Western civilization only, cultural phenomena have appeared which (as we like to think) lie in a line having *universal* significance and value" [emphasis in the original].[3] Weber believed that a number of intellectual developments, though they emerged uniquely in the West, had global implications for the whole world. He proceeded to enumerate such things as systematic theology, specialized conceptions of the logic of proof, and the full development of the experimental method in science. Likewise, he saw the uniqueness of canon law as conceived by the Christian fathers, but especially the whole corpus of systematic Western law, including the merchant law of international commerce, as exceedingly important for creating political stability. This, in turn, gave assurance to entrepreneurs that they would have a predictable future in which to make rational investments. Clearly, such developments had important implications for economic growth in Europe and elsewhere.

In this same sphere of historical economics, Weber is most famous for arguing that modern industrial capitalism was a unique product of the West and that in that development, the Protestant ethic of diligence and hard work, based on the idea of a religious calling, added a special dynamic to economic activity. In this way, Weber placed a strong but not exclusive emphasis on the products of the mind. He noted, for example, that even the peculiar idea of harmonic music structured around point and counterpoint of classical music, played by a symphony orchestra, was a unique cultural product of the West. In this way, one sees in Max Weber's thought a hint that products of the mind have had a huge impact on the shape of the Western world and, so he believed, had a large impact on the global world order that had been unfolding for centuries.

Max Weber's broader canvas gives us some idea of the many factors that need to be considered as we think about "Why the West?" Yet, Weber's thinking about the place of science in the modern world, or better, the role that science played in creating the modern world, needs finer articulation. In Weber's time, the history of science as a discipline had yet to be established. Indeed, the very conception of the scientific revolution appears to be a twentieth-century idea, first articulated by the French historian of science, Alexander Koyré, in about 1939.[4] That seasoned and disciplined inquiry into the history of science needs to be

[3] Max Weber, Introduction to *The Protestant Ethic and the Spirit of Capitalism*, trans. Talcott Parsons (New York: Scribners, 1958). This essay was written in 1904–5, but Weber's introduction to the volume, inserted by Parsons, comes from Weber's 1920 introduction to his *Collected Essays of the Sociology of Religion*.

[4] Steven Shapin, *The Scientific Revolution* (Chicago: University of Chicago Press, 1996), p. 2.

brought to bear on the developmental history of the West and other parts of the world.

At the same time, we need to separate the history of science from technology. Science as an intellectual activity is a search for the underlying principles and properties of nature. Scientists try to find the secrets of nature, the hidden processes of the natural world, not how to build a better mousetrap. Historically speaking, scientists (usually known as natural philosophers) wanted to understand such things as the nature of light, how it travels, the structure of the eye, and how it makes vision possible – the nature of the forces that hold the objects and materials of our world together. From before Aristotle's time in the fourth century B.C. to Galileo in the seventeenth century, natural philosophers have wanted to understand the nature of projectile motion, the reasons why and how objects fly through the air, the forces that impart or impede motion, and the exact mathematical description of falling and accelerating objects. Science in this sense is a search for fundamental explanations of how the world works. That search for fundamental explanations differs sharply from questions like the following: can we construct a building in which the wind blows through it, setting off a cascade of wind chimes, and these in turn release a valve that allows water to flow from a cistern and perhaps starts a water clock ticking? Such questions are engineering and technology questions, not scientific ones, or matters of scientific explanation.

But neither is science just doing mathematics, solving an equation, or developing new ways of calculating results from piles of data, as important as such activities are. From a scientist's point of view, mathematics is useful only if it can tell us more about nature than we already know, more about how and why natural forces operate as they do. For example, early Western astronomy was a set of geometrical operations with combinations of circles designed to make predictions of daily, monthly, and yearly patterns of the sun, moon, stars, and planets. What the real shape of the cosmos was remained in the hands of natural philosophers – that is, until Copernicus, Galileo, and Kepler broke the intellectual mold.

Copernicus believed that he had been able to use mathematics to find a new order in our universe, the order represented by a sun-centered universe. But to ward off the attacks of philosophers who might say that his new system was absurd, he declared that his book was for mathematicians. He was confident that if an astronomer understood his daunting array of geometrical models, the models would reveal the sun-centered nature of our universe. In this way, Copernicus made the extraordinary claim that his mathematical calculations were not just about mathematical symbols but about the real shape of the world in which we live. Galileo understood this claim perfectly: there is a shape, a structure to the universe – it is not arbitrary, "for such a constitution exists, it is unique, true, real, and could not possibly be otherwise,"[5] he wrote.

This realist philosophy of nature has always had its critics, but it has also been an indispensable handmaiden for natural philosophers who, throughout the centuries, were determined to make real claims about the natural world. Such claims of scientists transcend the beauty of aesthetically pleasing mathematical models.

Consequently, the rise of what I shall call modern science that makes claims about the way the world is constructed, and how it works, must be seen as a social phenomenon – that is, a worldview – that under certain circumstances rivals, and even displaces, religious and traditional authorities. In that sense, we need to think about the impact of science on society, the kinds of transformations that science and the scientific way of thinking have wrought on the human mind. As we now see more clearly, modern science did not develop in other parts of the world because of the absence of supportive social and cultural conditions. But once modern science emerged – let us call it the Copernican-cum-Newtonian worldview – that set of intellectual tools, that very powerful conception of a world system mathematically articulated, could have enormous implications for the way we think about the world and attempt to bring changes to it. Indeed, Newton's late-seventeenth-century worldview was breathtaking in the range of forces it encompassed. It had a surprising number of the elements, however inadequately understood, needed for what a twentieth-century Nobel Prize–winning physicist, Steven Weinberg, would call "Dreams of a Final Theory." As Weinberg came to see it, Newton's prescience fed into the still persistent hope that modern science can achieve a *final theory* of all the forces of nature, which will even provide us with the tools for understanding the very creation of the world. Whether or not such a dream is realistic, it is an expression of the modern scientific mentality that was not to be found before the scientific

[5] In Stillman Drake, *Discoveries and Opinions of Galileo* (New York: Doubleday, 1957), p. 97.

revolution of the seventeenth century and for centuries outside of Europe. Newton wrote:

Have not the small particles of Bodies certain Powers, Virtues, or Forces by which they act at a distance, not only upon the Rays of Light for reflecting, refracting, and inflecting them, but also upon one another for producing a great Part of the Phenomenon of Nature? For it's well known, that Bodies act one upon another by the Attractions of Gravity, Magnetism, and Electricity . . . and it is not improbable but that there may be more attractive Powers than these.[6]

These extraordinary ideas, flowing out of his *Mathematical Principles of Natural Philosophy* of 1687, were articulated in the Latin edition of his work on optics published in 1706. They give us further insight into the nature of the intellectual transformation of the modern world that had powerful implications for human cognition and for the possibilities of transforming the natural world for human habitation; that is, a deeper understanding of the fundamental laws of nature hold out the promise of making our social and economic systems more productive and efficient, far beyond the wildest dreams of scientists when presenting the first drafts of their new ideas. In this sense, commentators and intellectual historians of the last half century have greatly underestimated the impact of the scientific revolution – some even denying that it happened at all[7] – on all aspects of the human condition. As it now becomes increasingly clear, the Industrial Revolution of the eighteenth century would not have been possible without the scientific revolution, at the heart of which one finds the Newtonian conception of the natural world.

Fortunately, in the middle of the last century, when the history of science as a discipline was emerging, the British historian of science Herbert Butterfield argued that the scientific revolution of the sixteenth and seventeenth centuries "outshines everything since the rise of Christianity and reduces the Renaissance and Reformation to the rank of mere episodes, mere internal displacements, within the system of medieval Christendom." It changed "the character of men's habitual mental operations" at the same time that it transformed "the whole diagram of the physical universe and the very texture of human life." From this point of view,

---

[6] As cited in Richard Westfall, *The Construction of Modern Science: Mechanisms and Mechanics* (New York: Cambridge University Press, 1977), p. 141ff.
[7] Steven Shapin, *The Scientific Revolution* (Chicago: University of Chicago Press, 1996).

Butterfield declared, it is "the real origin both of the modern world and the modern mentality."[8]

This is an extraordinary statement, and yet it has even more powerful implications than Butterfield suggests. It is unfortunate, however, that Butterfield was inclined to devalue the Protestant Reformation that Weber and all scholars since his time have acknowledged as a major factor in the making of the modern world.[9] Indeed, the sociologist Robert Merton published a classic study showing that the Calvinist strands of the Reformation, what he called the "Puritan spur" to science, were a major factor in the spread of the scientific movement in the seventeenth century.[10]

Furthermore, it is has long been recognized that a central component of the Reformation was its authorization and support of universal literacy across Europe in the sixteenth and seventeenth centuries. That central fact (to be explored in the epilogue) provides another clue about the great disparities between East and West; that is, Europe experienced a major rise in literacy in the centuries following the Reformation with the result that non-European societies and civilizations lagged far behind the West in the achievement of literacy. Indeed, the Muslim world would not achieve the levels of literacy attained by England in the 1830s (and less so those of continental Europe) until the end of the twentieth century. Similar lags in literacy prevailed in India and China.

In short, the singular encouragement that the Protestant Reformation gave to Europe regarding literacy was an extraordinary advantage compared to other parts of the world community that would persist for hundreds of years.

At the same time, the Gutenberg press contributed an independent stimulus to literacy in the Western world, and we shall have to consider that influence also. For the printing press was banned in the Muslim world until the nineteenth century, except for a brief period among the Ottomans in the 1740s. And while China had long used woodblock printing, that technology spawned no literacy revolution there during its long existence (more on which is in the epilogue).

---

[8] Herbert Butterfield, *The Origins of Modern Science 1300–1800*, rev. ed. (New York: Free Press, 1957), pp. 7–8.

[9] Philip S. Gorski, *The Disciplinary Revolution: Calvinism and the Rise of the State in Early Modern Europe* (Chicago: University of Chicago Press, 2003). This is a good corrective for those who have lost sight of the social and organizational impact of the Reformation.

[10] Robert Merton, *Science, Technology, and Society in Seventeenth Century England* (New York: Harper, [1938] 1970).

If we restate Butterfield's indispensable insight so that both the Reformation and the scientific revolution worked together to create the modern world, we have a far more powerful assertion that dramatizes the radical transformations of the sixteenth and seventeenth centuries that swept Europe. At the same time, we shall find it necessary to bring in both the medieval legal revolution and the indispensable role played by universities from the twelfth century to the present.

By placing Butterfield's claim in the context of Max Weber's belief that certain intellectual advances – social, legal, and political – occurred in the West and had universal implications, we have compelling reasons for thinking that the modern world and the modern mentality were born in the Western world in the seventeenth century. That transformation grew out of long-evolving trends; once they came together in the scientific and Industrial revolutions, they established the demarcation between East and West, developed and developing, that was to persist for hundreds of years.

The Path Ahead

The task at hand is to unfold the historical deep structures of these developments and to do so in a comparative framework so that we can grasp the cultural and intellectual starting points that obtained in China, Mughal India, and the Ottoman Empire on the very eve of the scientific revolution of the seventeenth century. We must come to understand the institutional background and the attitudes toward science and scientific inquiry that prevailed in those parts of the world at the moment when the revolutionary architects of modern science in Europe – especially Galileo, Kepler, and Newton – were fashioning their extraordinary new scientific achievements. These innovations include Galileo's revolutionary telescopic discoveries; his work on the science of motion, suggesting the idea of inertia; Kepler's launching of a new science of celestial physics; and, finally, Newton's synthesis of all these ideas into the new science of mechanics.

The Telescope's Significance

There is a completely unexplored window through which this comparative perspective can be carried out. It is found in the invention of the telescope and its transmission around the world in the first two decades of the seventeenth century. This episode is as revealing of the contrasting levels of scientific curiosity around the world as it is neglected by historians of science. Let us consider first the nature and significance of the *occhiale*, the "spyglass," that emerged from the status of a military and diplomatic toy to a scientific instrument during the early seventeenth century.

Three contexts are needed to understand the significance of the telescope. First, it is the emblematic instrument of the modern scientific revolution. It is *the* instrument of empirical observation most associated with the scientific revolution of the seventeenth century, considering Galileo's use of it in astronomy.[11] Taking it around the world in the early 1600s was an invitation to others to participate in this ongoing empirical revolution focused on the heavens.

Second, and most practically, the telescope transformed the practice of astronomy in the seventeenth century: it transformed astronomy from a plodding science into an active, exploratory inquiry that constantly looks for new discoveries. In that sense, the telescope was a newly invented *discovery machine*. With it, an observer could see all kinds of new stars, satellites, and other celestial objects not seen before. At the same time, the telescope is a *portable laboratory* that could be taken anywhere in the world and used to explore the heavens.

Of course, it is true that the fundamental principles of optics were built into the telescope; that is, to build a telescope, a person needs to know that two lenses placed near each other will work together to magnify an object (so that it looks much closer than it is) only if the focal lengths of the two lenses are so aligned that their focal points converge at the point where the eye looks through the second lens (called the ocular). Many other principles need to be considered as well, such as whether one lens is convex and the other concave and that the magnifying power of any particular lens is a function of its curvature. An extreme example of this was Antoni Leeuwenhoek's discovery in the 1670s that a small, polished spherical glass bead, mounted for viewing, could magnify the object seen by several hundred times its actual size.

If two convex lenses are properly aligned – as Johannes Kepler understood – the field of vision will be larger, but the image will be inverted[12] (see the figures in Chapter 2). That's not good for terrestrial observation, but it was not a problem for astronomers. Much of this was unknown

---

[11] Albert van Helden, "Galileo and the Telescope," in *Novista Celesti e crisis del sapere* (Pisa, Italy: Giunti, 1983), pp. 149–58.
[12] Albert van Helden, "The 'Astronomical Telescope,' 1611–1650," *Annali dell'Istituto e Museo di Storia della Scienza* 1 (1976–77): 14–35.

outside Europe because eyeglasses were invented only in Europe in the thirteenth century.[13] Those who believe in the power of reverse engineering would imagine that the problem could be solved by local experimenters, if not by simple trial and error. The fact remains that though eyeglasses were taken around the world in the fifteenth and sixteenth centuries, no telescopes were invented there. Europeans continued to have an advantage in the science of optics in the seventeenth century, despite the significant optical advances in the Middle East of the eleventh century.

The third context concerns the telescope as a precision instrument. With improvements of the telescope, and a fixed mounting of Galileo's discovery machine, it could be used to make amazingly precise observations and many new celestial discoveries. The most startling of such new telescopically enabled discoveries in the mid-seventeenth century was made by the Danish astronomer Ole Rømer regarding the speed of light. Rømer had been observing the satellites of Jupiter, when he realized that the time elapsed for observing one of Jupiter's moons is longer when Jupiter is farther away from earth. From that, he inferred that it takes longer for the appearance of those moons to register here on earth when the planet is farther away than nearer. That means the speed of light is finite – not instantaneous. Rømer's observational data were too weak and limited to reliably establish a value for the speed of light, but his work and the method he used persuaded other leading astronomers of the time that the speed of light, though very rapid, is indeed finite.[14] Newton likewise recognized this and attempted to estimate that speed. It would be pleasant to think that Muslim as well as Chinese scholars were also studying and thinking about this problem, but evidence to that effect has not been produced.

It should also be noted that with the use of the telescope, or a quadrant mounted with a telescopic sight, the number of stars accurately catalogued by European astronomers more than tripled through the use of the

---

[13] Edward Rosen, "The Invention of Eyeglasses," *Journal of the History of Medicine and Allied Sciences* 11 (1956): 13–46, 183–218; Charles Singer, "Steps Leading to the Invention of the First Optical Apparatus," in *Studies in History and Method of Science* (Oxford: Oxford University Press, 1917), pp. 395–413; Vincent Ilardi, "Eyeglasses and Concave Lenses in Fifteenth-Century Florence and Milan: New Documents," *Renaissance Quarterly* 29 (1976): 341–60.

[14] James H. Shea, "Ole Rømer, the Speed of Light, the Apparent Period of Io, the Doppler Effect, and the Dynamics of Earth and Jupiter," *American Journal of Physics* 66, no. 7 (1998): 561–69.

telescope during the seventeenth century. Many double and triple stars were found, along with other substellar objects. The telescopic discovery machine vastly expanded human vision in the seventeenth century.

This context of new discoveries brings us to the penultimate issue, that of scientific curiosity. Clearly, the telescope as a discovery machine opened up an extraordinary domain of visual appearances in the universe: anyone with a trained eye who looked up at the skies in the early 1600s with a telescope would see all kinds of celestial objects, new stars, satellites of other planets, multiple stars, and so on. These had never before been seen by the human eye. These possibilities set Europeans on fire with excitement for new discoveries. To be sure, using a Galilean telescope in the early days of its invention was a difficult task because of the narrow field of vision and poor lens quality, but Galileo's extraordinary persistence was a pioneering model for others. Consequently, secular and religious scholars, ambassadors, missionaries, and merchants all sought to acquire telescopes and join this new inquiry. In this way, the arrival of the telescope serves as a sort of Rorschach test of scientific curiosity. How could one resist using this new device to look at the astonishing new sights revealed in the heavens?

In other words, if the telescope arrived in the hands of Chinese, Ottoman, or Mughal intellectuals and scholars, how could they not be curious about the heavens and use the telescope to explore it? Would they not see important scientific implications for astronomical theory? And if we probe the cultural backgrounds in these alternative civilizations, we would be compelled to ask, did madrasa education prepare its students for scientific inquiry? Did it make them curious about the natural world of the heavens? Likewise, the same questions need to be asked in the context of Chinese education. These explorations will give us the comparative context within which to evaluate the contrasting effects of the educational systems of East and West as well as Butterfield's claim that modern science in Europe transformed mentalities and paved the way for all sorts of practical changes in economy and society.

These revolutionary developments in astronomy, some associated with the telescope, were far from being the only landmark discoveries of the seventeenth century. For that reason, Chapters 7 through 11 explore other pioneering scientific discoveries in Europe. Like those in astronomy, they were the product of the high levels of scientific learning established in European universities since their inception in the twelfth and thirteenth centuries but absent elsewhere.

Finally, in the last chapter, we explore how all these scientific and intellectual changes converged to give Europe unparalleled advantages in social, political, and economic development. As I have suggested, the unprecedented changes of the seventeenth century constituted a package of advantages that transformed the world into its modern order.

2

Inventing the Discovery Machine

Oh telescope, instrument of much knowledge, more precious than any sceptre! Is not he who holds thee in his hand made king and lord of the works of God?[1]

– Johannes Kepler, 1611

Across the world in 1600, the night sky was a spectacular array of bright stars. Before the invention of electricity and other forms of lighting, to step out into the air on a clear night was to enter into a wonderland of starry objects filling the sky in all directions. This was as true in Europe or North America as it was in India, Africa, or China. The sky was filled with thousands of fixed stars that appeared to be attached to a blue background that rotated daily around the earth. Against that tapestry, the five planets – Mercury, Mars, Venus, Jupiter, and Saturn – followed their regular paths, tracked by their proximity to constellations among the fixed stars.

In the lucidity of this unpolluted sky, the nighttime observer was likely to see shooting stars that had their own mystical significance. Even today, if one goes outside the dense urban areas of our planet, where most people live, that dazzling vista can be seen. In the rural parts of our world, for example, in northern Maine or other parts of New England, or southern

---

[1] Johannes Kepler, *Dioptrice* (1611), as cited in *The Sidereal Messenger of Galileo Galilei, and Part of the Preface to Kepler's Dioptrics*, ed. and trans. Edward Stafford Carlos (London: Dawsons of Pall Mall, 1959), p. 86 ("Oh you much knowing tube, more precious that any scepter. He who holds you in his right hand, is he not appointed king or master over the works of God?" [also cited in Max Caspar, *Kepler*, translated and edited by C. Doris Hellman, New York: Dover, 1993, p. 201]).

France, in the mountains and villages north of Aix-en-Provence, or in rural Tunisia, among many other places, the vast array of stellar objects visible to the naked eye suddenly comes into view. For today's urban dwellers, this is a wondrous experience.

Since the time of Ptolemy in the second century A.D., professional astronomers and astrologers had carefully catalogued 1,028 stars, many of which were cast into anthropomorphic patterns called constellations. Ptolemy identified forty-eight constellations such as the Great Bear (Ursa Major), the Little Bear, the Hare, the Bull, and the Sea Monster. Not all stars fit into these patterns, and Ptolemy often pointed out other stars near or surrounding the constellations. The Big Dipper, for example, is a cluster of seven stars within the Great Bear. Other clusters of stars were given names, such as Orion's Belt and the Pleiades (Seven Sisters), but are called *asterisms*.

What early-seventeenth-century observers probably could not imagine was that those visible stars were but the minor part of the starry firmament that was to be tripled by the end of the century through the invention and use of the telescope. But the invention of the telescope itself was a prolonged affair based on the invention of glass, then lenses, concave and convex. By the Middle Ages, eyeglasses appeared in Europe to correct failing vision, and then after centuries of optical experimenting, the "spyglass" or the "long eyeglass" (*la longue lunette*), as the French called it, finally gave birth to what we call the telescope.

From Silica to Transparent Glass

It seems natural for contemporary readers to imagine that the invention of eyeglasses would quickly lead to the invention of the telescope. After all, early telescopes were just two lenses placed in a tube. However, lenses come in two basic forms, convex and concave. Convex lenses are known as magnifiers as they enlarge objects seen at a distance but also make them fuzzy. Concave lenses show things in sharp focus but smaller. The question for would-be telescope makers is: Which combination of the two types of lenses will give the telescopic effect that we want? Historically speaking, the two types of lenses did not appear on the scene at the same time. In both cases, the production of workable convex or concave lenses depended on the production of good-quality glass. So the first problem on the way to inventing the telescope is just getting lenses of sufficient clarity and uniformity that they produce a clear image. That is easy today: just go to your hobby shop, or better still, go online, and you can get your pick.

The art of making glass, however, has existed since ancient times, perhaps 4,000 years ago. Archeologists have found evidence of manufactured glass from about 2000 B.C. in the Mesopotamia valley outside the ancient city of Ur, located between the Tigris and Euphrates rivers. The process involves heating silica, a major ingredient of sand, along with soda (sodium carbonate) to a very high temperature. When it cools, it attains a crystalline look of greater or lesser clarity. Man-made glass and its techniques of manufacture spread to other peoples in the Middle East, including the Phoenicians, Egyptians, and eventually the Romans, long before the Christian era. By the thirteenth century, Italy, both Florence and Venice, had high-quality glass manufacturers.

That is when spectacles first appear among Europeans. Inventors in Pisa and Florence vied for the honor of inventing spectacles as well as the profits to be made from selling these indispensable aids to human vision (Figure 2.1). By the middle of the thirteenth century, the English savant Roger Bacon had carried out various experiments with glass globes. He observed that objects seen at a distance through a section of a globe that served as a convex lens would appear magnified: "glasses (perspicua) can be so constructed that objects at a very great distance appear to be quite close at hand, and conversely."[2] By the time of his death in about 1294, Italians had fashioned spectacles for correcting far-sightedness (*presbyopia*) using convex lenses. Indeed, the first pair of spectacles – "two glass disks enclosed by metal rims centrally connected so as to be held before the eyes" – had been invented in Pisa in 1286.[3]

Convex lenses are also called converging lenses because if you hold one up to the sun while focusing it on a piece of paper, the rays of light will converge on a very bright spot. As all curious young boys know from playing with magnifying glasses, the spot gathers so much light in one place that it generates heat, which will set the paper on fire. Since antiquity, this has been known as the *burning point*.

On the other hand, telescopes with upright images need both a concave and a convex lens. Concave lenses are called diverging because the rays of light passing through them diverge from the central axis of the lens.

---

[2] As cited in Charles S. Singer, "Steps Leading to the Invention of the First Optical Apparatus," in *Studies in the History and Method of Science* (Oxford: Oxford University Press, 1921), p. 398.
[3] Vincent Ilardi, "Renaissance Florence: The Optical Capital of the World," *Journal of European Economic History* 22 (1993): 508. A longer-term and more conceptual outline of the components of the telescope's invention is in Rolf Willach, *The Long Route to the Invention of the Telescope* (Philadelphia: American Philosophical Society, 2008).

FIGURE 2.1. A cleric wearing rivet spectacles in a painting by Konrad von Soest from the altar piece in the Stadtkirsche of Bad Wildungen, Germany, 1403.

The effect of this is to make the image of an object look smaller than it is. Consequently, concave lenses could be used to correct myopia, or near-sightedness. Just when concave lenses were invented is still debated but, in any case, they were not widely available until the end of the sixteenth century. That means that the invention of a working telescope probably could not have been achieved much earlier, although concave lenses were available after 1450.

But inventing such lenses (convex and concave) with suitable purity is only one part of the problem. Another factor that doubtless delayed the invention of the telescope concerns what is called the *focal length* of a lens. That is the distance from the lens to the spot where the image made by the light passing through the lens is most clearly seen. Every lens has it own focal length that depends on its curvature. The focal length of a convex lens is easily established, as noted earlier, by holding the lens up to the light while projecting the image onto a white surface and measuring the distance. Using the same procedure with a concave lens is more difficult because the light rays are diverging, and one has to adjust the distance between the lens and the projecting surface while judging the distance at which the image is most clear. This point is not so sharply delineated as it is with the convex lens. Indeed, it appears that in Galileo's time, at the beginning of the seventeenth century, it was doubted whether the concave lens had a focal point at all.[4] On the other hand, spherical concave mirrors were famous for their effects as burning mirrors, which, according to legend going back to ancient times, were used by Archimedes and assorted others to burn enemy ships when they approached a harbor.[5]

But concave lenses are different. The question for concave mirrors was where the focal point and the burning point were located, a short distance from the mirror's surface or at the center of the lens's curvature. In Galileo's time, some students of optics got the right answer and others did not. These historical facts help to explain why Galileo's explanation of how he got the telescopic effect he wanted with his two lenses is so vague, as we will see.

Looking at Figure 2.2, we can see how the rays of light are refracted in the "Dutch" or Galilean telescope. The rays of light are refracted by

---

[4] Sven Dupré, "Ausonio's Mirrors and Galileo's Lenses: The Telescope and Sixteenth Century Practical Optical Knowledge," *Galilaeana* 2 (2005): 145–80.
[5] This story of magical mirrors and their powers leading up to Galileo's telescope has been told by Eileen Reeves, *Galileo's Glassworks: The Telescope and the Mirror* (Cambridge, MA: Harvard University Press, 2008).

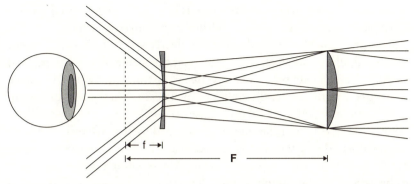

FIGURE 2.2. The Galilean or "Dutch" telescope using a convex objective and concave lens (ocular). The three rays at the center of the eye indicate the reduced field of vision seen with this lens system. Adapted from "Optics of the Galilean Telescope," http://www.pacifier.com/~tpope/Galilean_Optics_Page .htm#Hughes_1920, with permission of Jim Mosher.

the convex lens and then travel through the concave lens, where they are refracted again, yielding a smaller image. The green ray tracings (the three lines converging at the center of the eye) indicate the delimited bundle of rays striking the eye of the observer and also suggest the reduced image produced by this lens system. The most important feature of this arrangement, however, is the fact that the two lenses must be aligned so that the focal points of the two lenses converge where the eye looks into the instrument.

Figure 2.3 shows how light is refracted in the "astronomical" telescope comprising two convex lenses that was first proposed by Johannes Kepler

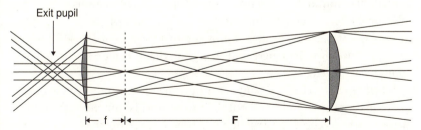

FIGURE 2.3. The Keplerian or "astronomical" lens system – using two convex lenses – developed by Johannes Kepler in 1611. The image seen is inverted, represented by the red and blue rays converging in the exit pupil, but is larger than in the Galilean system. Adapted from "Optics of the Galilean Telescope," http://www.pacifier.com/~tpope/Galilean_Optics_Page.htm#Hughes_1920, with permission of Jim Mosher.

in his *Dioptrics* of 1611. The ray tracings here in this modernized version of Kepler's diagram show that multiple rays converge at the exit pupil, where the image is magnified and most clearly seen. The convergence of rays – red, blue, and green – indicates that the image here is sharper and slightly larger than in the Dutch or Galilean instrument, though the image is inverted. Furthermore, the length of the telescope in the convex system is an additive function of the two lenses ($F + f$); that is, in the Galilean telescope, the length of the tube is determined by the difference between the focal lengths of the convex and concave lenses, as shown in Figure 2.2. But with the astronomical telescope, the mounting tube would have to be as long as the combined focal lengths of the two lenses, as shown in Figure 2.3.

When the Dutch invented their telescope and Galileo set out to build his own device based on rumors about the "Flemish" spyglass, very few of these optical details were clearly understood, and no manual existed for constructing telescopes. Without a clear understanding of the focal length of a concave lens and how two lenses worked in tandem, both spectacle makers and later telescope makers in the seventeenth century had to experiment to get the optical effects they sought. That is probably what Galileo did once he heard that a spyglass had been invented in Holland.

It has been estimated that in the first decade of the seventeenth century, convex lenses in the shops of European spectacle makers had focal lengths ranging between eighteen and twenty-one inches. For concave lenses, the focal lengths ranged between six and seven inches. Combinations of such lenses made it possible to create workable twelve- to sixteen-inch telescopes of low power. And prior to that, such lenses were even weaker in magnification so that experimenting with lens combinations to produce telescopic effects yielded unimpressive results, perhaps $1\times$ to $2\times$ powers of magnification. But, by the end of the first decade of the seventeenth century, lens power had increased enough so that greater telescopic effects could be achieved.

So we see that while contemporary optical experimenters can get fairly good quality lenses and carry out their experiments, not all of the optical elements of telescopic construction had been worked out in Galileo's time. Historically speaking, it took more than 300 years after the invention of eyeglasses for Europeans to solve these problems and invent a working "spyglass." Moreover, some students of the history of eyeglasses and the telescope think that the key to successful telescope making was stopping down the aperture of the ocular lens so that incoming rays enter only through the center of the lens with its better quality.[6] Galileo came to understand that; nevertheless, the story of the telescope and its transformation into a discovery machine has to begin in the Netherlands, where, by 1604, the art of lens and spectacle making had apparently been transmitted from Tuscany, the region of Italy that was home to Pisa and Florence.

Middelburg, Netherlands

In 1600, Middelburg was the capital of the province of Zeeland, on the south coast of the Netherlands. The province was a composite of peninsulas and islands jutting out like three fingers into the North Sea. In the 1580s, its population was about 10,000 but was still one of the twelve largest cities in the Netherlands. With the conquest of Antwerp by the Spanish in 1585, however, Middelburg's fortunes and its population began to rise. Merchants and artisans from Antwerp immigrated to Middelburg so that its inhabitants were three times as many in 1630.[7] In later centuries, a causeway would be built from the island to the mainland. Nevertheless, the center of the city was set back from the sea with a mile-long canal allowing ships to find shelter closer to the city, where dockworkers could unload cargoes.

Soon Middelburg would be the location of an important branch office of the Dutch East India Company, the Vereenigde Oostindische Compagnie (VOC). It was founded in 1602 to carry out the monopoly of Dutch trading activities in Asian colonies. The company was founded in no small part in reaction to the chartering of the British East India Company in 1600. Like its British counterpart, the VOC began to establish "factories" in India but also the famous Dutch colony of Batavia on the island of Java, then part of what were called the Spice Islands. In mid-sixteenth century, 50 percent of Middelburg's imports consisted of French wine as its merchants had won an imperial monopoly from the Hapsburg Netherlands.[8] Toward the end of the sixteenth century, Middelburg was also one of the centers at which tulips arrived from Constantinople and were

---

[6] Willach, *Long Route.*
[7] Anne Goldgar, *Tulipmania: Money, Honor, and Knowledge in the Dutch Golden Age* (Chicago: University of Chicago Press, 2007), p. 24; and Jonathan Israel, *The Dutch Republic, and Its Rise, Greatness, and Fall 1477–1806* (Oxford: Oxford University Press, 1995), p. 309.
[8] Goldgar, *Tulipmania*, p. 24; and Israel, *Dutch Republic*, p. 117, who estimate approximately 60%.

cultivated in backyard gardens before being sent elsewhere as the mania swept Europe.

But this small town of nearly 30,000 inhabitants also had a thriving glass-making industry that supplied lens blanks to lens grinders and spectacle makers. For decades at the end of the sixteenth century, various individuals claimed to be able to combine concave and convex lenses in an exact manner so that "you will see both distant and near objects larger than they would otherwise appear and very distinct."[9] It seems doubtful that the Italian Giovanni Della Porta, whose words claimed all this, actually made such a device in 1589, when a revised edition of his book called *Natural Magic* appeared. He was one of several optical experimenters who claimed to know what lenses so combined could do for vision. In the 1570s, the Englishmen Leonard Digges and John Dee, and later William Bourne in 1578, all claimed to have knowledge of "perspective glasses" or other optical devices that could enlarge objects seen at a distance.[10] Some of these devices involved concave mirrors and were claimed in legends to enable observers to see the details of landscapes many miles away. Sometimes the concave mirrors were combined with a convex lens that would serve to enhance vision even more.[11]

So it fell to a deeply religious, God-fearing man living in Middelburg to fashion the promise of this spectacle technology into a working device of far-seeing capacity. This was Hans Lipperhey, a resident spectacle maker, who presented to the Estates General in the Hague in October 1608 an actual device, "a certain instrument for seeing far," that was recognizably a "spyglass." If this device was truly what it claimed to be, then it had to be treated as a state secret, kept under wraps as a military enigma, and produced in no other country. That wish had slim chance of being realized.

Testing the device in the Hague gave the proof for which everyone hoped. This occurred in the presence of Count Maurice of Nassau, the chief executive and commander in chief of the armed forces of the United Provinces of the Netherlands. He was also the defender of the Protestant territories that were under siege by Catholic Spain. During the last

---

[9] Della Porta, in Henry King, *The History of the Telescope* (London: Charles Griffin, 1589/1955), p. 30; also in Albert van Helden, "The Invention of the Telescope," *Transactions of the American Philosophical Society* 67, part 4 (1977): 1–67.
[10] Van Helden, "Invention of the Telescope."
[11] Eileen Reeves, *Galileo's Glassworks: The Telescope and the Mirror* (Cambridge, MA: Harvard University Press, 2008). Reeves traces these legends from antiquity to the early seventeenth century.

few days of September of that year, Lipperhey demonstrated his device. According to a famous newsletter printed in the Hague in October:

From the tower of The Hague, one clearly sees, with the said glasses, the clock of Delft [about 8.6 kilometers away] and the windows of the church of Leiden [about 17.6 kilometers away], despite the fact that the cities are distant from The Hague one and a half and three and a half hours by road, respectively.[12]

This was impressive evidence that things far away could be seen distinctly with this new device. It was especially impressive news for the very slow-moving, marching troops constantly at war with each other in Europe at the time. Whether other officials from neighboring countries saw the demonstration, they learned of it almost instantly. Military jokes were reported, such as the count's remark that now "with these glasses" "the tricks of [the] enemy" could be seen in advance. The Spanish general Ambrosia Spinola is reported to have quipped, "From now on I [can] no longer be safe, for you will see me from afar," while the count replied, "We shall forbid our men to shoot at you."

With the military applications of this new device clearly evident, military and diplomatic representatives present at the time made strenuous efforts to procure these spyglasses for their countrymen. Pierre Jeanin, the official representative of the French king Henry IV, immediately approached Lipperhey, asking for one of these new instruments. When rebuffed, he appealed directly to the Estates General. He asked for and got two, one for his king and the other for the king's minister, the Duke of Sully.

Back in Brussels, others learned of the device and requested specimens. The Archduke of Austria, then in Brussels, obtained one of these new spyglasses, as did Guido Bentivoglio, who was the Pontifical representative from Rome. That spyglass would arrive in Rome by the end of April 1609,[13] perhaps before Galileo Galilei learned of the device's existence.

In the meantime, several other individuals, Jacob Metius and Sacharias Janssen, also spectacle makers living in Middelburg, put forth their claims to have invented some kind of far-seeing device. Consequently, though Lipperhey made a number of the devices for the Estates General, including sets of binoculars, he did not receive a patent for his device as he had requested. Yet, only Lipperhey's claim to have invented the device has been proven by scholars scouring the available documents.

[12] In van Helden, "Invention of the Telescope," p. 42.
[13] See the letter by Archbishop Guido Bentivoglio of April 2, 1609, in Engel Sluiter, "The Telescope before Galileo," *Journal for the History of Astronomy* 28 (1997): 226.

Still another Netherlander showed up at the book fair in Frankfurt that fall of 1608, offering a telescopic device for sale. By April 1609, at least one spectacle maker in Paris was offering spyglasses for sale. In just a short few months, telescopes were being made and circulated across Western Europe, from the Netherlands to France, Germany, and Italy. This suggests that European "contemporaries almost immediately recognized the importance of the new invention"[14] and set about exploiting its possibilities, commercial and applied.

Quite remarkably, the newsletter from the Hague indicates that the ambassador of Siam, now called Thailand, had been paying his respects to Count Maurice that fall in the Hague. When he returned to Thailand in Spring 1610, the worldwide dissemination of the telescope had begun. Meanwhile, the newsletter spread across Europe.

Padua and the University of the Late Renaissance

In 1609, the University of Padua, twenty-five miles from the provincial capital of Venice, was enjoying the last blush of its reputation as an outstanding Italian university. Since its founding in 1222, when renegade professors from the University of Bologna decamped to Padua, the university had risen to considerable fame, especially in medicine. In the middle of the sixteenth century, it had a broad-based faculty teaching the arts and medicine, surgery, anatomy, botany, and zoology. All students, those preparing for medicine especially, as well as theology, were taught Aristotle's physics, which included the science of mechanics and motion. Physics or natural philosophy was the ruling intellectual discipline in the northern Italian universities and had been since the Middle Ages, when Aristotle's so-called natural books became the mainstay of European university education. In the sixteenth century, students read Aristotle's major works, such as his *Physics*, but also books on plants and animals and his metaphysical tracts on the theory of knowledge. But Aristotle's great corpus included works on the causes of change and transformation, and these discussions implied that nature itself was a domain of unfolding causes independent of divine agency. Even more surprising to modern readers is the fact that the Aristotelian corpus included the study of meteorology, the ancient source of our contemporary science of weather analysis and prediction. In medieval times, this discipline concerned the "corruptible" sublunar domain just above the earth below the moon.

---

[14] Sluiter, "Telescope before Galileo," p. 227.

All this and many more Aristotelian subjects had been part of Galileo's university education in Pisa. But now he was a professor of mathematics in Padua, at the famous university that Copernicus attended, and before him the great pioneer of anatomy, Andreas Vesalius. In the sixteenth century, it was the medical faculty of Padua that became the most famous due to the presence of Vesalius. He published the illustrated and unsurpassed anatomical study, *The Fabric of the Human Body*, in 1543. In that extraordinary work, Vesalius revolutionized the teaching of anatomy and displaced after nearly a millennium and a half the teachings of the Roman physician Galen. The book was a product of the long use of dissection in Europe, extending back to the thirteenth century but which was forbidden in the Muslim world and Judaism.

Vesalius's pioneering anatomical illustrations laid the foundations for a number of subsequent medical discoveries, including the discovery that the heart serves as a pump pushing blood throughout the body. This was accomplished by the English physician William Harvey. He, too, had graduated with a bachelor's degree from Padua in 1597 and shortly thereafter made his discovery of the heart's function. Nevertheless, back in England, Harvey labored intensively for nearly three more decades, working out the invincible proofs of his theory before publishing it in 1628.

But 1543 was the very year that the equally revolutionary book, *The Revolutions of the Heavenly Spheres*, was published by Nicolaus Copernicus. Although Copernicus came from the German-speaking part of Poland, he had attended medical lectures at Padua in 1503 and probably witnessed dissections in the anatomical theater of the university. These two magnificent scientific treatises – the new astronomy of Copernicus and the new anatomy of Vesalius – stood as the northern and southern salvos marking the launch of the modern scientific revolution. Both were among the first books representing the new science printed on the newly invented printing press.

As for the shocking Copernican announcement that the sun, not the earth, is the center of our universe, it was another habitué of the University of Padua who rose to champion it. Galileo Galilei had been appointed to the university in 1592 to teach mathematics. At that time, professors of mathematics were regarded as lesser instructors than philosophers or teachers of natural philosophy because of the prevailing Aristotelian conception of natural science. Astronomy was taught by mathematicians, and this was mainly an exercise in geometry. What the world really looked like, whether the sun or the earth is at the center of our universe was a question of natural philosophy or metaphysics. That was the concern only of philosophers, not of mathematicians teaching astronomy.

That distinction was also reflected in the contrasting salaries of the two groups of academics. To compensate for that, Galileo did a lot of private teaching. While in Padua, Galileo took in student boarders but also made and sold military compasses for extra money. But it was the reduced status of mathematicians as intellectuals that provoked Galileo to aim higher. He wanted to address questions in cosmology, the shape of the world. As Galileo's intellectual capacities matured, so did his hauteur, inclining him to declare that he wanted to be both a philosopher and a mathematician. Yet, he did not want just an ordinary academic post; he wanted to be "philosopher and mathematician" to the Grand Duke of Tuscany.

This was a daunting and complicated request, not only because Galileo had failed to complete the degree requirements as a student at the University of Pisa in the 1580s. Even as a student at the University of Pisa, he acquired the label of "wrangler," suggesting that he was known for pursuing his own lights in a forceful way, missing classes and provoking his professors and fellow students.

He had enrolled to study medicine, but his true love was mathematics. His father, Vincenzo the musician, insisted that he enroll in the medical studies program, realizing, as parents often do, that being a physician would be more lucrative than being a teacher of mathematics. As Galileo would find out later, professors of philosophy or medicine earned six to eight times as much as professors of mathematics.[15]

Nevertheless, Galileo continued on his path of provocation, missing classes, finally landing in a crisis when university officials notified his father that Galileo was in danger of failing his program of studies. In the meantime, Galileo had found a model of mathematical enlightenment in the person of Ostilio Ricci, who served as counselor to the Medici court. The Medici court often moved its place of official business from Florence to Pisa, especially between Christmas and Easter. Professor Ostilio was a member of the famous Accademia del Designo in Florence (a unique academy of artists) and tutor to the Medici court. During that winter and spring of 1583, Ostilio was giving private lectures in Pisa to members of the royal court when Galileo managed to sneak into the lectures. He

---

[15] This is the estimate arrived at by Paul F. Grendler, *The Universities of the Italian Renaissance* (Baltimore: Johns Hopkins University Press, 2002).

found them especially captivating and probably convinced himself that mathematics was his real calling.

Back in Florence the following summer, Galileo arranged for his father to meet Ostilio, hoping that his father would agree to allow Ostilio to take Galileo under his wing and authorize the son to switch to mathematics. Vincenzo was not easily persuaded, with the result that Galileo bumbled along with his medical studies while secretly continuing with Ostilio's tutoring in Euclid, Archimedes, and other classical mathematical texts. Alas, the Galileo Galilei family ran out of money, and Galileo's half-hearted attempt to apply for a medical scholarship failed, forcing his withdrawal from the university without a degree.

Such unconventional training seemed unlikely to prepare Galileo for great things. Yet, his mathematical gifts stood out enough for him to be hired in 1589 as an instructor in mathematics at the University of Pisa for sixty florins a year, nearly half what was paid to the instructor he replaced. Galileo was a inveterate experimenter, boldly using the Tower of Pisa – at least, according to legend – to perform some of his experiments with falling bodies. It seems likely that if Galileo did use the Tower of Pisa, it would have been for dramatic effect, to demonstrate for others what he already knew, that light and heavy objects would fall at the same speed if not impeded by the resistance of the atmosphere.

News of the Spyglass in Padua

Galileo had moved to Padua in 1592 and gave his inaugural lecture at the university there in December. He continued his unconventional life by making a liaison in 1599 with a Venetian woman named Marina Gamba, with whom he sired three children out of wedlock. They were living separately, a short distance from Galileo's home in Padua, when the news arrived in July 1609 that "a Fleming" had invented a "spyglass." The newsletter of the journalist reporting from the Hague had been reprinted in Lyon that November 1608 and fell into the hands of Galileo's clerical friend Sarpi that same November. Paolo Sarpi was an official in the Republic of Venice, and he made further inquiries about this new instrument. Yet, it seems that Galileo did not hear of it until the spring or summer of 1609. Sarpi's inquiries must have convinced him that spyglasses had quickly become commonplace, that in the short weeks following the Dutch announcement of Lipperhey's invention, imitators were making some version of the spyglass in Paris, London, and other parts of Europe. Perhaps he and others thought this new device was just a toy.

When the idea of constructing a spyglass (*occhiale* in Italian) from two lenses came into Galileo's mind, Galileo quickly fashioned his own device and used it to explore the heavens. "Having dismissed earthly things, I applied myself to exploration of the heavens," he later wrote in the *Starry Messenger*. As far as we can tell, Galileo never saw an actual spyglass made by someone else before he set out to fashion his own. When he did that and aimed it skyward, he transformed the new device into a scientific instrument, indeed, a new discovery machine that would revolutionize the practice of astronomy in the seventeenth century.

For some time, historians of science have debated just how proficient Galileo was in the science of optics. Recent studies of his early optic knowledge and classroom teaching suggest that he knew the major works in the science of optics of his time. Indeed, he made a copy of a seminal work in optics that had until recently been lost from view by historians of science. This was Ettore Ausonio's *Theory of the Spherical Concave Mirror*, probably written in the 1560s. Ausonio was a Venetian mathematician and mirror maker whose optical ideas were seminal for students of optics in the sixteenth century and probably helped Galileo solve the problem of focal length. This was essential to solve the problem of how to align a concave and a convex lens to achieve telescopic vision.

Galileo claims to have first learned of the invention of the telescope after his recuperation from an illness in June 1609 that kept him bedridden in Padua. On July 19, he went to Venice, which is about twenty-five miles away, where he heard the news. According to Galileo's recollection in *The Starry Messenger* of 1610:

About ten months ago [which would be March/April 1609] a report reached my ears that a certain Fleming had constructed a spyglass by means of which visible objects, though very distant from the eye of the observer, were distinctly seen as if nearby. Of this truly remarkable effect several experiences were related, to which some persons gave credence while others denied them. A few days later the report was confirmed to me in a letter from a noble Frenchman at Paris, Jacques Badovere [one of Galileo's former students], which caused me to apply myself wholeheartedly to inquire into means by which I might arrive at the invention of a similar instrument.[16]

In Venice, however, Galileo learned that a stranger was in Padua with one of these new instruments. On August 3, he rushed back to Padua, only to find that the stranger had gone to Venice. At that point, Galileo

---

[16] Galileo, *The Starry Messenger*, in *Discoveries and Opinions of Galileo*, ed. and trans. Stillman Drake (New York: Doubleday, 1957), p. 29.

simply continued on with his own ingenuity. "My reasoning was this. The device needs... more than one glass.... The shape would have to be convex... concave... or bounded by parallel surfaces. But the last-named does not alter visible objects in any way;... the concave diminishes, and the convex, though it enlarges them, shows them indistinct and confused.... I was confined to considering what would be done by a combination of the convex and the concave. You see how this gave me what I sought."[17]

Here, Galileo the optical experimenter knows the contrasting effects of convex and concave lenses and quickly concludes that both kinds must be used to get the desired telescopic effect. Moreover, Galileo was a hands-on kind of person who knew how to grind lenses or quickly learned how to do it from local spectacle makers. As for constructing a telescope:

First I prepared a tube of lead, at the ends of which I fitted two glass lenses, both plane on one side, while on the other side one was spherically concave and the other convex. Then, placing my eye near the concave lens, I perceived objects satisfactorily large and near, for they appeared three times closer and nine times larger than when seen with the naked eye alone. Next I constructed another one, more perfect, which represented objects as enlarged more than sixty times. Finally, sparing neither labor nor expense, I succeeded in constructing for myself so excellent an instrument that objects seen by means of it appeared nearly one thousand times larger and over thirty times closer than when regarded with our natural vision.[18]

These results demonstrate both Galileo's genius and the fact that by the early seventeenth century, the workings of lenses in eyeglasses was broadly understood by scholars. Galileo, however, was able to figure out that the magnifying power of a telescope depends on the proportional relationship between the focal length of the two lenses. In mathematical terms, this would be $m = f_o/f_e$, in which f_o is the focal length of the objective and f_e is the focal length of the ocular.

It was known among students of optics that by combining convex and concave lenses, objects at a distance could be enlarged. As we saw earlier, the mathematician, writer, and dramatist Giovanni Della Porta from Naples had written much about optical phenomena in his book *Natural Magic*. It was widely known in many editions around Europe. When Johannes Kepler, the very clever young German astronomer, finally

---

[17] In Stillman Drake, *Galileo at Work: His Scientific Biography* (Chicago: University of Chicago Press, 1978), pp. 139–40.
[18] Galileo, *Starry Messenger*.

saw a copy of Galileo's *Starry Messenger* and the surprising powers of Galileo's telescope, he remarked that many people would see this as miraculous. "Yet it is neither impossible nor new." This kind of device had been announced, Kepler wrote, in Della Porta's *Natural Magic*, where the author speaks of the effects of combining convex and concave lenses. "When you put your eye behind the middle of the lens, you will see far away things so near that you seem almost to touch them with your hand." Such lenses can obviously be used to improve human vision. Della Porta had written, "if you know how to combine both types correctly, you will see remote as well as nearby objects larger and clear. To many of my friends, to whom distant objects used to look obscure and nearby objects blurred, I have given no small help, with the result that they saw everything with perfection."[19] William Bourne, the Englishman, had described such effects using two lenses as far back as 1578.[20]

But now, in August 1609, Galileo in Padua wrote to his friend Fra Sarpi in Venice, announcing his accomplishment. Sarpi knew that this was not a remarkable achievement in the light of such instruments that were already circulating in Europe, but he must have encouraged Galileo to return to Venice to show his new instrument to officials, especially since the stranger attempting to sell his spyglass to the Venetians asked too much (1,000 ducats). Sarpi advised the government against that purchase, probably hoping that Galileo would come up with a better device that would not be so expensive.

On August 21, Galileo went back to Venice, where he demonstrated his device at the top of various campaniles, especially the one in St. Mark's square. In a letter to his brother-in-law in Florence, Galileo recounted his experience on this occasion. His spyglass apparently was better than the one the Flemish stranger had demonstrated, with the result that he was called by the Signoria to demonstrate his instrument before the full Senate, "to the infinite amazement of all." It was such a novelty that "numerous gentlemen and senators, though old, have more than once climbed the stairs of the highest campanile in Venice to observe at sea sails and vessels far away that, coming under full sail to port," could not be seen without the spyglass for two more hours.[21] The effect of this instrument, Galileo claimed, makes "an object that is ... fifty miles away as large and near as if it were but five."

---

[19] As cited in Edward Rosen, trans., *Kepler's Conversation with Galileo's Sidereal Messenger* (New York: Johnson, 1965), pp. 15–16.
[20] Van Helden, "Invention of the Telescope," p. 30.
[21] Drake, *Galileo at Work*, p. 141.

Galileo then decided to give his telescope to the Doge. This brilliant donation led to his receiving a lifetime renewal of his appointment at the University of Padua, and with a much higher salary, rising from 520 to 1,000 florins per annum. For Galileo, this was both good news and bad. He wanted to return to his native Florence, but this appointment seemed to forestall that possibility. Galileo lamented, "I find myself here, held for life, and shall have to be satisfied to enjoy my native land sometimes during the summer months."[22]

But Galileo had been blessed by another event: his former pupil, Cosimo de' Medici, had been elevated to Grand Duke of Tuscany in February of that year. Cosimo, on hearing of Galileo's new device, asked that he be given one, and thus Galileo saw the opportunity to construct a more powerful spyglass for Cosimo, opening possibilities for engagement there, while encouraging rivalry between the various provincial rulers. Galileo, being proficient enough in optics to be able to grind his own lenses, had the key to making more powerful telescopes in the future – still called the *occhiale* ("spyglass") until the famous banquet of 1611, where he would be feted and the new device would be named the telescope.

Moon Watching in London: Thomas Harriot

News of the invention of the spyglass was rapidly spreading around Europe in October and November 1608. In November, the English ambassador in Paris was given a copy of the newsletter that announced the Dutch invention of the spyglass. It was given to him by the French journalist Pierre de L'Estoile. The latter had made an entry in his journal that November indicating that the newly invented spyglass could be used to see stars that otherwise were invisible. Unknown to Galileo, other Europeans were quick to seize on the spyglass's potential for astronomical purposes.

In England, the most notable of these heavenly explorers was "the greatest mathematician that Oxford has produced."[23] This was Thomas Harriot, born in Oxford, England, in 1560. He graduated from the University at Oxford in 1580 and was soon renowned for his solutions to seafaring and other mathematical problems associated with geography, mapmaking, and optics. He sailed to Virginia in the New World for

---

[22] Ibid.
[23] J. Fauvel, R. Flood, and R. J. Wilson, *Oxford Figures: 800 Years of the Mathematical Sciences* (Oxford: Oxford University Press, 2000), pp. 56–59, cited in http://www-history .mcs.st-andrews.ac.uk/References/Harriot.html.

FIGURE 2.4. Harriot's sketch of the moon made on July 26 (Julian calendar; August 5, Gregorian), 1609, four months before Galileo's observations. From the Lord Egremont Collection with permission of *Universe Today*. http://www .universetoday.com/2009/01/14/was-galileo-the-first/harriots-drawings/.

Walter Raleigh in 1585–86. His new world journal speaks of the frighteningly invisible spread of smallpox among the natives, who had no immunity to the disease.

But back in England, Harriot met legal difficulties brought against his patron, Sir Walter Raleigh. The latter had been accused of atheism and being involved in plots against the king. Harriot was temporarily imprisoned and thereafter sought to separate himself from Raleigh. He moved to work with the Duke of Northumberland, who had gathered scholars around him. The duke gave Harriot use of one of his estates in Syon (outside London). It was there that Harriot worked from 1597 onward on problems in optics. This led to his construction of a telescope ("perspectus") with which he looked at the moon on August 5, 1609, with a 6× power telescope. This was probably not the very first look at the moon with such an instrument, but Harriot's sketch of the moon at this time is thought to be the first of such efforts at *selenography*, or moon mapping (Figure 2.4).

Harriot never published his observations or writings, yet he was a silent competitor with Galileo, something that Galileo always feared, and rightly so. For while Galileo was busy making more powerful telescopes and angling for a means of returning to his native Florence, the gifted Englishman had already turned his spyglass skyward.

Galileo Begins His Moon Observations

In Fall 1609, Galileo set to work improving the power and design of his new instrument. He sent to Florence for more lens blanks so that he could grind his own lenses that would increase the power of his viewing tube. Although the exact date has not been definitively established, it appears that by September, he had a device that would magnify a distant object 20×. The fact that the Doge in Venice opened his purse so generously to Galileo when he gave the Doge his first *occhiale* doubtlessly encouraged Galileo to think the young Grand Duke of Tuscany would react in a similar fashion if Galileo could present him with an even more powerful instrument. But he had probably not yet dreamed that he would also have great heavenly discoveries to present to the grand duke.

Galileo turned his attention to the heavens that fall, perhaps as early as October, as he began observing the moon. This paid off handsomely when Galileo began to realize that the shading on the moon's irregular surface, as seen through the *occhiale*, revealed unaccountable unevenness, crags, and valleys. Galileo's actual recorded observations of the moon, painted in ink washes, appear to have been made between November 30 and December 18, 1609, with one remaining observation recorded on January 29, 1610.[24] But if Galileo set himself up to record such observations with paint and brush, then he must have made his discoveries earlier than late November, when he made the paintings (Figure 2.5).

When viewed during its early phases, the moon's surface casts shadows that Galileo thought were similar to shadows seen on earth when the sun comes up in the morning, first illuminating mountains and hilltops, then slowly illuminating the valleys below. Was Galileo's perception a product of his upbringing in painterly Florence? Was it the artist's gift to recognize that shadows in paintings appear on the opposite sides of

[24] Owen Gingerich and Albert van Helden, "From Occhiale to Printed Page: The Making of Galileo's *Sidereus Nuncius*," *Journal for the History of Astronomy* 34 (2003): 251–67, who follow Ewen Whitaker, "Galileo's Lunar Observations and the Dating of the Composition of 'Sidereus Nuncius,'" *Journal for the History of Astronomy* 9 (1978): 155–69.

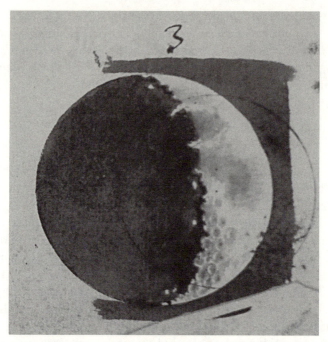

FIGURE 2.5. An image from Galileo's original watercolor sheet of ink washes. The terminator showing the division between the light and dark surfaces of the moon served to reveal those parts of the moon's surface that were raised, mountainous, or cratered, not smooth and flat. From *Galileiana* 48, f. 28, Biblioteca Nazionale Centrale, Florence. Courtesy of Owen Gingerich.

objects lit by a source of light? To Galileo's mathematical and scientific gifts, we have to add those of a watercolorist because of these seven watercolor sketches of the moon, which are his first and only record of his early moon observations. Only four of those paintings were ever published by Galileo, but even those printed engravings show the effects of embellishment.

As one art historian put it, "with deft brushstrokes of a practiced water-colorist, Galileo laid on at least a half-dozen different grades of washes, imparting to his images an attractive soft and luminescent quality."[25] Compared to the primitive sketches of Thomas Harriot's earlier attempt

[25] Sam Edgerton, "Galileo, Florentine 'Disegno' and the 'Strange Spottedness' of the Moon," *Art Journal* 44, no. 3 (1984): 229.

at graphic illustration, Galileo's gift was clearly something from the artistic world of Florence. Indeed, Galileo was familiar with the *designo* (design) perspectives of the famous Academy of Florence of that name – the first such school for artists founded in the mid-sixteenth century. Later, in 1613, Galileo would be admitted to the Accademia del Designo. One of the most outstanding painters of the academy of Galileo's time, Lodovico Cardi Cignoli, actually credited Galileo with teaching him all he knew about the geometry of perspective.

The four ink-wash drawings that were printed in Galileo's time appear to be embellished reconstructions serving Galileo's topographical intentions; that is, Galileo's arguments regarding the surface of the moon relied on assimilating the surface of the moon to the rough surface of the earth. In the morning, light comes over a mountain or hilltop, casting a long shadow over the valley below. The same thing happens on the moon, Galileo argued, as the sun's light strikes it at various positions in the sky visible to us. Consequently, Galileo's published image of the last quarter of the moon's surface appearing in the published engraving deviates markedly from both his original drawing and the actual surface of the moon (see Figure 2.5). Galileo used his images to illustrate what he saw as a theoretical observer rather than the strict realism of a photograph. He wanted to accent the shadings and contours of the moon's surface, indicating that it was not smooth and crystalline but rather rough, craggy, and pockmarked. At the same time, Galileo wanted to show that the line dividing the light and dark regions (the so-called terminator) stood out because of the mountainous surface of the moon and that the terminator shifted each succeeding day as the moon's angle with the sun changed. Furthermore, he seems to have exaggerated the crater placed just below the middle of the center to highlight the shadows created by the sun's light shining over the upper edge of the crater, creating the shadow below in the crater. Despite Galileo's use of reconstructive artistic techniques, his use of graphic devices made him a pioneer of drawings as scientific illustrations. Galileo's use of geometrical reasoning and his drawings persuaded him that some of these moon mountains were about four miles high, a remarkably good estimate. For him, all this evidence of the moon's rough surface was decisive in rejecting the idea that the moon is a crystalline body, thereby urging a rethinking of our cosmological assumptions. But before Galileo could finish his moon drawings and fully digest their scientific significance, his new *perspiculum* (as the telescope was called in Latin) revealed still more startling discoveries.

Jovian Moons

In Fall 1609, Galileo may have been exhausted from his efforts to discover the secret of the Dutch spyglass about which he heard only rumors. That may explain why he was slow to point his newly constructed but more powerful spyglass heavenward. He did not know that Thomas Harriot, who did not publish any of his astronomical findings, was busy outside London, constructing his own improved spyglasses and looking at the moon. Galileo had been told by his friend Sarpi that these new optical devices were to be found in many places around Europe, but he could scarcely imagine who had one and might use it to look at the sky. Between October 1608 and June 1609, no fewer than five other individuals in Italy alone – a cardinal, a nobleman, and gentlemen scholars in Milan, Rome, Naples, and Siena – had low-powered telescopes with which they had viewed the moon.[26] Indeed, the Pope received a spyglass – probably in April – from the Netherlands, and the Jesuits quickly used it for celestial observation.

No doubt Galileo had heard rumors of such possibilities, and this must have encouraged him to work in secrecy, carefully guarding his discoveries. The race to achieve scientific originality through discovery was well entrenched in the European community at that time. Galileo could be pretty sure that in addition to his ability to figure out the scientific problem of the relationship between the focal length of a lens and its power of magnification, he knew uniquely how to grind his own lenses that would suit his purpose. The combination of interest in telescopic exploration of the heavens and the lens-grinding skills that Galileo possessed were rare at the time; only Thomas Harriot in England has been identified as having them. Nevertheless, Galileo surely knew that rewards for originality go only to the first discoverer, not to those who later replicate and confirm discoveries. Furthermore, because Galileo was angling for the lofty position of "philosopher and mathematician" to the Grand Duke of Tuscany, he could hardly apply for such a job if he were second in line, either with a new spyglass or with new astronomical discoveries.

Unexpectedly, on January 7, 1610, before Galileo could finish his sequence of moon observations, he brought Jupiter into focus with his instrument. In doing so, he transformed Jupiter from a small point of light in the sky into a planet having its own shape, color, and texture.

[26] Mario Biagioli, *Galileo's Instruments of Credit: Telescopes, Images and Secrecy* (Chicago: University of Chicago Press, 2006), p. 93n53.

On that particular evening, Jupiter was only about two degrees away from the moon and must have made a tempting observational target. Once he got the bright planet in focus, what came into his view was not only the planet itself but three previously unseen starlets. Two of the little planets were seen to the east and one to the west of Jupiter. These objects did not scintillate as stars do, but it was when Galileo noticed that they moved, unlike fixed stars, that he realized they were indeed tiny "planets." (Only later did Kepler coin the word *satellite* to refer to these small captive moons.)

Galileo continued watching these little moons of Jupiter, observing on the next night that they were lined up on the west side of Jupiter. It was cloudy on January 9, preventing any observations. On January 10, only two of the planets appeared to Galileo, this time appearing on the east side. On January 11, still another configuration of the starlets appeared: two on the east side and only one, quite distant from the mother planet, on the west, looking like this:

$$\text{(East)} ** O * \text{(West)}$$

(Just these little notations were used by Galileo in his handwritten notes of those dramatic days.) See Figure 2.6.

From this point on, Galileo was convinced that he had seen something truly new in the heavens and that the evidence supported a new astronomical conception. For Galileo, this configuration of little planets was evidence of the Copernican hypothesis, that our universe is sun-centered and that all planets need not revolve around the earth. If a planet such as Jupiter could have small moons circling around it, then it was certainly possible to imagine a universe in which the earth had its moon and yet circled around a larger body such as the sun.

On the 13th day, after I had fixed the instrument very well, there appeared very close to Jupiter four stars in this formation ⋆ ⊕ ⋆⋆⋆ or better like this: ⋆ ⊕⋆⋆⋆ , and all appeared of the same size. The space between the 3 western ones was not larger than the diameter of Jupiter, and with respect to each other they were noticeably closer than the other evening; and they were not precisely in a straight line as before, but the middle one of the three western ones was a bit higher [elevated], or rather, the westernmost one was somewhat lower. These stars are all very bright, although very small, and other fixed stars that are the same size are not as brilliant.

FIGURE 2.6. An entry from Galileo's journal recording his observations of Jupiter's moons. From *Le Opere di Galileo* (1890–1910), 3, pp. 427–28, translation by Owen Gingerich and Albert van Helden, with permission.

With the new Jupiter observations in hand, Galileo declared that he "had decided beyond all question" that these little stars were "wandering about Jupiter as do Venus and Mercury about the sun."[27] And though he had surmised that there were four moons revolving around Jupiter, it was not until January 13 that he actually saw all four of the moons aligned with Jupiter, in still another formation:

(East) * O * * * (West)

Finally, on January 15, Galileo saw all four stars lined up on the west side of Jupiter:

(East) O * * * * (West)

At this point, Galileo was galvanized into action: it was imperative that he rush his new findings into print. Within two weeks of these events, Galileo went to Venice, where he engaged a printer to publish his findings. They were to bear that celebrated title, *Sidereus Nuncius*, the *Starry Messenger*. Scholars have debated whether the title should be translated as "message" from the stars (one meaning of *Nuncius*) or "messenger" of the stars; the latter has prevailed.

Galileo gave the printer the first part of his findings, which concerned the surface of the moon, including the four moon illustrations to be engraved, but held back his Jupiter results, which were to be included later. Although Galileo mentioned in his introduction to the essay that he had discovered four new planets, he did not mention the fact that these planets revolve around Jupiter.

In the meantime, Galileo continued his observations of Jupiter and sent a confidential communication to the grand duke's secretary in Florence, Baldessari Vinta. What name, he asked, should be given to these newly discovered heavenly objects? Should they be called planets or stars? Should they be called "Cosmian," after Cosimo de' Medici, or "Medicean," after the family name of the grand duke? Galileo sent his query on February 17 and got a reply back on February 26. The secretary's advice was to use the term "Medicean"; otherwise, people would not recognize that the starlets had been named after the Medicis of Florence. Typesetting had gone so far that "Cosmica Sydera" already appeared on one page, with the result that printed labels had to be prepared to paste over the incorrect title in all 550 copies.[28]

---

[27] Galileo, *Starry Messenger*, p. 53.
[28] Ibid., p. 262.

On March 12, 1610, Galileo's revolutionary little book, *The Starry Messenger*, came off the printing press in Venice. The substance of the book can be divided into two parts: the first on the moon (including the four engravings based on Galileo's ink washes) and the second devoted to Galileo's Jupiter observations. The first part had been given to the printer in early February and the last, containing Galileo's last-minute observations of Jupiter, was delivered only four weeks before it was published. Still, Galileo must have sworn the printer to secrecy between February 16, when he received the Jovian material, and the dedication date of March 12. A middle section of several pages fell in between the lunar engravings and the Jovian material. Galileo filled that with additional observations of the Milky Way and his quickly gotten observations of still more invisible stars spotted in the constellation of Orion. He sketched Orion's sword and belt, adding the new stars, as well as the Seven Sisters (Pleiades) containing many new stars that hitherto had been unseen. Galileo's Milky Way discussion rejected the traditional view that it was a misty vapor, claiming instead that it was composed of many hundreds of discrete stars that could only be seen with the aid of the new *occhiale*. This conclusion applied equally to other nebulae, cloudy clusters of thousands of stars invisible to the unaided eye.

By any standard, this was a remarkable parcel of discoveries. Galileo, now aged forty-six, was launched on a new career. He had just the gift that would cinch his position with Cosimo de' Medici II, the Grand Duke of Tuscany. That was his *Starry Messenger* and the telescope with which he made his unsurpassed discoveries. He also had a powerful batch of observations that he believed were uniquely supportive of the Copernican theory of the universe. For astronomers across Europe, Galileo's telescopic discoveries were revolutionary, pregnant with implications for planetary theory. Other observers were combing the sky, seeking confirmation or refutation of Galileo's remarkable claims. Soon ambassadors, merchants, traders, and missionaries would be taking the telescope around the world.

3

The New Telescopic Evidence

We are here... on fire with these things.
– Sir William Lower, June 21, 1610

During the year following the publication of the *Starry Messenger*, Galileo was thrown into a maelstrom of argument, debate, and more discoveries. Those without principled reasons for opposing Galileo's discoveries were enchanted and began to imagine all kinds of new things. An English astronomer, Sir William Lower, who had been a student of Thomas Harriot's, reacted enthusiastically to Galileo's news. He wrote to Harriot on June 21, 1610, "We are here... on fire with these things."[1] For him, Galileo's discoveries were more startling than Magellan's trip around the world. He and Harriot both wondered whether the planets Saturn and Mars might have hitherto unseen moons revolving around them. They were right: both do have satellites, but they would not be found for many years.

Becoming Mathematician and Philosopher

Galileo now pressed forward with his plan to become mathematician and philosopher to the Grand Duke of Tuscany. With his new book of discoveries in hand and his improved *occhiale*, Galileo had much with which to impress the grand duke.

[1] William Lower, as cited in John J. Roche, "Harriot, Galileo, and Jupiter's Satellites," *Archives Internationales d'Histoire des Sciences* 32 (1982): 16.

Around Easter, the grand duke's court made its headquarters in Pisa to avoid the heat of Florence. In a letter sent on May 7 to the grand duke's secretary, Belasario Vinta, Galileo laid out the present circumstances of his employment in Padua, a salary of 1,000 florins per month, his scientific instrument making for extra money, plus boarders. He hoped that the grand duke would grant him employment while easing his duties and the need for supplemental income. He made bold his claim to "particular secrets, as useful as they are curious, and admirable, I have in plenty."[2] He was certain that he would do "great and remarkable things," and he wanted from the grand duke only the leisure and financial security to pursue his plans. These appear to be larger than life:

The works I must bring to conclusion are these. Two books on the system and constitution of the universe – an immense conception full of philosophy, astronomy, and geometry. Three books on local motion – an entirely new science in which no one else, ancient or modern, has discovered any of the underlying properties which I demonstrate to exist in both natural and violent movement; hence I may call this a new science and one discovered by me from its foundations. Three books on mechanics, two relating to demonstrations and its principles, and one concerning its problems; and though other men have written on this subject, what has been done is not one-quarter of what I write. . . . I have also various little works on physical subjects such as *On Sound and Voice, On Vision and Colors, On the Tides, On the Composition of the Continuum, On the Motions of Animals*, and still more.[3]

Galileo was exuberantly confident of the path before him. In the end, he did fulfill most of his claims. His discovery of the law of free fall and the idea of inertia would be revolutionary, but it would take some time to work out the details. On the other hand, he did not write a general treatise on optics, color, sound, or the voice.

Not everyone was immediately persuaded regarding the reliability of the new astronomical observations reported in the *Starry Messenger*. Within three months of the appearance of that book, a young scholar by the name of Martin Horky, from Bavaria, published in Modena a short commentary attacking it. Others wrote letters directly to the Tuscan court advising against accepting Galileo's claims. In May, Galileo gave three

---

[2] In Stillman Drake, *Discoveries and Opinions of Galileo* (New York: Doubleday, 1957), p. 62.

[3] As cited in Stillman Drake, *Galileo at Work* (Chicago: University of Chicago Press, 1978), p. 160, May 7. The translation is modified, replacing "laws of nature" with "underlying properties."

public lectures at the University of Padua on his new discoveries. They were successful in bringing his colleagues around.

Seeing the little moons of Jupiter with this new telescopic device, however, could be difficult. Some observers were not able to do it. Watching for the Jovian satellites in the early hours of the morning can be tricky: waiting for Jupiter to appear above the horizon, fumbling around while trying to bring Jupiter into focus, waiting for one's eyes to adjust to the little objects. Furthermore, the field of vision of Galileo's telescopes was very narrow, less than half a degree across. In addition to that, Galileo often stopped down the lens with an inner ring that served to reduce distortion and sharpen the image by constricting the light to the center of the lens.

Giving advice to a Medici official, Antonio de' Medici, Galileo wrote, "The instrument [must] be kept steady and therefore, in order to avoid the trembling of the hand which is caused by the motion of the arteries and by respiration itself, it is good to fix the tube in some stable place."[4] Just as quickly as the moons may appear to the observer, they can disappear as the rising sun brightens the morning sky, hiding the light of the little moons. Watching at night was different, but this was possible only during certain periods of the annual celestial rotations.

Seeing new and unexpected things is always problematic, for scientists and laymen. In April 1610, Galileo set out for the summer residence of the grand duke in Pisa. On the way there, he stopped in Bologna to show his colleague Giovanni Magini and other astronomers the moons of Jupiter. One scholar refused to look through the telescope at all, and the others seemed unable to see the little moons. Magini was Galileo's rival, and Magini's assistant, Martin Horky (who came from Bavaria), being less restrained than his mentor, went out of his way to attack Galileo. He wrote a letter to Kepler reporting on the meeting of April 24 and 25, claiming that the demonstration had been a failure and that Galileo left Bologna sad and dejected.[5]

Horky's report of what he saw with the telescope repeats a frequent occurrence when scientists see entirely new and unexpected things: they often deny what they have seen until intellectual grounds have been prepared for seeing the new thing. Horky claimed to have stayed up all

[4] *Le Opere di Galileo Galilei,* as cited in Albert van Helden, "The 'Astronomical Telescope,' 1611–1650," *Annali dell'istituto e Museo di Storia della Scienza di Firenze* 1 (1976): 18n26.

[5] As cited in Drake, *Galileo at Work,* p. 160.

night that April in Bologna, using Galileo's telescope to view the sky. He thought the telescope was marvelous for terrestrial observations, but "in the heavens it deceives." He went on, "I observed with Galileo's spy-glass the little star that is seen above the middle one of the three in the tail of the Great Bear, and I saw four small stars nearby, just as Galileo observed about Jupiter."[6] He noted, "Other fixed stars appear double." Instead of treating these as possible new discoveries, Horky blamed the instrument of revelation. Clearly, Horky was not prepared to see what was there, that there are both double and triple stars that can be seen with the telescope. Like Pleiades (Seven Sisters) or Orion's Belt observed by Galileo, these constellations have more stars in them than the unaided eye had been able to detect since the beginning of human observation. Horky hastily discarded as false observations what should have alerted him to new discoveries. But worse, he was a mischief maker who sought to make as much trouble as he could for Galileo. His duplicitous letters to Kepler led Kepler to renounce him: "Since the demands of honesty have become," Kepler wrote, "incompatible with my friendship for you, I hereby terminate the latter."[7] In the meantime, Horky secretly made wax molds of the lenses in Galileo's instrument, hoping he would then be able to replicate them. He had no success with that. Horky was eventually put in his place. Even his mentor at the University of Bologna ended up chasing Horky out of his house.

In a letter to Secretary Vinta that summer, Galileo claimed success in all his endeavors, assuring his correspondent that everyone had been won over to his side and his opponents given up. He could also report that he had received a strong letter of support from Johannes Kepler, the imperial mathematician in Prague, though Kepler had yet to gain access to Galileo's new astronomical instrument. That would not happen until August of that year.

By the end of May, Galileo's appointment as "mathematician and philosopher" to the grand duke had been settled. On June 5, Vinta sent word to Galileo that he was to have the title of "Chief Mathematician of the University of Pisa and Philosopher to the Grand Duke." Galileo, how-ever, wanted the two words *mathematician* and *philosopher* conjoined in his title linked to the grand duke, Galileo's former student. This was

[6] As cited in Albert van Helden, ed., trans., *Sidereus Nuncius, or The Sidereal Messenger of Galileo Galilei* (Chicago: University of Chicago Press, 1989), p. 93.
[7] Kepler to Horky, August 9, 1610, as cited in Arthur Koestler, *Watershed: A Biography of Johannes Kepler* (New York: Doubleday, 1960), p. 196.

done on July 10, and Galileo's position was sealed for life, beginning in October.

More Startling Discoveries in Padua

Still living in Padua, Summer 1610 brought Galileo another unsettling discovery. Toward the middle of that summer, he chanced on the most distant planet, Saturn. In a letter to Vinta, the grand duke's secretary, Galileo decided to attach the first report on his observations of Saturn using his improved telescope. "I began," he reports,

on the 15th of the month again to observe Jupiter in the morning in the East, with his formation of the Medicean planets, and I discovered another very strange wonder, which I should like to make known to Their Highnesses and Your Lordship, keeping it secret, however, until that time when my work is published. But I wished [*sic*] to inform Their Serene Highnesses of it in order that, if others should discover it, they would know that no one observed it before me. I am quite sure that no one will see it before I have pointed it out. This is that the star of Saturn is not a single star, but is a composite of three, which almost touch each other, never change or move relative to each other, and are arranged in a row along the zodiac, the middle one being three times larger than the two lateral ones, and they are situated in this form.[8]

For contemporary readers, this diagram (Figure 3.1), perhaps looking like a cup with handles, is an odd revelation of what we now know as the rings of Saturn. With only the experience of Jupiter and its satellites discovered earlier in the year as a guide, Galileo was prepared to find Saturn with circulating moons attached. With his new telescopic device, Galileo was looking into a sky more wondrous than the 1,028 stars

---

[8] Galileo, *Opere*, X: 409–10, as cited in Albert van Helden, "Saturn and His Anses," *Journal for the History of Astronomy* 5 (1976): 105.

FIGURE 3.1. Saturn and its handles as seen by Galileo in 1616. From *Le opere di Galileo Galilei*, ed. Antonio Favaro (Edizione Nazionale, Florence: G. Barbera, 1890–1910).

catalogued by Ptolemy. It revealed all kinds of new objects neither imagined nor seen before by the human eye. Celestial objects were supposed to be either perfectly formed planets or stars; the idea of satellites revolving around heavenly objects other than the earth was preposterous. Double stars, triple stars, and supernovas had hardly been imagined. If Galileo's discovery of the little moons of Jupiter was the revelation of a typical pattern of celestial arrangements, then looking for other satellites circulating around Saturn, or possibly Mars, was the most likely expectation. But "rings" composed of star dust, ice, and stellar debris had never before been seen or imagined.

Consequently, Galileo saw these Jovian handles as evidence of two smaller globes circulating very close to the larger planet. Even Galileo's own telescopes gave him conflicting readings of the shape of Saturn, which he reported in November. There he cautioned observers that "if you look at [Saturn and its presumed companions] through a glass that does not multiply much, the stars will not appear clearly separate from one another, but Saturn's orb will appear somewhat elongated, of the shape of an olive, thus, o."[9]

When Galileo found little change in the positions of Saturn's mystifying "companions," he suspended his watch. But the question of just what the strange images of Saturn were remained a puzzle that would take nearly half a century, and improved telescopes, to resolve. On some occasions, Saturn would appear to have handles, whereas at other times, it would appear to be an elongated oval, and at others, it appeared to be a tilted disc with a protruding outer rim (Figure 3.2). Seen with today's slightly more powerful and much clearer lenses, Saturn is a showpiece: it can easily look like a brightly domed disc surrounded by a flat ring that fully conforms to our imagined flying saucers in the celestial regions.

Even Galileo's report of his puzzling observations was wrapped in an enigmatic anagram:

smaimrmilmepoetaleumibunugttauiras.

It was in this format that Galileo informed colleagues at the Roman College and the grand duke's brother Giuliano, the Tuscan ambassador

---

[9] As cited in Edward Stafford Carlos, ed., trans., *The Sidereal Messenger of Galileo Galilei and Part of the Preface to Kepler's Dioptrics* (London: Dawsons of Pall Mall, 1959), p. 91.

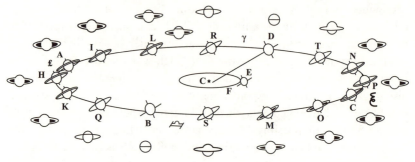

FIGURE 3.2. The changing appearance of Saturn. From Galileo's first sighting of the odd shape of Saturn in 1611 until 1656, European astronomers
reported many curious shapes that the planet took. Only in 1656 did Christiaan Huygens provide a plausible account showing that Saturn's handles were
actually the components of a flat ring "which does not touch him anywhere
and is inclined to the ecliptic." In his report in *The System of Saturn* (1656),
Huygens diagramed the path of Saturn around the sun and posted adjacent
to those positions the appearance that an observer would see at each position in that annual cycle. Reprinted in Huygens, *Oeuvres Complètes*, 1888–
1950.

in Prague, of his discovery. The letter, sent in early August to Prague,
was meant to inform Kepler of Galileo's latest discovery. Shortly thereafter, it was Kepler who first published news of Galileo's new discoveries.
Kepler's attempt to decrypt Galileo's anagram led to the mistaken guess
that Galileo had discovered something related to Mars: "Hail, twin com-
panionship, children of Mars."[10] Not until November did Galileo reveal
to his astronomical colleagues the real nature of his Saturnian observations. By then, the emperor in Prague was becoming impatient for
Galileo to reveal his secret. When Galileo was so informed of this irri-
tation by Kepler, Galileo sent a letter in reply, informing the reader that
his anagram actually spelled out *Altimissimum planetam tergeminum
observavi*, or "I have observed the most distant of the planets to have a
triple form."[11] This turned out to be a false clue since the companions
were really a ring. But whether Saturn was a solitary body, something
with handles, or a silk oval would remain unsolved. The larger message
was that the heavens contain far more than academic philosophers had
imagined.

[10] Ibid., p. 88.
[11] Ibid., p. 90, letter of November 15 to the ambassador in Prague.

Back to Florence

With his appointment as the royal mathematician and philosopher to the Grand Duke of Tuscany sealed, Galileo prepared to move back to his home country of Florence. This must have been a major undertaking, packing all his books, papers, and instruments for the horsedrawn trip back to Florence, about 150 miles away. Galileo was a bit of an entrepreneur, taking in boarders at "Hotel Galileo," tutoring private students, and designing unique new models of a geometrical and military compass. Added to that was the sale of a book called *Operations of the Geometrical and Military Compass*. This item was for sale only at his home. It was sold mainly to students of mechanics and mathematics, who also bought the special compass that was crafted by Mercantonio Mazzoleni and retailed by Galileo with a small markup. Mazzoleni lived with his wife in the extended Galileo household. With all these supplemental sources, Galileo "more than tripled his annual university salary,"[12] as he hinted in his letter to Secretary Vinta while applying to become chief academician of the Tuscan court. And then there was Galileo's lens-grinding machine requiring transport to Florence. Galileo would need such a machine for the indefinite future, as his career as a spyglass user and maker was just taking off. All these possessions had to be loaded into a horsedrawn wagon for the long, stiff journey to Florence.

To those unfamiliar with the regional variations in Italian culture, Galileo's preference for Florence over Venice might seem puzzling. Florence was and is the unsurpassed city of great Italian art with its own unique cultural ambience. Venice, too, has its unique charm and attractions such as canals, gondoliers, and unique architecture influenced by the styles of the Muslim Middle East. However, Galileo was employed by the University of Padua twenty-five miles away. By moving to Florence, Galileo was leaving a republic that was governed by semidemocratic procedures, whereas Florence was the capital of an absolutist state. Galileo's friend Giovanfrancesco Sagredo, in a long letter to Galileo sent from Venice in Fall 1610, warned Galileo of this. He spoke openly of the "freedom here and the way of life of every class of person [that] seems to be an admirable thing, perhaps unique in the world."[13] He raised all the questions that would loom large in Galileo's life: the fickleness of monarchs,

---

[12] Mario Biagioli, *Galileo's Instruments of Credit: Telescopes, Images, Secrecy* (Chicago: University of Chicago Press, 2006), p. 7.
[13] Drake, *Discoveries and Opinions*, p. 70.

the distractions of events and foreign affairs, and the potential threat of the Jesuits, who played a significant role in the court of Florence but not in Venice.

It would only take a few months before rumblings would be heard from surprising quarters about the large philosophical and cosmological issues that Galileo's discoveries would produce. Galileo, supremely confident in his intellectual abilities, probably underestimated the harm that could be done by less than stellar thinkers. He was bemused by some of his more creative opponents such as the one who suggested that the moon may have mountains hidden below but that there was actually a smooth and invisible covering all across the moon's surface. If that were true, Galileo opined, then perhaps on top of that smooth, invisible membrane were mountains ten times as high as what he observed, but invisible to human sight!

But Galileo's challenge to the underlying assumptions of Aristotle's physics, and to the Church's commitment to an immobile earth at the center of the universe, was not something that could be dismissed without serious consideration. Whether justified or not, many laymen and officials found biblical support for the geocentric view. It was only fitting that everyone take such cosmological speculation seriously. But that would mean that some would charge Galileo with heresy.

In the meantime, Galileo continued to make more discoveries that seemed to him incontrovertible evidence supporting the heliocentric hypothesis. Whatever Galileo's fate, Pandora's box had been opened: scientists, laymen, ambassadors, traders, and amateurs would be following Galileo's telescopic trail, and some would push the theoretical frontiers much further. Johannes Kepler had done that in his *New Astronomy* of 1609, but Galileo overlooked its significance.

That neglect was not done out of perverse rivalry with Kepler but rather because Galileo and Kepler had entirely different scientific styles. Kepler was a truly grand thinker who believed passionately in the mysterious unity of the cosmos and that physical causes of celestial movements existed and could be found by human beings, just as he had found them in his *New Astronomy*. For Kepler, the extraordinary calculator, the mathematical laws of planetary motion he described proved both the exquisite harmony of the heavens and God's hand in creating such a divine edifice.

Galileo, on the other hand, wished to be a pure rationalist, bracketing the possible design of God. On a practical level, Galileo was an experimentalist and a practitioner of what sociologists call *theory of the middle range*. He focused on the smaller pieces of the puzzle, studied them in great detail, using the precision of mathematics when possible, but leaving the big picture for later assembly. "I prefer to find," he wrote, "the truth even in unimportant things than to argue for a long time about [the] greatest problems without attaining any truth."[14]

But, of course, the "greatest problems" could not be avoided, as when Galileo inferred from his small truth (the existence of the satellites of Jupiter) the correctness of the Copernican system. In reality, Kepler's grand mathematical achievement was a better foundation for proving that Jupiter and its satellites belonged to a sun-centered system of eccentric orbits than telescopic sightings of Jupiter's moons, though both were necessary. Eventually, Galileo would be unable to resist the greatest problem by setting out a very complex set of arguments about the shape of the universe in his famous *Dialogue Concerning the Two Chief World Systems* in 1632. This was Galileo's way of handling the greatest questions of cosmology, the question of whether the Copernican or the Aristotelian system was the correct one. He would do so even under interdiction of the Church. But that is to get ahead of our story. For now, our small truth is that Galileo and Kepler had different approaches to cosmology, to understanding the real shape of our universe and the practice of science. A rich vein of new telescopic discoveries had now come Galileo's way, providing astronomers everywhere with hitherto unexpected evidence of the details of the heavens. Neglecting Kepler's proof that the orbits of the planets were elliptical, not perfectly circular, and centered approximately on the sun might be forgiven. As Kepler put it in his *Conversation with the Starry Messenger*, "instead of reading a book by someone else, he [Galileo] has busied himself with a highly startling revelation."[15]

A Telescope Comes to Kepler

The shortage of good-quality telescopes with magnifying power of twenty or thirty times the actual size of the object prevented many would-be observers from confirming Galileo's discoveries. In a letter to the Tuscan court in March 1610, Galileo claimed to have made sixty telescopes, but not all of them were usable. He certainly wanted his discoveries to be recognized far and wide; toward that end, he sent his instruments

[14] *Le Opere*, IV: 738, as cited in I. N. Veselovskii, "Kepler and Galileo," *Vistas in Astronomy* 18 (1975): 250.
[15] Johannes Kepler, *Kepler's Conversation with Galileo's Sidereal Messenger*, first complete translation, with an introduction and notes by Edward Rosen (New York: Johnson, 1965), p. 9.

to "great princes, and in particular to the relatives of the Most Serene Grand Duke."[16] This was consistent with Galileo's angling for royal patronage and because many of these high officials patronized scientists and astronomers. In that fashion, it aided the cause of the new astronomy. Nevertheless, it is surprising that Galileo did not send one of his telescopes to Kepler, though the Tuscan ambassador to Prague had advised Galileo to do so.

In one sense, Galileo's strategy did pay off. For in that letter in Spring 1610, Galileo mentions that he had already received a request for one of his instruments from the Duke of Bavaria, Elector of Cologne. With dispatch, Galileo supplied him with a telescope. On a trip from Vienna to Prague that summer, the very same Duke of Bavaria lent his telescope to Kepler, starting on August 30. Finally, the most knowledgeable astronomer and optician of Europe would have a chance to see the skies for himself with the new instrument. The experience at last of using the telescope sent Kepler into raptures: "Oh telescope, instrument of much knowledge, more precise than any sceptre! Is not he who holds thee in his hand made king and lord of the works of God?"[17]

Because of the skepticism surrounding Galileo's discoveries, Kepler had become increasingly alarmed by the fact that no witnesses had emerged in public who could verify Galileo's moons of Jupiter. In a letter to Galileo, he offered the possibility that "we are dealing not with a philosophical but a legal problem: did Galileo deliberately mislead the world by a hoax?" For, Kepler continued, "I do not wish to hide from you that letters have reached Prague from several Italians who deny that those planets can be seen through your telescope."[18] This alarmed Galileo enough to spur him to reply to Kepler, though Galileo still did not send a telescope.

With the Duke of Bavaria's telescope in hand, Kepler set to work at the end of August examining the skies for evidence of Galileo's discoveries. Kepler did indeed treat this problem of scientific verification as a problem of truthful witnessing. He had a young mathematician by the name of Benjamin Ursinus and other guests join him while observing Jupiter with the spyglass. Each of the guests was instructed to independently observe Jupiter and then mark down on chalk slates what he saw. Only later

---

[16] As cited in van Helden, *Sidereus Nuncius*, p. 92.

[17] Johannes Kepler, *Dioptrice* (1611), as cited in Carlos, *Sidereal Messenger*, p. 88. Or "Oh you much knowing tube, more precious that any scepter. He who holds you in his right hand, is he not appointed king or master over the works of God?" Also cited in Max Caspar, *Kepler*, trans. C. Doris (New York: Dover, 1993), p. 201.

[18] Kepler to Galileo, August 9, 1610, as cited in Koestler, *Watershed*, p. 197.

were the reports compared with each other, with great satisfaction. These observations were reported in a little tract that Kepler wrote bearing the title *Johannes Kepler's Narration about the Four Wandering Satellites of Jupiter Observed by Him*. In the form of a letter, this tract was transmitted to Galileo on October 25, 1610, while its official publication occurred in Florence in 1611. This event, coupled with the April release of his *Dissertatio* (*Conversation with the Starry Messenger*), marked the turning of the tide in Galileo's favor insofar as the reliability of the telescope was concerned. This little work on the Jovian moons was also the one in which the term *satellite* was first coined by Kepler. Although Kepler's "conversation" with the *Starry Messenger* was written without the aid of a telescope, Kepler's analysis suggested that Galileo's report was fully consistent with Kepler's astronomical experience as well as the laws of astronomical optics about which he had published a major tract in 1604. But now Kepler's *Narration* contained the telescopic confirmation.

For the next four months, in Fall 1610, independent reports confirming the existence of the satellites of Jupiter began to emerge. In England, Thomas Harriot reported seeing them in October; Joseph Gaultier de La Valette saw them in Aix-en-Provence; and Nicolas-Claude Fabri de Peiresc witnessed the satellites elsewhere in southern France in November. By early December, Father Christoph Clavius, the head mathematician at the Roman College, confirmed seeing the moons of Jupiter as well as many previously unseen stars in the Milky Way. At least, these sidereal objects were no longer in dispute.

The Phases of Venus

December 1610 was full of portentous discoveries and excited correspondence. Astronomers and educated laymen all over Europe finally got a chance to confirm Galileo's discoveries. But more surprises came from Galileo himself.

The most theoretically important of these new discoveries concerned Venus, the evening star. It shines so bright in the evening sky that Kepler often referred to it as the planet made of gold, shining brightly no matter how it turned. Close students of the implications of the Copernican system could visualize how different some of the planets would appear to observers on earth as they revolved around the sun or perhaps circled the earth. If Venus revolved around the sun instead of the earth, then its angle to the sun would change as it moved, and Venus would appear horned – that is, only partially illuminated like our moon. If, on the other

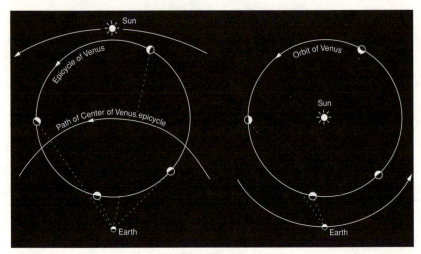

FIGURE 3.3. The phases of Venus. In the Copernican system (right), the full range of Venus's phases can be seen, whereas in the Ptolemaic system (left), the entire set of phases is never seen. From Owen Gingerich, *The Great Copernican Chase* (New York: Cambridge University Press, 1992), p. 100, with permission of Owen Gingerich.

hand, Venus revolved around the earth, then this change in Venus's illumination would display a different pattern (Figure 3.3). Galileo and Kepler had this astronomical visual sense, and so did Galileo's friend and former student, Benedetto Castelli. He was a friar who had joined the Benedictines but remained closely associated with Galileo during the time when Galileo first made his telescopic discoveries.

In a letter sent to Galileo on December 5, 1610, Castelli wrote, "Since (as I believe) the opinion of Copernicus that Venus revolves about the sun is correct, it is clear that she would necessarily be seen by us sometimes horned and sometimes not, although the said planet is at equal distances from the sun, at those times, that is, when the smallness of the horns and the effusion of rays do not impede the observation of this difference."[19]

This was not news to Galileo but, if it were true, then telescopic observations would reveal the phases of Venus mimicking those of the moon. If those phases were observed, they would provide the first direct observational evidence of the heliocentric hypothesis. So Castelli asks, "Now I would like to know from you if you, with your wonderful spyglass, have

[19] As cited in van Helden, *Sidereus Nuncius*, p. 106, and in Stillman Drake's careful chronology of these events, "Galileo, Kepler, and Phases of Venus," *Journal for the History of Astronomy* 15 (1984): 206.

noticed such an appearance, which, without doubt, will be a sure means of convincing any obstinate mind" of the veracity of the Copernican model.

Just when Galileo began his observations is not clear. On November 13, he wrote to Giuliano de' Medici, the Tuscan ambassador to Prague, that nothing much was going on with the planets. Recent study of celestial records suggests that only toward the end of that month did Venus become an interesting planetary object.[20] Shortly thereafter, on December 11, Galileo sent a letter to the Tuscan ambassador in Prague that contained an encrypted message that foretold Galileo's new discovery:

Haec immatura a me jam frustra leguntur, o y.

The message not only had two extra letters that did not fit into Galileo's anagram, but it was also a double message. The translation of the surface meaning of the Latin phrase could be rendered as "these [observations] are at present too young to be read by me"[21] (or "these unripe matters are brought together by me in vain").[22] Read this way, the message suggests that Galileo was not yet sure of what he had seen and that reporting on his observations now was premature. But the transposed message, whose real meaning would not be sent to Kepler and the ambassador in Prague until January, was as follows:

Cynthiae figuras aemulatur mater amorum

or "the mother of love [Venus] emulates the figure of Cynthia [the Moon]." The planet Venus has phases just like those of the moon. This was just as predicted by Copernican theory.

By the end of the month, Galileo's observations of the phases of Venus were finally complete, and he was sure of what he saw. Now he could reply to Castelli: "Know, therefore," he wrote on December 30, 1610,

that about 3 months ago I began to observe Venus with the instrument, and I saw her in a round shape and very small. Day by day she increased in size and maintained that round shape until finally, attaining a very great [angular] distance from the Sun, the roundness of her eastern part began to diminish, and in a few days she was reduced to a semicircle. She maintained this shape for many days, all the while, however, growing in size. At present she is becoming sickle-shaped,

---

[20] Owen Gingerich, "Phases of Venus in 1610," *Journal for the History of Astronomy* 15 (1984): 209–10.
[21] Gingerich, "The Galileo Affair," *Scientific American* 247, no. 2 (1982): 136.
[22] Van Helden, *Sidereus Nuncius*, p. 107n66.

and as long as she is observed in the evening her little horns will continue to become thinner, until she vanishes. But when she reappears in the morning, she will appear with very thin horns, again turned away from the Sun, and will grow to a semicircle at the greatest digression [from the Sun]. She will then remain semicircular for several days, although diminishing in size, after which in a few days she will progress to a full circle. Then for many months she will appear, both in the morning and then in the evening completely circular but very small in size.[23]

In these observations, Galileo saw many consequences for astronomy. The first was that the light of Venus, bright as it is, comes not from the planet itself but from the reflected light of the sun. The second and greater implication was that Venus, as well as Mercury, revolves around the sun, not the earth, and this was strong evidence that the Ptolemaic system was wrong.

But earlier in the 1570s, the Danish astronomer Tycho Brahe had modified his new system to form a so-called geoheliocentric system (Figure 3.4). According to it, the solar system had the planets revolving around the sun, but the sun, with this great entourage of planets, revolved around the earth. This configuration – a massive sun circled by five planets (some larger than the earth) in turn revolving around a much smaller earth – seemed quite implausible to Galileo.

Such dynamics were not compatible with the principles of mechanics as Galileo understood them to work on earth. Still, Brahe's system did retain much of the heliocentric hypothesis. Deciding between it and the Copernican system was not something that could be done easily with a few new observations. Indeed, it would take a century or more for definitive observational evidence to emerge. The universe was much larger than anyone thought, with the consequence that virtually no observations could be made with the telescopes available that would definitively decide between the two systems. When the Jesuit missionaries went to China, they discovered that the Tychonic system could be very useful. It did not violate the official prohibition against spreading the Copernican view, yet it retained elements of heliocentrism and was compatible with the new astronomical data. Observations of solar and lunar eclipses, the satellites of Jupiter, Venus's phases, and the elliptical orbit of Mars were all compatible with the Tychonic system, as the Chinese astronomers set out to show.

---

[23] As cited in van Helden, *Sidereus Nuncius*, pp. 107–8. Drake, "Galileo, Kepler, and Phases of Venus," p. 207.

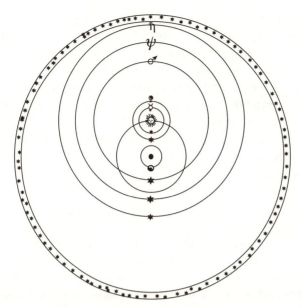

FIGURE 3.4. Tycho Brahe's geoheliocentric system. The five planets are arranged in orbits around the sun, while the sun revolves around the earth. From http://upload.wikimedia.org/wikipedia/commons/9/91/Tychonian.png.

Two days later, on New Year's Day 1611, Galileo sent a letter to the ambassador in Prague that unscrambled for Kepler the anagram mentioned earlier, revealing the moonlike phases of Venus. It was in the form of these letters to Prague that the world would learn of these latest telescopic discoveries of Galileo. Kepler would publish them in the preface to his book on optics, the *Dioptrics* of 1611. In the letter from Galileo that Kepler published, Galileo spelled out the consequences of his discoveries for astronomy. The observed phases of Venus indicated that it must orbit

around the Sun, as do also Mercury and all the other planets – something indeed believed by the Pythagoreans, Copernicus, Kepler, and myself, but not sensibly proved as it now is by Venus and Mercury. Hence Kepler and other Copernicans may glory in having believed and philosophized well, though this is but a prelude, and will continue a prelude, to our being reputed by the generality of bookish philosophers as men of little understanding and practically as fools.[24]

This last disparagement of the "bookish philosophers" might have been true, but it was hardly politic to state it so bluntly. For the gulf between the two camps, those willing to explore nature in all its richness with the

[24] As cited in Drake, *Galileo at Work*, p. 164.

telescope and those who insisted on preserving Ptolemaic assumptions within Aristotelian orthodoxy, would become increasingly pronounced. Galileo's harsh statement would allow his enemies to treat him with even more contempt, ending finally with the Inquisition.

In Galileo's hands, the Dutch-invented spyglass had now become the new discovery machine. In a matter of months, Galileo's improvements to the toylike spyglass, combined with his unsurpassed tenacity as an astronomical observer, had turned the device into an indispensable scientific instrument. For the moment, it was time for Galileo to bask in the glory of scientific success.

Celebrating Success: Rome

As was pointed out earlier, the new year of 1611 began with Galileo's triumphant revelation of the phases of Venus. That had followed other successes, especially the confirmation of his Jupiter observations by other astronomers and especially members of the Jesuit College. Now was a time for celebration.

Galileo had been planning a trip to Rome for some time for the purpose of discussing his discoveries with scholars there. But he became ill in Florence in January and retreated to the Ville del Selve, owned by his friend Filippo Salviati, on the western outskirts of Florence. By March, he had recovered and prepared for the trip. This month and the next two, April and May, would bathe Galileo in four celebratory events that would not be repeated for the rest of his life. But for those fleeting moments, Galileo could justly feel celebrated as a savant for the ages.

The grand duke provided for his transportation while Galileo made regular evening observations with his telescope at a half dozen small cities and towns on the way to Rome. Arriving on March 19, Galileo was put up in the Tuscan Embassy at the Palazzo Firenze. From that moment on, Galileo was swept up in a whirl of activities and events that celebrated his achievements.

After visiting with a cardinal with close ties to the Tuscan court on the first day, Galileo's next stop was at the residence of Christoph Clavius at the Roman College. He also met with two professors, Christoph Grienberger and Odo Maelcote, and their students. There seemed to be elation among all the participants, reveling in the wonders of Galileo's discoveries. "I found," he wrote later in the month, "that all these fathers having finally recognized the truth of the new Medicean planets, have been observing them here continuously for two months and continue to do so. And we have compared [their observations] with mine and found that they agree."[25] This meeting doubtless paved the way for the grand celebration that the members of the Roman College planned for Galileo later in May.

The Malvasia Banquet and the Academy of Linceans

Two weeks later, Galileo was taken by another party of admirers to a banquet at the hillside villa of Monsignor Malvasia just outside the Porta San Pancrazio. This was located on the Roman passageway up the hillock known as the Janiculum. This party of eight had been arranged by the son of a leading Roman family with extensive ties to the church. This was Federico Cesi, the founder of the Accademia dei Lincei, the Lincean Academy. Cesi had founded this scientific society when he was only eighteen, in 1603. From an early age, he had been drawn to observing the plants and wild animals that lived in the hills surrounding the family estate in Aquasparta, a village in Umbria about 560 miles south of Rome. In addition to these living natural wonders, there were mysterious fossils. With three friends, Cesi founded what was to be the first modern scientific society in Europe. Its goals were "not only to acquire knowledge of things and wisdom, and living together justly and piously, but also peacefully to display them to men, orally and in writing, without any harm."[26]

The goals of the academy were truly commendable: "The Lincean Academy desires as its members philosophers who are eager for real knowledge, and who will give themselves to the study of nature, and especially to mathematics. At the same time, it will not neglect the ornaments of elegant literature and philosophy, which like graceful garments, adorn the whole body of science."[27]

It was difficult to say exactly what the underlying program of the Linceans was. They were fascinated by the problem of classifying the natural world in all its riches, but the mysterious fossils of Umbria raised still other questions about heat and light. "What exactly were those fossils – animals, vegetables or minerals? Why did they glow in the dark, or heat up

---

[25] Galileo to Vinta, March 30, as cited in James M. Lattis, *Between Copernicus and Galileo: Christoph Clavius and the Collapse of Ptolemaic Cosmology* (Chicago: University of Chicago Press, 1994), p. 187.

[26] As cited by van Helden, "Accademia dei Lincei," *Connexions* (http://cnx.org).

[27] As cited in James Reston, Jr., *Galileo: A Life* (New York: HarperCollins Publishers, 1994), p. 115.

when dipped in cold water, and what was the nature of the subterranean fires that burned beneath them?"[28]

The three friends joining Cesi, "prince of the Lincei," in founding the order included the Dutch physician Johannes Eck, Francesco Stelluti, and Angelo de Fillis. Eck was a charismatic and mercurial intellect who was held in a Roman prison on charges of murder. Stelluti had been given the nickname, the "Slow One" (*el Tardigrardo*) and de Fillis the "Eclipsed One" (*l' Eclissato*), and both were retained as "chief counselors" to Cesi.

Perhaps the Academy of Linceans (the *Lynx Eyed*) has been overrated because of Galileo's association with it. Although the members of the academy did help Galileo publish two of his important essays – the *Letters on Sunspots* and the *Assayer* – there was little of scientific value in their work apart from Galileo's publications. Its membership never exceeded thirty, and it failed to attain a weighty institutional presence in Italy. Yet, Cesi was a self-confident, open-pursed, and ready dispenser of noblisse oblige who could bring older scholars to join his enterprise.

At the time of its founding, Count Cesi's paranoid father quickly set about disbanding the group before the end of its first year. Believing that Johannes Eck was a Protestant heretic bent on converting his son, he persecuted Eck across Italy, driving him out of the country. The others found the environment so hostile in Rome that they made off for Naples. In that southern city, Cesi eventually met up with the notorious writer, magician, optician, and dabbler in various arts, Giovanni Battista della Porta, of whom we heard earlier. Born in 1535, della Porta was seventy when Cesi met him, but in 1610, Cesi invited della Porta to join the Lincean Academy. Della Porta was later to become controversial for his elusive claim to having invented the telescope before Galileo.

By Spring 1611, Federico Cesi and the Linceans had reconstituted themselves back in Rome. Cesi's father had died, and his mother left a significant inheritance that allowed him to continue his scientific pursuits and other benefactions.

The naturalistic agenda of the Linceans, focused on natural history and local flora and fauna, seems remote from astronomy and mathematics, yet Cesi was quick to grasp the importance of Galileo's *Starry Messenger*. That is why Cesi brought Galileo and the six other guests together for the sumptuous banquet in the villa of Monsignor Malvasia in the Roman hills on the evening of April 14. It was at this meeting that Cesi invited

---

[28] David Freedberg, *The Eye of the Lynx* (Chicago: University of Chicago Press, 2002), p. 69.

Galileo to become the sixth member of the academy. It was also at this celebrated meeting that Galileo's new instrument was given the name "telescope."

At that meeting, Galileo set up his telescope so that the guests could enjoy the wonders of his device. According to one report on the meeting, late in the afternoon, Galileo focused his instrument first on the palace of the Duke of Altemps in the Tuscan Hills. Although it was sixteen Italian miles away, "we readily counted its each and every window, even the smallest." Following that, the telescope was focused on the inscription on the lintel of the Church of the Lateran of St. John Lateran. It had a Latin inscription placed over it by Pope Sixtus V that the guests could clearly see, "so clearly that we distinguished even the periods carved between the letters, at a distance of at least two miles."[29] After dusk, the party saw Jupiter and its satellites. With all these showy successes, Galileo must have been very pleased.

Report of the Jesuit Mathematicians

It has often been thought that the Jesuits as a religious order were hostile to Galileo and his inquiries. Yet, as we saw, when Galileo arrived in Rome, many of the scientifically trained scholars at the Roman College had been following his discoveries with enthusiasm. Indeed, some had been using telescopes to observe the heavens, looking for Jupiter and its newly found satellites for months. In all likelihood, a telescope acquired in the Netherlands in 1608 had been received by Church officials in Rome before Galileo even had one.

Furthermore, one of the most important officials in Rome, Cardinal Robert Bellarmine, who would preside over the Church's censure of Galileo in 1616, decided to have the best mathematicians and astronomers at the Jesuit College examine Galileo's discoveries reported in the *Starry Messenger*. On April 19, while Galileo was still in Rome, Bellarmine sent a letter to the mathematicians at the Roman College seeking their opinion on five questions: Are there many invisible stars out there, unseen by the unaided eye, especially in the Milky Way and other nebulas? Is Saturn really a triple planet with companions on its sides? Does Venus change shape, becoming half lit, horned, and full like our moon? Is it true that

---

[29] Julius Caesar Legalla, *Lunar Phenomena*, as cited in Edward Rosen, *The Naming of the Telescope* (New York: Henry Shuman, 1947), p. 53.

the surface of the moon is rough and ragged? Does Jupiter really have four little moons revolving around it episodically?[30]

Bellarmine, like many other Europeans, needed answers to these questions. "I want to know this because I hear various opinions spoken about these matters, and Your Reverences, versed as you are in the mathematical sciences, will easily be able to tell me if these new discoveries are well founded, or if they are rather appearances and not real."[31] The scholars, Christoph Clavius, Christoph Grienberger, Odo Maelcote, and Giovan Lembo needed only five days to come back with their answers, for they had been using telescopes, some of their own making, since December 1610. According to a letter written by Grienberger in January 1611, Giovan Lembo had solved the problem of making telescopes powerful enough to see the satellites of Jupiter months earlier.[32] Clavius had also received a better telescope from Galileo's friend and Venetian merchant Antonio Santini, who was probably the first person after Galileo to see the satellites of Jupiter in Italy.

So there was not much doubt but that the scholars from the Roman College could confirm all of Galileo's discoveries: the rough surface of the moon, the satellites of Jupiter, the phases of Venus, and the existence of many tiny stars in the Milky Way, along with additional stars in the constellation of Pleiades. Of course, they could only confirm that Saturn had an odd shape, often described as a teacup with handles, not the existence of companions. Likewise, the panel of experts acknowledged the existence of thousands of tiny stars in the Milky Way and other nebula but could speak only with high probability that all nebula were composites of such stars invisible to the naked eye.

Regarding the elemental facts of observation, the Jesuit panel agreed with Galileo. What they could not confirm was the veracity of the Copernican sun-centered hypothesis; for that, more evidence of a different sort was needed.

Induction into the Lincean Academy

On the next day, April 25, Galileo met again with Count Cesi and the five existing members of the Academy of the Linx Eyed. This was the day of Galileo's formal induction as the sixth member of this historic but

---

[30] In Lattis, *Between Copernicus and Galileo*, p. 190.
[31] Ibid.
[32] Ibid., p. 186.

short-lived scientific society. Given the hostility that existed in some quarters in Italy, but outside the Roman College, the academy proved useful in getting some of Galileo's controversial works published. His observations on sunspots were letters written to the German nobleman Mark Welser, and the latter was instrumental in getting the letters into print. These observations by Galileo claimed that sunspots were actual irruptions on the surface of the Sun – not passing clouds but constantly changing solar manifestations. These ideas disturbed the view that celestial bodies, such as the sun, were perfectly circular and unchanging. So on this day, Galileo was sworn in as the sixth member of the Accademia dei Lincei. Although the Lincean Academy proved to be of short life, Galileo proudly included his Lincean membership among his titles listed in his books.

Another important connection with the Linceans was the Swiss scholar Johannes Schreck. He was a gifted mathematician and naturalist who attended Galileo's lectures back in Padua. He was inducted a few weeks after Galileo as the seventh member of the academy, but his membership in the academy lasted only a few months, until he joined the Jesuit order. He then changed his name to Father Terrentius. Seven years later, he was on board a missionary ship from Lisbon, bound for China. While in China, he kept up a one-sided correspondence with Galileo, who declined to respond.

Feted at the Roman College

The capstone of Galileo's triumphant visit to the capital city of Italy in 1611 was the large public gathering of the faculty and students of the Roman College celebrating Galileo's astronomical discoveries. This boisterous session of May 8 was presided over by the scholar from Brussels, Odo Maelcote, officially an "extraordinary lecturer" in mathematics at the college. Galileo had shown him sunspots earlier during his visit to Rome, and Maelcote had served on the expert panel of astronomers assembled by Cardinal Bellarmine to evaluate Galileo's discoveries. That report was suitably conservative as an official document. But now, in his festive lecture on "The Starry Messenger at the Roman College," Maelcote was enthusiastic. Although others on the panel remained skeptical regarding the rough, cratered surface of the moon, Maelcote was bold to claim that "the lunar body is bounded by a figure that is in no way perfectly spherical, but is a rather rough and uneven surface."[33] In a nod

33 Ibid., p. 194.

to other members of the panel, he acknowledged their reservations while averring that it was enough for him to describe what he had seen through the telescope while enlisting the opinion of others: "You be the judge on the outcome of the matter."

Maelcote included in his address both the telescopic revelations of *Starry Messenger* and Galileo's additional discoveries regarding Saturn and Venus that had been published by Kepler. Regarding Saturn, he could add little except to paraphrase Galileo's own descriptions. On the question of Venus, Odo Maelcote sided wholly with Galileo. Some members of the audience writing later about this historic session mention "demonstrations" that most probably showed the path of Venus around the sun and what should be seen of Venus at various points in its orbit. "Behold," he wrote in the lecture, "now it is clear to you that Venus is moved around the sun (and the same can indubitably be said of Mercury)." Pushing further, Maelcote joined Galileo in stating that "it is undoubtable that the planets shine only by the light borrowed from the sun, which I do not judge to be true of the fixed stars."[34]

One can easily imagine the jubilation that arose from this gathering of volatile students and bemused faculty. This was truly an unsurpassed moment in the history of astronomy in Italy. With this extraordinary intellectual celebration of Galileo's telescopic discoveries at the famous Gregorian University in the heart of Rome, Galileo's findings were as accredited as any such discoveries become in the early phases of scientific research. His insistent championing of the Copernican system, however, would bring him into fateful clashes with resistant Church authorities. But for posterity and the world, Galileo's celestial discoveries with the telescope were there for all to see and accredited by the members of this famous college.

New Recruits for China?

One of Maelcote's auditors at that singular scientific ·celebration was Adam Schall von Bell, a German student from Cologne studying at the Roman College. He would join the Jesuit order in the same year. And in a few short years after that, he would accompany Schreck (known as Terrentius) on the ship from Lisbon destined for the China mission in April 1618. Like Schreck, he was taking cutting-edge astronomical knowledge to the Far East. That knowledge was not primarily of the

[34] Ibid.

Copernican system, now three quarters of a century old; it was rather the technical and practical knowledge of hands-on astronomy. It was knowledge of the most recent telescopic discoveries and firsthand experience with telescopes. For knowledge about the construction of telescopes was not difficult to find among the faculty members at the Roman College. Beyond that, Schall, Schreck, and their recruiter returned from China, Nicolas Trigault, knew a good deal about *Kepler*'s new astronomical works, especially his work on astronomical optics. Still shrouded in mystery is the fact that the returning recruits had in their possession a "Keplerian" telescope sent to Schreck by Cardinal Borromeo of Milan.[35] Between 1611 and the 1618 departure for China, Schall, Schreck, and other students of astronomy had occasion to examine and use multiple versions of Galileo's new scientific instrument. This was clearly scientific knowledge not available in other parts of the word. Consequently, all this firsthand knowledge, coupled with Schall's scientific and managerial gifts, would propel him to the directorship of the Chinese Bureau of Mathematics and Astronomy by midcentury. Such an appointment seems extraordinary to us, but it carried its own perils.

[35] Pasquale D'Elia, *Galileo in China* (Cambridge, MA: Harvard University Press, 1960), p. 28.

4

The "Far Seeing Looking Glass" Goes to China

One Adam having driven us out of Paradise; another has driven us out of China.[1]

The Jesuit Mission in China

The earliest certain transmission of the telescope to Asia occurred in 1613, when a Dutch sea captain brought it to Japan.[2] The question of whether the representatives of the king of Siam took a spyglass back to Thailand in 1610 when they returned is still unanswered, as their ship was wrecked in a storm somewhere along the coast of Indonesia. Nevertheless, telescopes were taken to Thailand by the Jesuits soon thereafter.[3] As we shall see later, the British ambassador, Sir Thomas Roe, brought a telescope to the Mughal court of Jahangir in 1615. In the same year, however, Chinese scholars could read a preliminary account of Galileo's celestial discoveries written and translated into Chinese by a Portuguese Jesuit.[4] By 1619, a "Keplerian" astronomical telescope arrived in China with a new batch of missionaries. The Jesuit scientists Johannes Schreck (known among

---

[1] A joke attributed to Jesuit officials banished to Macao when Adam Schall was arrested by Chinese authorities in the Rites Controversy of 1665. As cited in Jonathan Spence, *To Change China* (London: Penguin Books, 1980), p. 22.

[2] Shigeru Nakayama, *A History of Japanese Astronomy* (Cambridge, MA: Harvard University Press, 1969), p. 100n49; Peter Abraham, "The History of the Telescope in Japan," http://home.eurpopa.com/~telescope/tsjapan.txt.

[3] S. J. Peitro Cerutti, "The Jesuits in Thailand – Part 1 1607–1767," http://www.sjthailand .org/english/historythai1.htm.

[4] Joseph Needham, *Science and Civilisation in China* (hereafter *SCC*; Cambridge: Cambridge University Press, 1959), 3:444–45.

Jesuits as Terrentius) and Johann Adam Schall had arrived in China with firsthand experience using the Dutch or Galilean telescopes in Europe at the moment of Galileo's discoveries. But the Jesuit mission in China had already been launched before Matteo Ricci arrived in 1583.

For more than three decades, Ricci and his followers had been laying the groundwork for bringing European science and astronomy to China. That task, as it turned out, was far more complicated than anyone imagined. It was more complex than transmitting the telescope and related parts of Western astronomy to other parts of the world. Long-distance spying, as could be done with the Dutch invention, would surely raise issues in the Muslim world as well as in China. But China's intellectual walls were anchored in unique and highly articulated ancient patterns of thought that were always ready to be recovered and reimposed.

The longer the Jesuit missionaries were in China, the more they realized that bringing Western ideas – not only Christianity but also Western science itself – to China entailed a colossal civilizational encounter. This could be negotiated only by undertaking a Herculean metaphysical struggle. Only through such underworker efforts as philosophers can contemplate could intellectual space be made within Chinese thought that might allow the insertion of a worldview quite foreign to the Chinese outlook.

Matteo Ricci, who died in China in 1610, grasped this early on in his mission to China. His mission, naturally enough, was to bring Christianity to China, but he quickly discovered that the imposition of the spiritual life of Christianity could be aided by transmitting and translating works of Western science and philosophy, as well as mathematics, into Chinese. That was why, only five years after the appearance of Galileo's *Starry Messenger*, a Chinese-language publication appeared containing the main ideas of Galileo's discoveries along with a drawing of Saturn and its handles, without mentioning the name of Galileo himself. Western science was to be the vanguard tool that Ricci and his followers hoped would enable them to convert the Chinese to Christianity.

Ricci's Plan
Ricci arrived in China in 1583 and spent the rest of his life setting up the Jesuit mission there. Just getting from Macao to the mainland took him a number of months. To get from Canton to Beijing in 1601, it took him nearly nineteen years. From then onward, Ricci's grand plan to convert the Chinese coalesced around the idea of importing and translating a huge library of Western scientific, mathematical, and philosophical writings.

During his early years, Ricci had been more successful making eclipse predictions than his Chinese hosts. Ricci's command of mathematics, learned at the Roman College under the tutelage of Christoph Clavius, greatly impressed Chinese officials, who sent their sons to him for tutoring.[5] Ricci had also mastered Chinese so that he could hear confessions in Chinese and write missionary tracts in the language of China. By the time Ricci reached Beijing, he could write:

Because of my world-map, my clocks, spheres and astrolabes and the other things I do and teach, I have gained the reputation of being the greatest mathematician in the world. And although I have no book on astronomy, I am able with the aid of certain Portuguese calendars and periodicals, to predict the eclipses more accurately than they do.[6]

Ricci had been most fortunate in meeting the brilliant Chinese scholar Xu Guangqi in 1601. Xu converted to Christianity in 1603 and thereafter became known as Dr. Paul. In the same year, he passed his last Civil Service Examination, granting him the highest scholarly title, the *jinshi*, in 1604.[7] Shortly after, he began studying Euclid's *Elements* with great intensity. Xu became convinced that Western mathematics was far more systematic and logically constructed than what was then known of ancient Chinese mathematics. As a result, both Ricci and Xu Guangqi hit on the idea of translating foundational works of Western science and mathematics into Chinese. This corpus would become a device that could be presented as an authoritative statement to persuade the Chinese that Christianity and its teachings were correct and superior to Chinese religious ideas, just as Western math and science were presented as demonstrable truths by Ricci and Xu Guangqi. The strategy worked. Despite many setbacks during the seventeenth century, the missionaries fell back repeatedly on astronomical demonstrations to prove the greater accuracy and value of Western science and astronomy. Even Chinese scholars who objected to adopting the "new" or "Western" method recognized its superior predictive as well as explanatory power.[8]

---

[5] Peter M. Engelfriet, *Euclid in China* (Leiden, Netherlands: Brill, 1998), pp. 64–67; Jonathan Spence, *The Memory Palace of Matteo Ricci* (New York: Penguin, 1984).
[6] John Dunne, *Generation of Giants* (Notre Dame, IN: University of Notre Dame Press, 1962), p. 210.
[7] Catherine Jami, Peter Engelfriet, and Gregory Blue, *Statecraft and Intellectual Renewal in Late Ming China: The Cross-Cultural Synthesis of Xu Guangqi* (Leiden, Netherlands: Brill, 2001), p. 403.
[8] John B. Henderson, "Ch'ing Scholars' Views of Western Astronomy," *Harvard Journal of Asiatic Studies* 46, no. 1 (1986): 121–48.

Another Chinese scholar who had passed the Palace Examination became a great admirer of Ricci's, though he did not convert to Christianity. In a preface to Ricci's famous tract, *The True Meaning of the Master of Heaven* (1604), Feng Yingjing (1555–1606) claimed that "Ricci traveled 80,000 *li* [to China]. He measured the heights of the nine heavens and the depths of the nine oceans without the slightest error. He has already fathomed forms and figures [in the sky] which we never have. Having such reliable evidence in these matters, [his teachings about] divine principles . . . ought to contain no falsehoods."[9]

With that goal in mind, Xu Guangqi took up the task of translating Euclid into Chinese. The project was completed (the first six chapters) and the work published in 1607. Both Ricci and Xu wrote prefaces for it.[10] In the meantime, Ricci had begun the project of translating a textbook on astronomy. This was the thirteenth-century text by Sacrobosco (d. 1256) that presented the universe as a set of nested spheres. Such a work was essential in Ricci's view because the Chinese were still located in a flat, rectangular world with a canopy overhead. Furthermore, *On the Sphere* was the Ptolemaic model still used by Europeans and had been revised to some degree by Ricci's teacher, Christoph Clavius, the new "Euclid" and leading mathematician at the Jesuit College in Rome.[11]

With the completion of Euclid's *Elements* in Chinese, it was thought advisable to rework Ricci's translation of *On the Sphere*. This was taken up by Manuel Diaz (1564–1659), who had arrived in China in 1602. But, in addition to retranslating Clavius's revision of *On the Sphere*, Diaz added an appendix explaining the new telescopic discoveries of Galileo that had been reported in Galileo's *Starry Messenger* of 1610. Although the Jesuits in Rome had, by April 1611, confirmed most of Galileo's telescopic discoveries, as reported in Chapter 3, Galileo's championing of the Copernican system led to intense controversy and, after 1616, official rejection. Consequently, Diaz did not mention Galileo by name but did discuss clearly Galileo's findings. He also provided a diagram showing the handles of Saturn. These were the *rings* of Saturn, but not for several decades would Christopher Wren and Christiaan Huygens (in 1656) correctly identify the handles of Saturn as rings. Galileo referred to these objects as the companions of Saturn, believing, as he did then,

---

[9] As cited in Willard Peterson, "Learning from the West," in *The Cambridge History of China* (New York: Cambridge University Press, 1998), p. 807.
[10] Engelfriet, *Euclid in China*; Jami et al., *Statecraft*, p. 34.
[11] Engelfriet, *Euclid in China*, chap. 4.

that Saturn, like Jupiter, had satellites circling it. (It did, but they were not visible in Galileo's telescope.)

Diaz's discussion of the odd shape of Saturn signifies how well informed many of the missionaries were because the first knowledge of Galileo's sighting of Saturn and Venus was published in 1611 by Kepler as letters he had received from Galileo.[12] As we saw in Chapter 3, Kepler had been so impressed by Galileo's telescope that he set about writing a new work on optics, explaining the geometry of telescopic vision published under the title *Dioptrics*. It was there that Kepler explained the optics of convex lenses used in his new version of a telescope.[13] In the preface to that work, Kepler published several letters by Galileo in which his discoveries regarding Saturn as well as the phases of Venus were first revealed, although in anagrams. Galileo had intended to publish a new edition of the *Starry Messenger* that would reveal these discoveries, but it remained for Kepler to do the honors.

Diaz's discussion of the telescope and Galileo's findings in Chinese have been known to historians and Sinophiles for at least half a century (Figure 4.1). Two pages of the pamphlet on Galileo's discoveries were translated into English by Joseph Needham and are well worth printing here:

Most of the above observations (on lengths of days and nights, eclipses, etc) were made with the naked eye. But such vision is short: how could it attain to even one ten thousandth part of the mysterious principles of the vast Heavens? Quite recently, however, a famous scholar of the West, particularly learned in calendrical science and himself an observer of the sun, moon and planets, deploring the weakness of his eyes, has made an ingenious instrument which helps them. With this device an object one foot in size can be seen at a distance of 60 *li* as clearly as if it was in front of one's eyes. The moon, seen through [this telescope] seems a thousand times larger than usual.

[Left hand page of the Chinese text]:

Venus appears as large as the moon; its light increases and decreases just like the moon's. Saturn as shown in the above diagram, seems to have a rounded shape like that of an egg, with two small stars on each side, but whether or not they are attached to it we do not know. Jupiter can be seen always surrounded by

---

[12] Edward Stafford Carlos, *The Sidereal Messenger of Galileo Galilee* (London: Dawsons of Pall Mall, 1959).

[13] Albert van Helden, "The Invention of the Telescope," *Transactions of the American Philosophical Society* 67, pt. 4 (1977): 1–67.

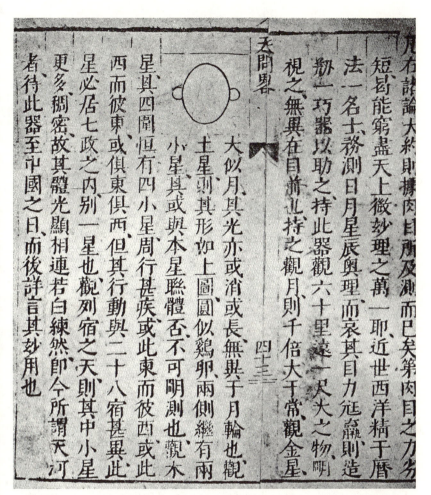

FIGURE 4.1. A Chinese brochure of 1615 advertising the Diaz's appendix describing Galileo's discoveries with a diagram showing Saturn's handles. Owned by the author.

four small stars, continually revolving around it at great speed, one at the west and one at the east, or vice versa, and some times all on the west side or all on the east side – (in any case) their movement differs greatly from that of the 28 *hsiu* [heavenly lodges]. For these stars must remain in the regions of (each of) the seven planets, and form (indeed) a special sort of star. Then, when one looks (with the telescope) at the great constellations in the firmament one sees an immense multitude of small stars closely crowded together, hence the light from their bodies seems to form a white stream, which we call the Milky Way.

As soon as one of these instruments arrives in China we shall give more details on its marvelous uses.[14]

Considering their great distance from Europe, it is remarkable that just three years later, in April 1618, a boatload of new missionary recruits was on its way from Lisbon to China and, in all likelihood, had a Keplerian telescope on board that had been sent to Schreck by Cardinal Borromeo of Milan.[15] This telescope ended up in the hands of the emperor, but it seems to have led an underground existence for a number of years. Some of those who used it reported seeing an inverted image that is the signature pattern of the Keplerian telescope. It was because the whole science of astronomy was shrouded in secrecy by the emperor and his officials that instruments like the telescope were not supposed to be in the hands of unauthorized persons, much less used to make observations of the heavenly domains. Later on, in the 1630s, this charge of illicit use of the telescope would be leveled against the missionaries and Li Tianjing, the director of the Bureau of Mathematics and Astronomy who succeeded Xu Guangqi.[16]

Send Me an "Astrologer"

From the outset, Ricci had known that he needed missionary astronomers to carry out his grand plan. He repeatedly petitioned Rome to send him an "astrologer," as astronomers were often termed at the time. In 1613, the head of the Jesuit mission, Niccolò Longobardo, had sent Nicholas Trigault back to Europe to procure new recruits trained in astronomy and mathematics who could explain Western scientific ideas to the Chinese. Trigault also set about collecting a great library of scientific and philosophical writings for the China mission. He first went to Rome and then canvassed Europe for books and donations for the China mission.

---

[14] Needham, *SCC*, 3:445; M. Pasquale D'Elia, *Galileo in China* (Cambridge, MA: Harvard University Press, 1960), p. 18.

[15] D'Elia, *Galileo in China*, p. 28.

[16] Ibid., p. 28. Keizo Hashimoto reports in *Hsü Kuang-chi'i and Astronomical Reform: The Process of the Chinese Acceptance of Western Astronomy 1629–1635* (Kansai, Japan: Kansai University Press, 1988) that the Chinese who used the telescope spoke of the inverted image, the distinctive feature of a Keplerian telescope. Also Keizo Hashimoto, "Telescope and Observation in Late Ming China," *Journal of the Division of Social Sciences, Kansai University* 19, no. 2 (1988): 91–101. Li Tianjing, in his memorial of December 19, 1634, says that "this instrument, brought by [Giacomo] Rho and [Johannes] Schall from their kingdoms, was afterwards decorated for presentation to you." D'Elia, *Galileo in China*, p. 49.

When Trigault departed on his return trip to China, he had twenty-two recruits, four among whom were outstanding mathematicians: the Italian Giacomo Rho, the Bohemian Wensceslaus Pantaleon Kirwitzer, the German Johann Adam Schall von Bell, and the Swiss Johannes Schreck.[17] The latter was the outstanding scientist and scholar who had become the seventh inductee, following Galileo, into the famous Accademy dei Lincei, the Academy of the Linx Eyed, back in May 1611.[18] While in China, Schreck carried on an extensive correspondence with Europeans in an effort to get better planetary tables for eclipse predictions. Galileo offered no help, while Kepler did supply Schreck with parts of his new *Rudolphine Tables*. These were very complex compilations of the coordinates of the sun, moon, and planets that enabled astronomers to calculate the future position of the celestial bodies and hence to make predictions of eclipses and other stellar conjunctions. Later *Supplements* to Kepler's tables were received by the missionaries and held in their Beijing library. Most probably they were also used to make solar and lunar predictions later in the century. For throughout their work in China, the missionaries assigned to the Bureau of Mathematics and Astronomy had to work out the daily and monthly positions of the stars and planets – that is, ephemerides. For that they needed planetary tables available only from Europe.

With the arrival in China of the rich cargo of scientific and philosophical books, the missionaries began publishing works in Chinese fully explaining the telescope, other mechanical devices, and European astronomy. There has been considerable debate regarding how much the missionaries revealed about the new Copernican system in the early days, but by the middle of the century, the Copernican system had been presented in one form or another, especially in Kepler's *Epitome of the Copernican System* (1627), which was the first full exposition of the Copernican system in the early seventeenth century.[19] Consistent with their missionary

[17] Ibid., p. 25.
[18] Edward Rosen, *The Naming of the Telescope* (New York: Abelard-Schuman, 1947); David Friedberg, *The Eye of the Lynx* (Chicago: University of Chicago Press, 2002); Richard Westfall, "Galileo and the Accademia Dei Lincei," in *Novita Celesti e crisi del Sapere*, ed. P. Galluzi (Florence, Italy: Giunti Barbera, 1984).
[19] See Johannes Kepler, *Epitome of Copernican Astronomy* (1618–21). Nathan Sivin, in "Copernicus in China," *Colloqui Copernicana* 2 (1973): 66–122, claims that knowledge of the Copernican system was too incomplete to be understood for anything more than a short period of time. But this seems implausible given all the books on astronomy translated by the Jesuits. See the very short list of such works in Peterson, Note 81. Hashimoto's *Astronomical Reform* also suggests a different conclusion than Sivin.

enterprise, the European missionaries, working with highly qualified Chinese scholars, set about providing all the basic tools that were needed to put Chinese astronomy on the same observational footing as European astronomy of the early seventeenth century.

The Chinese Bureau of Mathematics and Astronomy

The Chinese Bureau of Mathematics and Astronomy (hereafter for convenience, Bureau of Astronomy) was located within one of the great bureaucratic divisions of the Chinese government called the Ministry of Rites. That ministry had a mandate to administer state ceremonies, rituals, and sacrifices but also the civil service examinations used to recruit scholars to government service. Within that large organization, the Bureau of Astronomy was responsible for all astronomical activities, which included astronomical observation, the determination of planetary movements, calendar making, and divination. This latter activity involved determining lucky and unlucky days for a whole range of human activities such as taking a trip, buying or selling property, arranging a marriage, and scheduling a burial and its siting.

The whole cosmological system of the Chinese was embedded in a moralistic outlook that was grounded in a metaphysical view linking the organic evolution of the heavens and human events, especially the activities of the royal family. Signs in the heavens, it was thought, could indicate what would happen on earth, whereas the misbehavior of Chinese officials, especially the emperor, could result in famines, earthquakes, droughts, or hail storms. The mere occurrence of such events suggested that things on earth were out of order and were displeasing to the spirit world. Similarly, anomalies in the heavens were taken as portents of the future; hence, astronomers looking at the sky might divine future events that should only be known to the emperor.

From this point of view, various celestial events – eclipses, shooting stars, conjunctions of planets and stars – could be seen as heavenly warnings. As such, they could be interpreted as heaven-sent reports on the state of human affairs. For these reasons, there was a powerful inclination to keep sky watching confined to the official bureaucracy whose personnel were required to keep a constant vigil on the skies and to report all unusual events to the emperor himself.

The main product of the Bureau of Astronomy was the annual calendar: some were designed personally for the emperor, others for the general Chinese population located in the various regions of the country.

Those calendars required making meteorological predictions all across the country, which could be a dangerous thing to get wrong. Each year, the presentation of the calendar was celebrated in a very public ceremony in front of the emperor and all his officials, as the Jesuit astronomers were to witness firsthand as creators of those calendars.

Within this context, the Chinese were interested in improving predictions of obvious events such as solar and lunar eclipses and the conjunctions of planets and stars. This was clearly an astrological enterprise. Searching for the true scientific or cosmological theory of the universe was an altogether secondary consideration.

The most severe Chinese critics of the new or Western astronomy based their attacks on the clear priority to preserve Chinese astrological traditions. They opposed the very idea of importing something that they considered foreign and alien. This was so even though they acknowledged that the Western system yielded more accurate predictions than the Chinese methods.

It was only late in the day, after the Christian astronomers gained access to the Bureau of Astronomy, that they realized how deeply embedded their work was in the Chinese traditions. Those traditions were seen by the Jesuits as unalterably heathen and superstitious as well as out of step with Enlightenment values then emerging in Europe. Consequently, both parties felt themselves compromised by the web of contrasting goals and desires that they tried to satisfy by working together. Despite all their misgivings about Western science and culture, the Chinese had known for some time that their system badly needed reform, and the Jesuit astronomers, by the 1620s, had the knowledge and personnel to carry out the reforms.

On the other side, the Jesuits found the transmission of Western science, especially mathematical astronomy, an essential tool for garnering the attention of the Chinese while diverting them toward Christianity but also helping them to reform the Chinese calendar. In the end, however, Chinese calendar making was above all an astrological enterprise centered on making predictions that came down to designating lucky and unlucky days for the whole daily panoply of Chinese culture. That necessarily involved the actions of the emperor down to the ordinary village peasant. In other words, whereas the Jesuits were concerned with improving the scientific predictions of Chinese astronomy, the Chinese wanted accuracy of auspicious days, accompanied by meteorological prescience for all the regions of China. If the Jesuits were to bring about reform of the Chinese system, they had to participate in this superstitious calendar making that involved not only the selection of auspicious days but also the location and siting of imperial rituals.

By midcentury, the Jesuit officials assigned to help out in the Bureau of Astronomy tried to separate themselves from those activities, but it was not always possible, as Adam Schall would learn. A misstep along those lines of forecasting and site selection could have deadly repercussions for the very men who found all of it to smack of heretical ritual.

Explaining the Telescope

When the newly recruited missionaries arrived back in China (in Macau, actually, in 1619), the status of the Jesuits had fallen dramatically from the time of Ricci's death in 1610. At that time, Ricci and the missionaries were seen by Chinese officials as experts who could be used to reform Chinese astronomy. Ricci and the Chinese officials knew that the system needed to be reformed, and they frequently appealed to him to help carry out the reform project during the two decades before Ricci arrived in Beijing. Because Ricci had made such a strong impression on various officials in southern China with his mathematical skills, when he tried to tell them that he was not capable of managing the reforms, he reported, "They do not believe me."[20]

In December 1610, seven months after Ricci's death, Xu Guangqi arranged to test the predictive powers of the European and Chinese astronomical systems. The result was a defeat for the astronomers in the Chinese bureau of astronomy, as their predictions were off by nearly half an hour.[21] Xu Guangqi then petitioned the emperor to entrust the calendar reform project to the missionaries. The emperor agreed, with the result that the Jesuits Sabatino de Ursis and Diego de Pantoia, assisted by Xu Guangqi, immediately began to work on the reform. The starting point was the translation of the books then existing in China explaining European planetary theory and other Western ideas. This astronomical defeat of the Chinese officials began the long battle of opposition to adopting Western astronomy, mathematics, and science. By 1616, the

---

[20] John Dunne, *Generation of Giants* (Notre Dame, IN: University of Notre Dame Press, 1962), p. 210. On the need for reform, see Willard J. Peterson, "Calendar Reform Prior to the Arrival of Missionaries at the Ming Court," *Ming Studies* 21 (1986): 45–61; Thatcher E. Deane, "Instruments and Observation at the Imperial Astronomical Bureau during the Ming Dynasty," *Osiris* 9 (1994): 127–40; and B. Elman, *On Their Own Terms, 1550–1900*. (Berkeley: University of California Press), 2005, pp. 76–77.
[21] Dunne, *Generation of Giants*, p. 114; D'Elia, *Galileo in China*.

Jesuits had come under fire, with several arrested and carted off to jail in cages. This was the result of the persecution of the missionaries by the Chinese official Shen Que. Although he was only a provincial official, his antimissionary campaign was unrelenting and successful in temporarily displacing the Jesuits.[22]

Consequently, when Trigault arrived back in Macau (in 1619) with the new recruits, he and they discovered that it was almost impossible to get into China proper and had to resort to subterfuge to gain access to the mainland. Shortly thereafter, three groups of two each secretly entered China from Macau between 1619 and 1621. They followed a path to a safe haven in Hangzhou.[23] Schall and Schreck were among those who met with Niccolò Longobardo, the head of the mission, and then traveled secretly on to Beijing in 1623. There they began providing technical assistance to the missionaries who had finally been released from prison and allowed back into the city.

Along with the twenty-two new missionary recruits that Trigault brought back to China on the 1618 ship from Lisbon, there was an extraordinary collection of European books on science, philosophy, technology, and other areas that Trigualt had collected during his return visit to Europe. The number of these books has been given as 7,000 by many sources.[24] This rich cargo was at the heart of the Jesuit plan, originating with Matteo Ricci, to import and translate into Chinese a very substantial portion of the Western tradition of science and philosophy. When the Jesuits regained their footing, a major effort was made to carry out that program. Foremost for present purposes were those books on astronomy and the newly authored works explaining the telescope, its construction, and use. Schall and Schreck were the primary authors of many of these early works of the 1620s, designed to explain the telescope and other mechanical devices from Europe.

Seven or eight years after the missionaries reestablished their standing in China and became fully ensconced as advisers to the emperor and the Chinese Bureau of Astronomy, Schall presented some of the newly imported European instruments (and possibly the new Keplerian telescope) to the emperor. The latter was also presented with an index of

---

[22] Dunne, *Generation of Giants*, chap. 8; Elman, *On Their Own Terms*, pp. 92–93; Jacques Gernet, *China and the Christian Impact* (Cambridge: Cambridge University Press, 1990), p. 61.

[23] Engelfriet, *Euclid in China*, p. 335.

[24] D'Elia, *Galileo in China*, pp. 94–95.

Western astronomical writings.[25] During the next decade or so, the missionaries and their Chinese collaborators published a number of books explaining the telescope, what is seen with it, how to construct a telescope, and what the new telescopic observations tell us about the heavens and planetary orbits.

In addition to the writings of Schall and Schreck, the Chinese convert Wang Cheng (known as Dr. Philip) and Giacomo Rho published important works. Adam Schall's small book of 1626, *Treatise on the Telescope*, was the first exposition on the device itself: he lauds the virtues of the telescope and tells the reader what can be seen with it and how to make one. In all these works, the idea that the telescope is the most important astronomical discovery device ever invented is repeatedly stressed. According to Schall:

It is easy for all to see near and great objects, but it is difficult to see those which are little and distant. Now with the telescope there is no longer either small object or distant object. In short, with the telescope both heaven and earth become part of our visual field. In the mountains or on the sea, with this, one can see ahead of time incursions of brigands or pirates. It is besides of great use to everybody. It is an instrument which unexpectedly renders sight acute, and it is the joy of the [scholar].[26]

Schall then reviews more details of Galileo's discoveries: the appearance of the moon's surface, its mountains and valleys. He describes the phases of Venus as seen through the telescope as well as the oval appearance of the sun in the morning that he ascribes to refraction caused by "the fine dust which rises in the air." Then he mentions sunspots. Jupiter's satellites are mentioned, along with Saturn's appearance, the many new stars found in the constellation of Pleiades ("not just seven but more than thirty"), and the Milky Way, which can be seen with clarity using the telescope. Schall also includes a picture of a lovely Chinese-style telescope, probably the first such depiction in China (Figure 4.2). There are several more illustrations in the book depicting Saturn, Orion, the cluster known as Praesepe, the phases of Venus, the Great Bear, and so on. Pushing on, Adam Schall discusses lenses and the construction of telescopes in nine sections.[27]

---

[25] Engelfriet, *Euclid in China*, p. 335.
[26] In D'Elia, *Galileo in China*, p. 34.
[27] Ibid., pp. 36–37; also see Needham *SCC* 3, Figure 186, and Hashimoto, *Astronomical Reform*, p. 175.

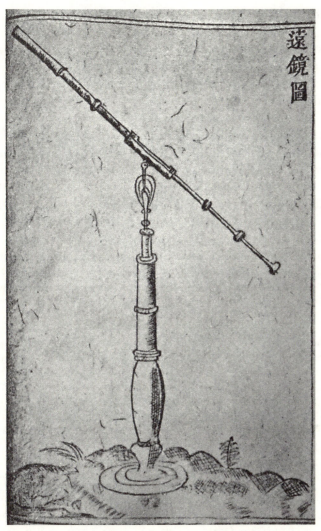

FIGURE 4.2. The first Chinese picture of the telescope, the "far-seeing optick glass" that was depicted by Schall in his book *The Telescope* in 1626. Note that the device appears to be made of many different interlocking tubes that can be adjusted for length by screws, according to Wang Cheng.

In the following year, 1627, the convert known as Dr. Philip (Wang Cheng, 1571–1644), working in collaboration with Schreck, published an impressive little book called *Diagrams and Explanation of the Marvelous Devices of the Far West*. Dr. Philip was another exceptional scholar who had passed his Metropolitan Examination in 1622, and in his small treatise, he draws on Schall's earlier description of the telescope, its components, and its construction. However, at that time, in the 1620s and 1630s, the Chinese language did not have the technical terms equivalent to such Latin terms as *convexum* and *cavum*. For example, in the Latin edition of the *Sidereus Nuncius* (of 1610), Galileo spoke of one lens as *sphaerice convexum* (spherically convex) and the other as *cavum* (concave), which he placed at the two ends of his lead tube.[28] Nevertheless, Schall and the other missionaries were able to convey the shape and meaning of the two different lenses in their descriptions of the telescope. Consequently, Wang Cheng (Dr. Philip) gives the explanation that can be found in a number of later Chinese sources.[29] He tells the reader:

First place a lens that is made of glass which seems flat but in fact is not at the mouth of the tube. The lens is called mouth-piece, also called center-protruded lens [convex], or front lens; next place a lens that is a bit curved-in [concave], also named eye-piece, center-recessed lens, or rear lens, at the back of the tube; if the proportion of the distance between the two lenses corresponds, one can see things.

There are only two pieces of lens, but the number of tubes can be added as desired. One tube fits into the other, and the tubes can be shortened or lengthened. The tubes can be fastened with screws which allow free movement of the telescope up and down or left and right. Viewing is done using only one eye. An object of 60 *li* [miles] seems two hundred steps away. One can thus observe the moon, Venus, the sun, Jupiter, Saturn, and star constellations. When one observes the sun and Venus, one adds thereto a dark green lens. Alternately, place a piece of white paper under the telescope to observe the sun.[30]

---

[28] This page and its translation are contained together in van Helden, "Invention of the Telescope," p. 45.

[29] J. J. L. Duyvendak, "Comments on Pasquale D'Elia's, *Galileo in China*," *T'oung Pao*, second series, 38 (1948): 326–27.

[30] The passage quoted is from Fang Hao, *Studies in the History of the Relations between China and the West* (Peiping, China: Institutum Sancti Thomae, 1948), p. 293, who is quoting from Wang Cheng's book. This important discussion of Wang Cheng's work based on Father Fang Hao's research, published a year after D'Elia's original study, was added to D'Elia's account by the translators on p. 38 of *Galileo in China*. However, the translators mistranslated the key terms *chung wa ching* (glass with sunken center, or concave lens) and *chung gao ching* (glass with a protruded center, or convex lens). The result was the reversal of the proper lens arrangement in a Dutch or Galilean telescope, which must place the convex lens as the objective and the concave as the eyepiece (ocular). The same mistake was made by Keizo Hashimoto, *Astronomical Reform*, p. 183n58, who was probably following the translation in D'Elia. I am indebted to Chai Choon Lee for providing me with this revised and corrected translation of the Chinese text.

The technique of projecting the image of the sun onto a piece of paper was first discussed by Kepler in the context of using a pinhole or camera obscura device to study the patterns of light back in his *Optics* of 1604. Projecting light through a pinhole into a darkened room was a widely discussed technique in sixteenth-century European optics.[31] However, the use of different colored lenses may have worked, but no one before Newton, and hence in the 1620s or 1630s, knew about the multicolored composition of the spectrum of light and that different colors refracted at different angles, although the rainbow spectrum produced the same phenomenon.

In this treatise on the telescope, Wang Cheng mentions several uses of the telescope, including sea voyages, warfare, for painting, and as a *camera obscura*. He elaborates on the use of the telescope for wartime events, espying the enemy, counting horses, detecting fortifications, discharging cannons, and so on. "Nothing is more useful," he asserts, "than this instrument."[32] All these authors emphasize that the Milky Way is a vast expanse of stars, now seen clearly with the telescope and with ten times as many stars as seen by the naked eye. All this was new territory for the Chinese. For when the Jesuits arrived, Chinese craftsmen did not possess the ability to grind lenses for corrective glasses or telescopes,[33] nor did they possess the level of optical understanding of Ibn al-Haytham (d. 1038) in the Middle East.[34]

About a decade later, another of the recruits brought back by Trigault, the Italian Giacomo Rho, published a book on planetary theory in which he likewise mentions that "a famous mathematician of the regions of the Occident has manufactured a telescope with which one can see distant things as if they were near and small things as if they were great."[35]

---

[31] Sven Dupré, "Ausonio's Mirrors and Galileo's Lenses: The Telescope and Sixteenth-Century Practical Optical Knowledge," *Galilaeana* 2 (2005): 145–80, and Eileen Reeves, *Galileo's Glassworks: The Telescope and the Mirror* (Cambridge, MA: Harvard University Press, 2008).

[32] D'Elia, *Galileo in China*, p. 38.

[33] "By 1600 Europe was already ahead of Asia in producing basic machinery such as clocks, screws, levers and pulleys that would be applied increasingly to the mechanization of agriculture and industrial production. China in the period from 1600 to 1800 did not possess optical-lens makers, horological gear-wheel cutters, and men who could make accurate microscrews." Donald Lach, *Asia in the Making of Europe* (Chicago: University of Chicago Press, 1977), pp. 397–400, as cited in Benjamin Elman, "Jesuit *Scientia* and Natural Studies in Late Imperial China, 1600–1800," *Journal of Early Modern History* 6, no. 3: 228.

[34] Needham, *SCC* 4:78.

[35] D'Elia, *Galileo in China*, p. 39.

Finally, in about 1628, Schreck wrote a *Brief Description of the Measurement of the Heavens* that was presented to the throne in 1630. In this work, Schreck extols the virtues of the telescope and spells out what some of Galileo's observations meant for planetary theory. Schreck refers only to "a celebrated Mathematician of the kingdom of the Occident" when he refers to Galileo.[36]

One should not make too much either of this oblique reference to Galileo or the omission of his name. As I noted in Chapter 2, telescopes were all over Europe within a matter of months after their invention; Galileo had not invented the telescope himself, though he fashioned his own and transformed it into a discovery machine. Dozens of European observers had corroborated his observations regarding the moon, Venus, Jupiter, and Saturn. But, on the other hand, Galileo was clearly an extraordinary astronomical observer, unequalled for such observations during his lifetime. He could grind his own lenses, giving him a technical advantage over all others. Furthermore, when Schreck refers to Galileo as a "celebrated" mathematician, he means that literally, as we saw on the occasion when Galileo was feted by students and faculty of the Jesuit College in Rome.

In his *Brief Description*, Schreck recalls the powers of the telescope in the context of Galileo's observations of Venus and in the context of ancient theories of the universe that Schreck found implausible. But he continues that the

celebrated mathematician in order to observe Venus has constructed a lens which permits one to see afar. [With this instrument] sometimes [the star] is obscure, sometimes it is completely illuminated, sometimes it is illuminated either in the superior quarter or in the inferior quarter [see Figure 3.3]. This proves that Venus is a satellite of the sun and travels around it. When the planet is distant, contrary to what happens with the moon, a half of it is illuminated, which proves that then it is above the sun. When it is full it appears very small. When it is not seen this means that it is under the sun. When, finally, it is on the sides of the sun, one has the two quarters. The same may be said of the very small Mercury which is exceedingly close to the sun and of which one does not easily discern the luminosity or obscurity; also it moves like Venus and one measures it in virtue of the same principles.[37]

The missionaries' reluctance to discuss the Copernican heliocentric hypothesis was due not only to the official Church's interdiction of the idea but also because decisive evidence in favor of the theory had not

[36] Elman, *On Their Own Terms*, p. 99.
[37] D'Elia, *Galileo in China*, p. 40.

yet been brought forth. Of course, Galileo and his supporters thought the Venus observations were decisive for the Copernican theory, but others were not so convinced.

It is striking that Schall, Schreck, and others discussed at all the claim that the planets Venus and Mercury revolve around the Sun. This was consistent with Tycho Brahe's geoheliocentric system that had not been condemned by Church officials; but it also was consistent with the Copernican hypothesis. In that light, Tycho's system granted the missionaries considerable latitude to discuss various accredited astronomical observations revealing at least partially the sun-centered nature of our universe.

For these reasons, Schall was the first to introduce into China Tycho Brahe's geoheliocentric system in his book on the telescope. According to Brahe, the five planets – Mercury, Mars, Venus, Jupiter, and Saturn – revolve around the sun, but it in turn revolves around the earth (see Figure 3.4). Observations such as those mentioned confirmed that at least two of the five planets revolve about the sun and thus correspond reasonably to Tycho's system. Galileo and Kepler both rejected that system in favor of the Copernican view, but they clearly were in the vanguard of those trying to establish the new system. When the calendar reform program finally got going after 1629 under the direction of Xu Guangqi, astronomical discussion was not just philosophical. It was linked to more precise astronomical observations made possible by the arrival of the telescope. Official rejection of Copernicanism did not prevent the missionaries from discussing the implications of actual observations supporting the sun-centered paths of Venus and Mercury. The Church's interdiction of the Copernican system did require that the missionaries in China adopt the next best system, and that was Tycho's system. But, as Joseph Needham pointed out, from an observational point of view, there was nothing between which to choose regarding the two systems, Copernican or Tychonic. On a calendrical level, "the geocentric and the heliocentric hypotheses were in strict mathematical equivalence; whether the earth or the sun was at rest, lengths and angles were identical, and similar triangles had to be solved."[38] Transporting the Tychonic system to China was no more of a problem there than it was in Europe, where both systems had supporters. With the invention of the telescope and its arrival in China, a new and more precise observation-based astronomy was possible.

[38] Needham, *SCC*, 3:446.

Launching the Reform

Xu Guangqi was put in charge of the reform of the Chinese calendar in 1629. The precipitating event that led the emperor to give the reform project to the missionaries was the eclipse challenge of 1629. This was the culmination of the early-seventeenth-century series of eclipse predictions participated in by both the missionaries and the Chinese scholars. On the day preceding June 21 of that year, proponents of all three systems – Chinese, Muslim, and European – were asked to make predictions in writing for the solar eclipse expected the next day. "The traditional mathematicians foretold that it would start at 10:30 and end at 12:30, lasting two hours. Instead the eclipse occurred at 11:30 and lasted only two minutes, just as the missionaries predicted."[39] This experiment, as explained by Xu Guangqi, finally convinced the emperor that the problems with the Chinese calendar predictions were not mathematical errors but rather an inherent defect of the whole Chinese calendrical system.

Xu Guangqi was an exceedingly able astronomer and mathematician as well as a Christian convert. He was an ideal candidate for the job, a person who had passed twenty-four additional examinations after his *jinshi*, all of which enabled him to teach in the prestigious Hanlin Academy.[40] He fully understood the new Euclidean geometry recently introduced in his translation and the historically grounded problems of the Chinese calendar. For him, a major task of the reform was the translation of the great cargo of books on all subjects brought back from Europe by the missionaries. He was convinced that the reform could only be carried out if the foundations of Chinese astronomy, based on Western assumptions, were fully understood. Anticipating the day when some defect in the system would arise, he insisted on having all the Western sources available in Chinese so that such a defect could be adequately understood and remedied when it arose.

Xu Guangqi believed that only the Tychonic system would work because he had had very little knowledge of the Copernican system and because the newly acquired telescope seemed to present evidence consistent with Brahe's system while contradicting the Chinese system.[41] At the same time, Xu and the missionaries were certain that the Ptolemaic system had been vanquished. The observations of Venus in China were accepted

---

[39] D'Elia, *Galileo in China*, p. 4.
[40] Jami et al., *Statecraft*.
[41] Hashimoto, *Astronomical Reform*, pp. 166ff.

as refuting the Ptolemaic system of *The Sphere* and seemed to demonstrate that Venus and Mercury revolve around the sun, as that Western system predicted. In that context, it could be said that the Chinese astronomers in the Bureau of Astronomy had gone through two revolutions: from the flat world of traditional China to the spherical Ptolemaic world, and from that to the geoheliocentric cosmos in which all the planets except the earth revolved around the sun.

It was against this background that the project of translating fundamental works in mathematics, astronomy, philosophy, and the auxiliary sciences that the Jesuits did their best to provide the Chinese with the modern tools of astronomy. They brought the new telescope as well as Tycho Brahe's classic instruments with which he revolutionized European observational astronomy in the late sixteenth Century. But they also brought many new mathematical tools – for example, Euclid's geometry, logarithms, and trigonometry (in the 1650s) – as well as the philosophical and technical means for understanding those developments.[42] Most impressive of all, the missionaries brought the new cutting-edge optics of Kepler.

Optics and Observation

One of the most striking aspects of the missionaries' astronomical efforts in China is how assiduously they placed astronomical inquiry within sound *observational* theory and practice. The books by Schall, Rho, Schreck, and their collaborators repeatedly stress the importance of the telescope, what can be seen with it, and what those observations imply for astronomical theory. The arrival of the telescope and its incorporation into the reform project directed by Xu Guangqi marks the beginning of the transition from the age of naked-eye astronomical observation to the "telescopic phase" of inquiry in China. This is at least partially parallel to the European situation in which the invention of the telescope dramatically altered the *practice* of astronomy. As discussed in earlier chapters, the telescope transformed European astronomical practice into an exploratory science looking for new discoveries. Xu Guangqi realized that more precise observations in China were both necessary and possible with the telescope and other Western instruments used by the Europeans.[43]

---

[42] Jami et al., *Statecraft*; Engelfriet, *Euclid in China*.
[43] Hashimoto, *Astronomical Reform*, pp. 52ff.

Indeed, the major thrust of Xu Guangqi's efforts to reform Chinese astronomy rested on the assumption that scholars associated with the Bureau of Astronomy – more than fifty individuals – would construct new observational instruments, including telescopes, with which new observations could be made. These were to be tested against predictions derived from the traditional Chinese system (the *Datong* system) and from Western astronomy – that is, the Tychonic system.

By the 1630s, Adam Schall von Bell and Giacomo Rho had undertaken the publication of a book called the *Complete Treatise on the Measurement of the Heavens* (*T'se é Liang Ch'uan I*) that was "exclusively devoted to the explanation and use of the astronomical instruments [of Tycho's *Mechanica*, 1598] which were indispensable for the astronomical reform" in China.[44] The appendix of the volume, which was presented to the throne on August 27, 1631, contained a list of all the Tychonic instruments with a brief explanation. Among these one finds the quadrant, parallactic ruler, horizontal azimuth quadrant, bipartite arc, astronomical ring, astronomical sextant, great equatorial armillary sphere, zodiacal and equatorial universal armillary sphere, and so on.[45]

As part of this reform project, Schall and Rho, together or separately, published such books as *Treatise on Lunar Motion, Theory of Solar Motion* (1631), *Treatise on Eclipses* (1631), and *Treatise on the Motion of the Five Planets* (1634). There was also a volume on the motion of the fixed stars as well as a companion volume of *Tables of the Five Planets* (*Wu Wei Piao*, completed in 1634).[46] This formed the basis for the predictions that would test the Western and Chinese systems. Xu Guangqi's concern was to show that solar and lunar predictions as well as predictions of planetary positions based on the Western system (that of Tycho) were more accurate than those based on traditional Chinese methods.

After considerable preparation and the training up of Chinese students in the "new" system, a number of such predictions were made and tested

[44] Ibid., p. 206.
[45] The existence of these instruments in China has been known for a half century due to Needham's *SCC* 3:451ff., pioneering research, but they are usually associated with Ferdinand Verbiest, who succeeded Schall as the head of the Bureau of Astronomy in China in 1669. However, as is clear from the discussion, these became known in the 1630s, and probably early versions of them were manufactured before Verbiest arrived in China in 1657. For questions of accuracy, see Allan Chapman, "Tycho Brahe in China: The Jesuit Mission to Peking and the Iconography of European Instrument-Making Processes," *Annals of Science* 41 (1984): 417–43.
[46] Hashimoto, *Astronomical Reform*, p. 224.

against empirical evidence. Each time, the Tychonic predictions proved more accurate. Likewise, Xu's successor, Li Tianjing (1579–1659), who took over when Xu died in 1633, stressed the accuracy of telescopic observations. In a memorial sent to the emperor on November 4, 1634, he wrote:

This instrument can individuate two stars so close that the human eye does not distinguish their outline. It can see those stars which are so small that the eye discerns them only with difficulty. When two stars are distant the one from the other by a half a degree, that is, 30 minutes according to the new astronomy, and therefore cannot be observed either by measuring instruments or the eye, the telescope can observe both of them clearly, because it has a capacity of a little more than a half a degree, which constitutes its measure.[47]

With this and the other instruments, Xu Guangqi and his coworkers set out to get much more accurate observations that would clearly separate the virtues of the Western and Chinese systems.

Before considering those results, we should note also that the Jesuits brought to China a very sophisticated new science of optics based on Johannes Kepler's innovations that took astronomical observation still further beyond traditional Chinese assumptions and possibilities. Although the telescope could theoretically yield very precise observations, Kepler believed that there are physical limits to observation that are imposed by the laws of optics – the laws of refraction and the structure of the human eye. More important, Western astronomers back to Tycho and then Kepler realized that observations of the sun on the horizon are deflected so that the apparent position of the sun, or possibly the moon, is not where it appears to be. The process of refraction redirects the light so that an object can appear either ahead of or behind its actual position.[48]

However, Kepler's research in optics concerned the relative size of the sun and moon during a partial eclipse. It was this problem that Kepler took up when he observed a partial solar eclipse in July 1600 through a pinhole. This was an idea suggested to him by Tycho Brahe. During an eclipse, the moon's position between the earth and the sun is such that it looks as large as the sun as it blocks out the sun's rays. But Kepler's experiment revealed that the projected image of the moon through the pinhole was smaller than the sun, despite the fact that the moon's image was so large that it blocked the sun's rays. This paradox led Kepler to

[47] In D'Elia, *Galileo in China*, pp. 47–48.
[48] Alfred Joy, "Refraction in Astronomy," *Astronomical Society of the Pacific*, leaflet 220, June: 162–69.

take up his optical investigations that resulted in his new theory of human vision.

Kepler plunged into the existing theory of optics, especially the thirteenth-century work of Witelo. Part of what Kepler (re)discovered was that light radiates out as a cone of light. Each ray, he suggested, could be thought of as a "pencil of light," and these pencils of light, after passing through the apparatus of the eye, draw the image seen on the retina. In this way, Kepler worked out the process by which light is transmitted through the various parts of the human eye and then is projected upside down onto the retina.[49] This was a very new understanding of how the eye works and became the foundation for all succeeding generations of students of optics, such as Descartes and Newton. The results of Kepler's optical research later appeared in his landmark work on the optical part of astronomy (*Astronomia Pars Optica*, 1604). He also worked out some tables of refraction for celestial objects when they are at various positions in the sky. This was important because the process of refraction in the air can make objects appear to be located several degrees from their actual location. The implications for astronomers were obvious. Sightings of celestial objects may have to be corrected to account for the refraction of light passing through the atmosphere. Consequently, calculations could be used to correct the apparent position of a star or other planetary object under specified conditions. These ideas were transmitted to China.

It was Adam Schall and Giacomo Rho who introduced many of Kepler's optical ideas into Chinese. In the *Treatise on Eclipses*, they translated major parts of Kepler's master work (*Astronomia Pars Optica*). Similarly, the *Treatise on Lunar Eclipses* (1632) and *The Complete Treatise on Measurement of the Heavens* (1632) invoked Kepler's optical insights.[50]

Another device that Schall introduced from Kepler was a method by which solar and lunar diameters can be estimated. This is based on an instrument that Kepler constructed. It consists of a pinhole in a piece of wood mounted on the end of a pole mounted on an inclined plane. The light shining through the pinhole strikes a block of wood on which units of measurement are marked to enable accurate measurement. This

[49] Owen Gingerich, "Johannes Kepler," *Dictionary of Scientific Biography* 7 (1973): 289–312; David C. Lindberg, *Theories of Vision from al-Kindi to Kepler* (Chicago: University of Chicago Press, 1978); van Helden, "Galileo and the Telescope," in *Novita Celesti e crisi del Sapere* (1984).

[50] Hashimoto, *Astronomical Reform*, pp. 182–200.

instrument was brought to China and used by the Chinese and missionary advisors.[51] The principles of this device were then extended to the telescope so that another method of estimating solar and lunar diameters became available.[52]

The pinhole technique used by Kepler could also be used as a method by which the sun could be observed during an eclipse and the magnitude of the eclipsed part estimated. The treatise on eclipses explains that as with observing the sun, one places a white sheet of paper beneath the telescope, which has drawn on it "double concentric circles, the inner one of which had the line of the diameter divided into ten, each division of which had the length of about 2 inches as the unit. The magnitude of an eclipse could thus be determined by making use of these divisions."[53] This was a more sophisticated method than the five-finger tradition of Europe whereby one estimated how many fingers were covered by the shadow of the occluding planet.

With this device, a unit of measurement could be established to estimate the degree of solar or lunar eclipse, which was important for various purposes. It was especially important when the Chinese and Western missionaries conducted face-offs involving eclipse predictions and the proportion of the object that would be occluded. It became useful in September 1644, when Adam Schall staged an eclipse prediction for the emperor, challenging the traditional scholars and requiring an estimate of the time as well as the percentage of the sun that would be occluded.[54] In addition to these techniques, others were offered for the measurement of angular distance, starting with the moon as a reference point.

To some degree, the optical insights of Kepler were marginal techniques that should be used under special circumstances, but the fact that they were introduced suggests how seriously the scientifically trained Jesuits took their task of explaining the latest astronomical and optical advances of Europe that they knew.

Armed with all these insights, tables, and techniques, between 1632 and 1635, Xu Guangqi and his coworkers carried out a sequence of observations based on predictions drawn from Western astronomy and from traditional Chinese astronomy. These involved the positions of Saturn,

---

[51] Ibid., pp. 184–85.
[52] Ibid., p. 189.
[53] Ibid., p. 175.
[54] Fu Lo-shu, *Documentary Chronicle of Sino-Western Relations (1644–1820)*, 2 vols. (Tucson: University of Arizona Press, 1966), pp. 3–4.

Venus, Mercury, Mars, and Jupiter as well as solar and lunar eclipses.[55] Three particular predictions stand out. The first was the conjunction of Mars and Saturn in October 1634: "According to the prediction by the new method, Mars and Saturn were to be seen in the same degree in longitude and were to be separated by 1° 54' in the direction of latitude at the beginning of dusk on October 25, 1634."[56] But according to the traditional system, the Datong, this conjunction should occur about three days later.

The second prediction concerns Venus and Saturn, which were predicted to be in the same celestial position separated by 3° 31' of latitude at 6:30 in the morning of October 28. The traditional system predicted that the event would occur a day later.

The third crucial test concerned Venus and Mars, which also would experience a conjunction separated by 1° 30' of latitude in the morning of November 1, 1634. The Datong system predicted that the event would take place eight days later. In all these cases, the predictions of the missionary astronomers, based on ephemeridean tables from European scholars, proved correct. Likewise, when the eclipse predictions were tested, all those based on the new system were found to be correct. Most of the observations, designed to compare the predictive accuracy of the two systems, were carried out between 1632 and 1635 but some later.[57]

In short, the reform under Xu Guangqi was based on creating a whole research program that was systematically carried out. First, the Chinese scholars working with the Christian missionaries created the theoretical and documentary basis of the new astronomy. Then they designed and constructed new instruments, including telescopes, that were used to test the predictions against the theoretical values derived from the two systems of astronomy. The whole enterprise could be seen as an extraordinary research and training operation involving more than fifty individuals (Chinese and European). Even the missionaries themselves must have experienced this as an extraordinary training exercise that they had never before experienced in such a hands-on way during their university studies in Europe a decade or two earlier. Indeed, it would be difficult to say when modern graduate students in astronomy first had such a complete hands-on training exercise in a university, testing one theory against another.

---

[55] Hashimoto, *Astronomical Reform*, chap. 4-ii.
[56] Ibid., p. 225.
[57] Ibid., p. 226.

As each of the parts of the crucial mathematical and astronomical works was finished, it was formally presented to the emperor. The result was, as Joseph Needham put it, "a monumental compendium of the scientific knowledge of the time."[58] During the next century or so, this massive collection of scientific and philosophical writings became the centerpiece of Chinese astronomical thinking, called "The Complete Studies on Astronomy and Calendar," published in 1723.[59]

Platform for a Scientific Takeoff?

In somewhat understated terms, this extraordinary astronomical and cosmological reform can be summarized as Benjamin Elman has done: by 1630, Chinese specialists "had available to them a rich tool kit of new computational techniques, more accurate observations, and a view of the cosmos that modified the 'spherical heavens' system." They also had "the latest precision instruments that went beyond those available to the Gregorian reformers in Europe a generation earlier."[60] In addition to that, they now had Kepler's new optical theory and, by the 1650s, the new trigonometry of Europe as well as logarithms.[61]

Beyond that, all those associated with the Chinese Bureau of Astronomy had the practical experience of building and testing new instruments and then using those instruments to test two different theories of observational astronomy. This was a remarkable transformation that Ferdinand Verbiest would characterize as a "revolution" in astronomy, though he was thinking of the 1669 restoration of the new Western astronomy rather than the original accomplishment of 1631–34.

Given all these innovations and aids to understanding the new astronomy, along with the telescopic discovery machine, one might have expected the presumed excellence of the Chinese in science to have propelled them forward in the seventeenth century, making significant discoveries and laying the foundations for a unified celestial and terrestrial physics. One might also have expected them to make innovations with regard to the telescope such as the invention of longer and more powerful telescopes produced by Europeans from the 1640s onward. Some Chinese

---

[58] Needham, *SCC*, 3:447.
[59] The compilation has been republished in many versions, usually titled as *Si Ku quan shu*. It is available online for Chinese readers. Also see Needham, *SCC*, 3:448.
[60] Elman, *On Their Own Terms*, p. 99.
[61] Engelfriet, *Euclid in China*; Jami et al., *Statecraft*.

scholars took a great interest in the new Western mathematics and astronomy, as historians have reported.[62] Some minor celestial discoveries were
made, but no significant astronomical innovations were forthcoming, nor
were any innovations made regarding the telescope.[63] The main drift of
the scholars close to the emperor was toward severely restricting the use
of the new natural studies and attempting to prove that underneath the
Western successes, there was really a Chinese past.

Schall at the Helm: The Rites Controversy

From 1629 to 1664, the assumptions of the new Western-based system of
astronomy held center stage, with minor incidents. Once the official impri-
matur of the emperor was given in fall 1629, the calendar reform went
forward under the leadership of Xu Guangqi and other Chinese scholars,
guided in crucial ways by the missionary scientists. They produced dozens
of small treatises laying out the technical foundations of Western astronomy. The new or Western calendar was officially announced in 1634 and
remained in place until 1665. For those three decades, the new system
flourished, and new calendars were regularly produced. They contained
the expected predictions for planets, for lunar and solar eclipses, and
for lucky and unlucky days. Astrological divination was an odd business
for the Christian spiritualists, but it was the compromise thought necessary to prepare for the conversion of the emperor and all his subjects to
Christianity.

There had always been sharp differences of opinion, even within the
Jesuit order, regarding the idea of accommodating Christian belief and
practice to Asian ways. Yet, the very success of Ricci's mission in China
rested on the notion that the missionaries learn Chinese and adopt the
everyday costumes of their Chinese hosts. Educationally and psychologi-
cally, the Jesuits identified with the literary elite of China, the mandarins,
not the village peasant. Furthermore, it became obvious that China was
a top-down society and that only by converting sections of the literati,
the bureaucratic leaders, and the emperor himself could China be won

---

[62] John Henderson, "Ch'ing Scholars' Views of Western Astronomy," *Harvard Journal of Asiatic Studies* 46 1 (1986): 121–48.

[63] Zhang Baichun, *The Europeanization of Astronomical Instruments in Ming and Qing China* (English abstract, 2001), and Minghui Hu's review of Zhang Baichun, "The Europeanization of Astronomical Instruments in Ming and Qing China," *Isis* 95, no. 4 (2004): 699–700.

over to Christianity. When it came to religious and philosophical issues, the Jesuits attempted to use Chinese terms such as *Shangdi* (Lord-on-high) and *tian* (heaven), Chinese equivalents to Christian ideas. Deeper probing into this metaphysical abyss led the Jesuits to assert that China had a "natural religion" in Confucianism, which would be acceptable to Christian theology, but the task was to round out this indigenous religion with more explicit Christian doctrines. Matteo Ricci attempted this in his early years by writing theological tracts in Chinese.

On the other hand, the few mendicant missionaries (Dominicans and Franciscans) in China saw those Chinese terms, *Shangdi* and *tian*, as tainted. As the numbers of European missionaries in China grew, so did the tensions over these doctrines. None of this could escape the great power conflicts of the seventeenth century between the ascendant French, the Spanish, and the Portuguese with their crumbling empire, all of whom tried to manipulate the Holy See in Rome to win greater jurisdiction for their nationals within China.

Given a three-decade run with the new system in place, one might have supposed that the Jesuits were at last victorious, both within Catholic Christianity and within China, and that the Chinese scholars had been converted to the new astronomical system and its metaphysical assumptions. But without transforming the educational system and teaching the new astronomy locally, there really was not much hope that China could be converted to the new system, however superior its predictive powers were demonstrated to be. The Achilles heel of the Jesuit program was precisely its deep involvement in Chinese divination, siting, and the selection of auspicious days.

Adam Schall von Bell, a German missionary born in Cologne in 1591, had been trained in philosophy and theology – and no doubt much more – in Rome, where he joined the Jesuits in 1611, prior to his embarking on the mission to China in 1618. As we saw earlier, Schall and Father Terrentius (Schreck) became collaborators in the effort to write new scientific texts containing the best of European astronomy and mathematics, while rendering them into Chinese. It is possible that Scheck tutored Schall in mathematics during their sea voyage to China, but it is apparent that Schall was a very capable person in his own right.

In 1630, he had been called back from a regional posting in China to Beijing to replace Schreck when the latter died. As one of the technically most proficient astronomers among the Jesuits, Schreck had been assigned to Xu Guangqi's great project of reforming the Chinese calendar.

Back in Beijing to replace Schreck, Schall's outstanding competence as an astronomer soon put him at the head of the Jesuit team working on the reform of the Chinese calendar.

When Xu Guangqi died in 1633, his assistant carried on for a time, but Chinese expertise built up through Xu's big enterprise soon faded. The other leading missionary scientist besides Schreck was Giocomo Rho, who died in 1638. With Schreck's demise, and then Rho's death, Schall was appointed director of the Board of Mathematics. Prior to that, he had been providing the astronomical calculations needed for the annual calendar, the prediction of eclipses and planetary positions, and auspicious days in Chinese calendars. The Jesuits later disputed over whether Schall only provided calculations for these astrological dates or whether his deep involvement in traditional Chinese cosmology was too compromising for a Christian missionary.

In 1638, Schall was given the official title of Mandarin, though he claims to have many times refused the honor. This was a great honor for a Westerner, but for Christian missionaries attempting to be *in* the world but not *of* the world, it suggested an intimate involvement in un-Christian superstition. His reputation grew among the Chinese, and when the declining Ming empire was under assault by the Manchus in the 1640s, Schall was pressed into service constructing cannons. The effort was so futile that the last Ming emperor walked out into his garden and hung himself.

With the launching of the Qing Dynasty in 1644, Schall's reputation and indispensable expertise remained intact, resulting in his appointment under the new rulers as director of the Bureau of Mathematics and Astronomy. As head of the Astronomical Bureau within the Ministry of Rites, Schall's duties included mathematical astronomy, astrology, and time service. The later activity involved choosing auspicious times, dates, and locations for marriages and imperial burials. Throughout China, there were district schools of divination whose purpose was the training of officials who could divine auspicious dates and choose appropriate sites based on topography and orientation.[64] Such officials during the Ming and Qing dynasties were appointed to serve in the Department of Time Service of the astronomical bureau. It might be anticipated that such individuals, trained according to Chinese metaphysical assumptions, would

[64] Huang Yi-long, "Court Divination and Christianity in the K'ang-his Era," *Chinese Science* 10 (1991): 1–20.

eventually clash with the new astronomical principles from the West. Indeed, Schall dismissed several of these Chinese workers, later resulting in enmity among the Chinese.

The depth of this divinatory belief in properly chosen sites and times of burial, above all for royal family members, is revealed in the writings of the famous consolidator of Neo-Confucianism, Zhu Xi (1130–1200). He had been involved with the siting and burial of an emperor centuries earlier. According to Zhu Xi:

By the word "burial" . . . we mean "safely storing away" . . . that is how one safely stores away the bodily remains of ancestors. When descendants wish to store away the bodily remains of ancestors, they must find a way to fully express their solemn and sincere feelings, in order to attain abiding peace and security. If the corpse remains intact and the spirit finds peace, the descendants will be prosperous and ceremonial offerings will be uninterrupted; this is a natural principle. . . . But if by chance the divination is not expert and the place not auspicious, then there are bound to be underground streams, winds, or insects to disturb it underground. The spirit thus moved will not be peaceful, and the descendants' line will be wiped out – a matter greatly to be feared. And it may be that, even if a fortunate location is chosen, but the funeral is not a generous one and [the body is] not stored deep, in the chaos of war it will be abnormally dug up and exposed. This is also an important matter for reflection.[65]

The message was clear: the siting and burial of royal ancestors was a momentous event with serious consequences for all those involved. For a foreigner such as Adam Schall and his associates to be involved in such things could be risky, not to mention heretical from a Catholic point of view. In the late 1650s, Schall found himself embroiled in a Rites Controversy that dragged on for a decade.

Saved by an Earthquake

The episode occurred during the transition of imperial power in the first quarter century of the Qing Dynasty. At the heart of the controversy was a charge that Schall was responsible, as head of the Bureau of Astronomy, for selecting an inauspicious time for the burial of a young prince who died prematurely. This had occurred back in 1657, but the case was so serious that the burial did not take place for two years, and the dispute lingered into the 1660s. It was propelled along by the fiercely anti-Christian Chinese scholar Yang Guangxian (1597–1669).

[65] As cited in ibid., p. 3.

Yang was a traditional scholar, steeped in the ancient rites of hemerology (the choosing of auspicious days) but not trained in mathematical astronomy. He admitted that he "knew only the *li* (the ultimate principle) of the calendar but nothing about the methods of calendrical calculations."[66] His concern was to preserve the traditional ways of Confucian China. For many years, he had been an anti-Christian crusader, even sending memorials to the throne protesting the presence of the foreigners. These had started in 1659 and continued into 1665. In 1659, he published a tract on siting, burials, and auspicious days claiming that the date, time, and location of the prince's burial lacked any auspiciousness. This resulted in a serious charge of dereliction of duty on Schall's part. Until 1664, the Manchu officials sent the memorials back; but, with the realignment of political power in the palace after the death of the Shunzhi Emperor in 1661, things changed.

If the hemerological principles that Yang claimed to know were not fully accredited even by Chinese scholars, based as they were on watching the mysterious filtering out of ashes sealed in a vessel, it was easy for him to paint the Jesuits as a band of foreigners bent on spying and subverting the empire.[67] In a memorial sent by Yang and his associates in 1664, they claimed:

The Westerner Adam Schall was a posthumous follower of Jesus, who had been the ringleader of the treacherous bandits in the Kingdom of Judea. In the Ming dynasty he came to Peking secretly, and posed as a calendar-maker in order to carry on the propagation of heresy. He engaged in spying out the secrets of our court. If the Westerners do not have intrigues within and without China, why do they establish Catholic churches both in the capital and in strategic places in the provinces?[68]

With the change in imperial rule after 1661, the stability of the arrangements in the Bureau of Astronomy unraveled. The new emperor, Kangxi, was only eight years old when his father died, requiring that a caretaker official rule in his stead. The new officials in Beijing now gave Yang Guangxian's memorial a hearing. The result was that Schall, the newly arrived Ferdinand Verbiest, and other Jesuits were rounded up, bound with "nine long and thick chains of iron, all with iron locks; three around

[66] As cited in Pingyi Chu, "Scientific Dispute in the Imperial Court: The 1664 Calendar Case," *Chinese Science* 14 (1997): 9.
[67] Huang Yi-Long and Chang Chih-ch'eng, "The Evolution and Decline of the Ancient Chinese Practice of Watching for the Ethers," *Chinese Science* 13 (1996): 82–106.
[68] Ibid., as cited in Spence, *To Change China*, p. 21.

the neck, three on the arms, and three on the feet," and carted off to jail.[69] In the meantime, Yang Guangxian submitted still another memorial claiming that Schall, through his choice of an inauspicious date, was responsible not only for the premature death of the prince but also his mother and the emperor himself, who died of smallpox in 1661.

Schall was partially paralyzed by a stroke precipitated by these events and had to rely on the Flemish Jesuit Ferdinand Verbiest for his defense. An investigation was undertaken and, on April 24, 1665, Schall and all the others were judged guilty: Schall was to be executed by dismemberment. Others involved were to be exiled after receiving forty blows with the bamboo.[70]

The next day, however, an earthquake rocked Beijing, leading all concerned to believe that perhaps an injustice had been done. The Princess Dowager intervened, absolving the Jesuits. The Jesuits, except for Schall, were released. Schall was placed under temporary house arrest, two non-Christian officials were pardoned, while several Chinese Christian converts, including Schall's assistant Li Zubai, were beheaded for treason.[71] Schall died the following year.

The charges on which the Jesuits were sentenced were those of sedition because the judge and the Council of Deliberative Officials could not determine which astronomical system was correct. Yang admitted that he knew nothing about mathematical astronomy, while relying on the computational skills of his Muslim assistant, Wu Mingxuan. The judge wrote that "the methods of calendar-making are profound and subtle; it is very difficult to tell the difference (between the two systems)."[72] They knew that changes had been made to the old system, but now the officials of the Manchu reign rejected the changes introduced by the missionary scientists.

Verbiest's Rise and His Daring Tests

Verbiest had arrived in China only in 1659, just as the Yang Guangxian–inspired Rites Controversy was about to unfold. By 1669, he was fully appraised of the strengths and weaknesses of the Jesuit plan, which relied

---

[69] Letter of Manuel Jorge, as cited in Liam Brockey, *Journey East: The Jesuit Mission to China 1579–1724* (Cambridge: Belknap Press, 2007), p. 125.

[70] Brockey, *Journey East*, p. 129.

[71] Ibid., p. 129; slightly different accounts are in Spence, *To Change China*, p. 22; Pingui Chu, "Scientific Dispute in the Imperial Court"; and Huang, "Court Divination," p. 13.

[72] In Pingyi Chu, "Scientific Dispute in the Imperial Court," p. 16.

on Western science as the tool of conversion. This is the theme he stresses repeatedly in his famous *Astronomia Europaea*, published in Europe in 1687 – the year of Newton's *Principia Mathematica*. Verbiest wanted to proclaim a revolution in astronomy in China also, which was the result of the Jesuit success in guiding the Chinese Bureau of Astronomy to adopt the Western system. Although Verbiest (and many others) saw the landmark date as 1669, when Verbiest played his part in the daring intercivilizational astronomical test, it had all been played out before based on the Tychonic system introduced between 1631 and 1634 by Xu Guangqi and his missionary advisors.

But now, in 1669, following four humiliating years of house arrest, Verbiest and the others were released from confinement. When the young Kangxi Emperor became aware of the calendar dispute, he told his advisors to search the country for experts in astronomy who could correct the errors in the traditional calendar. Finally, Verbiest and his accomplishments came to the attention of the emperor, who summoned him and his companions to a royal audience to discuss the problems. When the emperor asked whether there was an empirical test that could determine what the correct arrangements of the heavens is and could be witnessed right in front of everyone's eyes, Verbiest said there was. This was the observation of the length of a gnomon's shadow in broad daylight, based on a prediction of the length of the shadow cast by the sun. Liking this idea, the emperor commissioned Verbiest to immediately make preparations for such an observation the next day. When this was done, Verbiest's predictions proved exactly correct. To make sure that this was not a lucky outcome, the emperor asked for two more observational predictions to be witnessed by high officials, including those from the Astronomical Bureau. Verbiest again proposed several observational tests, and his calculations proved to be correct.[73]

Yang Guangxian was so incompetent as an astronomer that he had twice refused to take the position of director in the Bureau of Astronomy offered him by the Manchu officials after Adam Schall had been ousted. Finally, he was forced to do so. That meant that he and his computus had to come up with a new calendar. Once they did, the young Kangxi Emperor, now only sixteen but making his move to gain control of the throne, gave a copy of the new calendar worked out by Yang and Wu Mingxuan to Verbiest. This was shortly after the success of the

[73] Verbiest, *Astronomia Europaea*, ed. and trans. Noel Golvers (Nettetal: Steyler, 1993), pp. 62–64.

gnomon tests. Verbiest quickly found mistakes in it. He noticed in particular that Wu Mingxuan mixed up Arabic and Chinese characters, which he interspersed in the tables, hoping thereby to call it a "Chinese-Arabic" calendar.[74] There was also an extra month that did not fit in Verbiest's calendar. This was actually acceptable according to the Chinese lunar-solar calendar that required the addition of an additional month every few years. The exact date when such an extra month should be inserted was a matter of judgment. The approximate rule for adding an intercalary month was seven additional ("leap") months in a nineteen-year cycle. The point of doing that was to bring the (shorter) lunar calendar in line with the solar year. Verbiest clearly did not warrant that system.

The resulting dispute led the emperor to declare that the matter should be settled by a public witnessing of observations using instruments at the observatory. Once again, the missionary scientists, now firmly under the leadership of Ferdinand Verbiest, staged another astronomical showdown designed to prove the greater accuracy of the Western system. Verbiest proposed a sequence of five observations that would test the new Western system and the traditional Chinese system, taking place over several days. Because Verbiest had seen the calculations worked out by the Muslim computus who worked for Yang Guangxian, he chose positions that showed the greatest departure from the calculations based on European ephemerides.

In his communications with the emperor regarding the difficulties of evaluating the two calendrical systems, Verbiest hinted that if officials were to follow the lead of Yang, things in the heavens might be out of order; that is, Yang's calendar might describe one set of heavenly arrangements, but the sky itself might reveal another. This audacious suggestion implied that the throne itself might be in jeopardy.

Each party was required to submit predictions in advance. Verbiest's challenge included specifying "the degrees and the minutes of the zodiac which they [sun, moon, and planets] should reach in the heavens on a given day of the month and at a given minute of the day," in February 1669.[75] These positions included the appearance of the sun in Aquarius on February 3; the position of Jupiter at night of the same day as well as the position of Mars that night; the position of the moon on the night of February 18; and the appearance of the sun in the sign of Pisces on February 18 at noon. When the days of observation came and went,

74 Ibid., p. 65.
75 Ibid., pp. 67–69 and 195.

the values of the celestial objects determined by Verbiest proved correct, whereas those of Yang and Wu Mingxuan failed.[76]

The upshot was the sacking and banishment of Yang to his home province, during which trip he died. Verbiest was then immediately appointed head of the Bureau of Astronomy, though he claims to have rejected the assignment four times before accepting. The new calendar was reimposed, while Verbiest recalled the faulty calendar with the extra month that Yang and Wu Mingxuan had already sent out for use by local astrologers. It was during this time that Verbiest was also commissioned to make a new set of instruments for the Beijing observatory. These were to be based on the models of Tycho Brahe (in his *Astronomia Instauratae Mechanica*, 1598), which had been described thirty years earlier by Schall and his associates. These instruments were eventually placed on the top of the great observatory platform in Beijing, long famous because of Joseph Needham's discussion of them in his landmark study.[77] (See Figure 4.3.)

Aftermath

The young Kangxi Emperor (r. 1661–1722) became a devoted student of mathematics and the new astronomy. Until Verbiest's death in 1688, he regularly requested tutorial sessions with the missionary scholar. After his death, other Jesuits prepared lectures for the emperor, some of them based on a textbook then in use in Europe.[78]

With restoration of the new astronomy and calendar, Kangxi requested that Verbiest explain to him each of the dozens of theoretical and applied topics that were addressed in the *Astronomical Compendium* (*Chongzhen lishu*) that had been assembled by Xu Guangqi and later revised by Adam Schall. According to Verbiest's account in the *European Astronomy*, the emperor was

very well versed in mathematics [and] was very happy when, measuring the height and length of objects by means of an instrument and drawing chorographical maps, he noticed that his calculations so closely approximated the very reality of things and the actual distance between places.... Thereafter from these earthly measures he moved on to the higher heavenly spheres, and he assiduously examined the magnitude of all planets, the distance between them and from the earth;

---

[76] Noël Golvers provides more details of the observations and errors in calculations in his comments in Verbiest, *Astronomia Europaea*, p. 199n27.

[77] Also see Chapman, "Tycho Brahe in China," and Needham, *SCC*, 3:451ff and plates LXVI and LXVII.

[78] Jami et al., *Statecraft*.

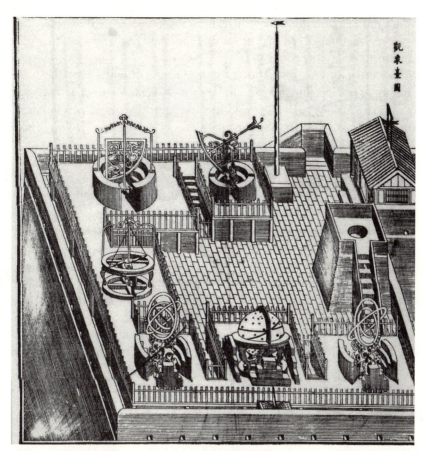

FIGURE 4.3. The Observatory in Beijing and the New Instruments designed by Verbiest. An engraving made by Melchoir Haffner for Verbiest's *Astronomia Europa*, 1687.

moreover, he wanted their movement, orbits and rotations, and the whole theory to be demonstrated by means of various instruments, even paper (plani-) spheres. He had impressed upon his mind the whole system of fixed stars, including their names, the order and the position of each of them (he even spent nights studying these things) so that, when he raised his eyes to the heavens, he could immediately indicate any star with his finger, and call it by its proper name.[79]

This tutoring of the Kangxi Emperor is a notable episode in intercivilizational cultural encounters. It is consistent with the view that Kangxi was the most pro-science Chinese emperor of any dynasty. True as that is,

[79] Verbiest, *Astronomia Europaea*, p. 100.

it did not result in the transformation of the Chinese scientific enterprise into one that paralleled that of Europe. The null result was not just that all this tutoring was confined to the royal household.

During the course of the seventeenth century, dozens – probably hundreds – of competent Chinese scholars and students were trained according to the new system of astronomy and mathematics by Xu Guangqi and his missionary friends in the 1630s and by Adam Schall in the 1640s, 1650s, and 1660s, when he was in charge of the bureau. Xu's enterprise in itself employed more than fifty workers. Likewise, in Verbiest's era from 1669 to his death in 1688, dozens more were trained in the technicalities of the new calendar that had received the full support of the Kangxi Emperor. Most important, all this training and immersion in the new astronomy was not carried over to training for the Civil Service Examinations. Indeed, after Verbiest's departure, the study of astronomy was increasingly restricted to the palace and the Bureau of Astronomy. It was forbidden to place questions on the subject in the Civil Service Examinations.[80] Thus "natural studies" were carefully sequestered from the general public of scholars-in-training outside the court.

Verbiest fully understood that there was a problem of educational training for Chinese scholars insofar as Western philosophy and science were concerned. The traditional Chinese path to educational enlightenment did not emphasize natural philosophy, and there was nothing quite like Aristotle's natural books in Chinese philosophy (more on which is in Chapter 6). Verbiest had grasped this and proposed a new curriculum that would present a well-ordered sequence of Aristotelian philosophy as he understood it from his training at the university in Coimbre (Portugal). A number of Aristotle's books had already been translated, such as *De Anima*, *De Coelho* (*On the Heavens*), his metaphysics and meteorology, and other works.[81] Verbiest, however, wanted to create a systematic corpus and hoped that it would be incorporated into the Chinese educational system. After the emperor saw the tract, he turned it down, forbidding its printing in China.[82]

[80] Elman, "Jesuits *Scientia* and Natural Studies," p. 22; Elman, *On Their Own Terms*, pp. 167–68.

[81] See Willard Peterson, "Western Natural Philosophy Published in Late Ming China," *Proceedings of the American Philosophical Society* 117, no. 4 (1973): 295–316; and "Learning from Heaven," in *The Cambridge History of China*, vol. 8, pt. 2, eds. Denis Twitchett and Frederick W. Mote, 789–839.

[82] Noël Golvers, "Verbiest's Introduction of Aristoteles Latinus (Coimbra) in China: New Western Evidence," in *The Christian Mission in China in the Verbiest Era: Some Aspects of the Missionary Approach*, ed. N. Golvers, 33–51 (Leuven, Netherlands: Leuven University Press, 1999).

Taking Stock

This leaves us with three measures of the impact on the Chinese of the works that Needham classified as "a monumental compendium of the scientific knowledge of the time."[83] Let us call them instrumental, observational, and theoretical. The first indicator would be modifications of the telescope such as the Europeans carried out during the seventeenth century. The second concerns star charts and catalogues: Which improvements or advances were made there? And third, what can be said of progress in developing a unified physics of celestial and terrestrial motion?

Our sketch of developments in seventeenth-century Chinese astronomy has shown that the missionaries had imported the full panoply of analytic tools, including the telescope, advanced forms of mathematics (trigonometry and logarithms), planetary tables, and the Tychonic system of the universe. Although this system was a geoheliocentric system, it was still a system with an important sun-centered component. It was a respectable system – indeed, the very same one that Europeans themselves had to work through on their way to proving the Copernican system. The real criticism of the Jesuits in this regard is not their failure to present Copernicus's original work – because of official interdiction – but rather to present Kepler's defense of Copernicus (in Kepler's *Epitome of the Copernican System* of 1627) and Kepler's laws of planetary motion of his *Astronomy Nova* of 1609. That said, the Chinese had all the basic tools and an amazing range of specialized treatises written in Chinese by the missionaries to explain each and every major aspect of the new astronomy from Europe. The link between the theoretical and observational parts was especially striking under the guidance of the missionaries.

With regard to the first indicator of innovative impulses concerning the telescope itself, all studies to this date indicate that the Chinese made no improvements on the telescope: they did not build larger, more powerful telescopes comparable to those in Europe, nor did they invent the micrometer for measuring small angles within the telescope. They did not make any improvements on the telescope, nor did they invent the reflecting telescope that relies on a deeper understanding of the principles of optics.[84] This was an invention made operational by Isaac Newton in 1669, if not before.

---

[83] Needham, *SCC*, 3:447.
[84] Baichun, *Europeanization of Astronomical Instruments*; Minghui Hu, "Review," pp. 699–700.

With regard to observational astronomy, Chinese astronomers had a centuries-long tradition of celestial observation and have been credited with knowing about sunspots before the arrival of Europeans and also the observation of novae explosions centuries earlier. As the Jesuit missionaries became more and more familiar with the Chinese past, they often emphasized the continuity with a presumed superior past of Chinese scientific inquiry. There was a widespread belief among the Chinese that a great treasure trove of lost wisdom had disappeared from the past. This idea was seized on by Xu Guangqi and many other Chinese scholars.[85] Eventually, the idea morphed into the notion that the lost wisdom had been transported to the West sometime long ago and that Western knowledge brought by the missionaries to China was that same Chinese wisdom from the past. One version of that mythology credited Cheng Ho, the Chinese sailor of the fifteenth century, with spreading Chinese scientific knowledge to the West. Corrections were made to that knowledge according to this legend, and it became the source of Western knowledge brought back to China later by the missionaries.[86] This is but one of many such fables.

The problem was that the new, Western astronomy was embedded in a very different metaphysics. The unity of heaven and the Chinese imperial throne was not in it, neither in Ptolemy's spheres nor in Tycho Brahe's modified system. It also abandoned the lunar calendar that had been used from ancient times. The absence of Chinese astrological assumptions in the new astronomy was a reason for rejecting it. From Yang Guangxian's point of view:

> It is better to have no good astronomy than to have Westerners in China. If there is no good astronomy, this is no worse than the Han situation when astronomers did not know the principle of apposition between the sun and the moon and consequently claimed that the solar eclipses often appeared on the last days of the month; still the Han dynasty enjoyed dignity and prosperity that lasted for four hundred years.[87]

Even the most gifted mathematician of the late seventeenth century, Mei Wending (1633–1721), had difficulty accepting the new system; he could

---

[85] Among others, see Engelfriet, *Euclid in China*; Pingyi Chu, "Remembering Our Grand Tradition: The Historical Memory of the Scientific Exchanges between China and Europe, 1600–1800," *History of Science* 41 (2003): 193–215; Minghui Hu, "Provenance in Contest: Searching for the Origins of Jesuit Astronomy in Early Qing China, 1665–1705," *International History Review* 24, no. 1 (2002): 1–36.

[86] George Wong, "China's Opposition to Western Science during Late Ming and Early Ching," *Isis* 54, pt. 1 (1963): 35.

[87] Ibid., p. 34.

not contemplate eliminating traditional Chinese cosmology. In a dialogue discussing the question of adopting the Western system of astronomy, he observed that getting the whole system right takes time, but "nowadays, the application of the new astronomy is to use its methods to supplement the incompleteness of the old; it is not the complete abolition of the old method in order to follow the new."[88] Adopting revolutionary scientific ideas was too much to ask.

On a strictly observational level, the Chinese had a rich tradition of celestial observation that was directly connected to calendar making and the activities of the Jesuits. This was the making of star charts and atlases. Those planetary tables could then be used to construct an *ephemerides*, the daily and monthly positions of sun, moon, and planets. These had to be worked out to predict forthcoming solar and lunar eclipses and astrological conjunctions that were thought to be vital signs of the unity of the heavens and the imperial court. These planetary positions were also needed for locating auspicious days.

The Star Catalogue found in the Chongzhen astronomical compendium (of 1631–34) is reported to contain 1,365 stars dating from the first year of the Chongzhen Emperor's reign.[89] This certainly is a respectable number of entries for that period of time. However, the Chinese scholars working with the missionaries also compiled a new atlas that contained 1,812 stars.[90] Studies of this compilation suggest that the extra stars came from European sources, especially Tycho Brahe (because they used his coordinates) and the compilation of the Roman College mathematician Christoph Grienberger of 1612. Some southern stars not visible in central China were also included, suggesting that they were either based on the observations of Schall and Schreck during their sea voyage to China or taken from other Western sources. According to Sun Xiaochun, these southern borrowings included twenty-three new constellations taken from the Bavarian astronomer Johann Bayer's *Uranometria* of 1603.[91] When Schall and the missionaries compiled all the astronomical data during the reform years, they also created an

---

[88] As cited in ibid., p. 37.

[89] Xiaochun, "On the Star Catalogue and Atlas of Chongzhen Lishu," in *Statecraft and Intellectual Renewal in Late Ming China: The Cross-Cultural Synthesis of Xu Guangqi*, eds. Catherine Jami, Peter Engelfriet, and Gregory Blue, 311–21 (Leiden, Netherlands: Brill, 2001), p. 314.

[90] Ibid.

[91] Ibid., p. 319.

extraordinary astronomical map of the heavens that has been overlooked by historians of science.[92]

This star catalogue (and the atlas) could be called a good start, and if the Chinese scholars had continued, they might have reached the same results as Europeans. By the mid-seventeenth century, European catalogues had increased the number to more than 2,000 entries, and by the end of the century, John Flamsteed, the British Astronomer Royal, had compiled nearly 3,000 entries using a telescopic sight with his sextants.[93] From what is presently known, Chinese astronomers in the seventeenth century did not undertake such a systematic study of the sky as was carried out by Flamsteed. Chinese observers using the telescope did make a few celestial discoveries of new stars, including the discovery that the star called Mizar in Ursa Major is a double star.[94] But nothing as dramatic as increasing their star catalogue by nearly half is found. Nor was there a major effort to explain the strange appearances of Saturn and its rings.

There was what may be called a *curiosity deficit*, with the result that nothing extraordinary was undertaken or fulfilled in astronomy by the Chinese in this period. This contrasts with the example of Europeans, especially Christiaan Huygens (and Christopher Wren), who resolved the odd appearances of Saturn by putting forth the ring hypothesis in 1656 (see Figure 3.2). Likewise, Ole Rømer undertook the startling idea of proving that the speed of light is finite, not instantaneous, based on his telescopic observations of the satellites of Jupiter. That act of creativity may have been serendipitous, but it was entirely of a piece with Isaac Newton's extraordinary achievement in his *Principia Mathematica* that brought all the objects of the universe under the domination of universal gravitation.

Still another example of impressive European scientific curiosity that we will meet up with later was the "weighing of air" in the 1640s undertaken by Torricelli, Blaise Pascal, and many others yet absent entirely in seventeenth-century Chinese science.

---

[92] Schall's Eight Panel Astronomical Map of 1634, housed in the Vatican, was placed on display by the Library of Congress. See http://www.loc.gov/exhibits/vatican/romechin.html.

[93] Allan Chapman, "Jeremiah Horrock, the Transit of Venus and the 'New Astronomy' in Early Seventeenth Century England," in *Astronomical Instruments* (Aldershot, UK: Ashgate, Variorum, 1990), chap. 5; Joseph Ashbrook, *The Astronomical Scrapbook: Skywatchers, Pioneers, and Seekers in Astronomy* (Cambridge: Cambridge University Press, 1984). For images of Flamsteed's sextant, see atschool.eduweb.co.uk/bookman/library/ROG/P36.JPG.

[94] Hashimoto, *Astronomical Reform*.

The third and most important gauge of Chinese scientific progress in the seventeenth century centers on the modern physics of motion. When Joseph Needham investigated the history of physics among the Chinese, he concluded that physics as a systematic field of science was underdeveloped. As best he could tell, "one can hardly speak of a developed science of physics."[95] It lacked the systematic thinkers that one encounters in the history of mechanics in the medieval West.[96] These included such names as Buridan, Bradwardine, and Nicole d'Oresme. These thinkers in the West had a long history of arguing with Aristotle's conceptions of motion. Indeed, the science of motion leading up to the time of Galileo was a great debate with Aristotle, leavened with ideas from Archimedes and others.[97] Absent such a history, Needham found none in China who would correspond to the so-called precursors of Galileo in the West, reflected by such names as Philoponus, Buridan, Bradwardine, and d'Oresme. Without the preparation that those pioneers accomplished in the science of mechanics, "no dynamics and no cinematics."[98] Without those theoretical foundations, it is difficult to see how one could arrive at Kepler's laws of celestial physics, much less Newton's synthesis, building on the new, seventeenth-century science of mechanics, along with Galileo's idea of inertia. Observationally, it required groundings in celestial motions, precise positions for Mars and the moon, which were needed by Kepler and Newton, respectively, for working out their physics of planetary motion.

The arrival of the telescope in China did not have the fructifying effect on scientific inquiry that it had on astronomy in Europe. As far as science is concerned, "the epistemological premises of modern science were not triumphant in China until the early twentieth century."[99] When assessing the significance of this outcome, it is not a question of the Jesuits losing out with their efforts, nor of whose system was best from an ethnic point of view.[100] Progress toward the new mechanics of a unified celestial and terrestrial physics was a matter for the whole world and required precise mathematical analysis. Without that, a Newtonian synthesis was

---

[95] Needham, *SCC*, 4/1:1.
[96] Marshall Clagett, *The Science of Mechanics* (Madison: University of Wisconsin Press, 1959).
[97] See Edward Grant, *A History of Natural Philosophy* (New York: Cambridge University Press, 2007).
[98] Needham, *SCC*, 4/1:1.
[99] Elman, "Jesuits *Scientia* and Natural Studies," p. 229.
[100] Benjamin Elman nicely titles his book *On Their Own Terms: Science in China, 1550–1900*. This seems appropriate, except that scientific progress is more like "the tide and the time" that waits for no one.

not possible. The Newtonian synthesis laid an indispensable foundation for the Industrial Revolution. Getting to that point required a revolution in scientific thought that did not emerge in China in the seventeenth or eighteenth century.

As Needham put it, progress toward modern science was an effort to produce "a body of incontestable truth acceptable to men everywhere. Without it plagues are not checked, and aircraft will not fly." Moreover, "the physically unified world of our own time" is something that only "happened historically in Europe."[101]

But let us consider what happened in other parts of the world when the telescopic discovery machine arrived there.

---

[101] Needham, *SCC*, 3:448–49.

5

The Discovery Machine Goes to the Muslim World

[Jahangir is] the greatest and richest master of precious stones that inhabits the whole earth.[1]

When the early models of the spyglass appeared in Holland, Europeans quickly recognized the importance of the new device for both military reconnaissance and celestial exploration. Shortly thereafter, missionaries, sea captains, and traders began taking the telescope around the world, first across Europe and then to Asia. In 1615, the British ambassador Sir Thomas Roe presented a telescope to the Mughal court of Jahangir. This occurred in the same year as Chinese scholars could read a preliminary account of Galileo's discoveries written in Chinese.

Mughal India

When Europeans began exploring India in the late sixteenth century, and more extensively in the early seventeenth century, they were stunned by the amount of wealth that was in the hands of the rulers of Mughal India. As one British official put it, Sultan Jahangir was "the greatest and richest master of precious stones that inhabits the whole earth."[2] Others noted the great disparity of wealth and power between Jahangir and "Christian kings," saying that it was so great as to be "incredible."[3]

---

[1] Edward Terry, chaplain to Sir Thomas Roe, in Ellison Banks Findly, *Nur Jahan, Empress of Mughal India* (Oxford: Oxford University Press, 1993), p. 64.
[2] Edward Terry, *A Voyage to East-India*, as cited in Findley, *Nur Jahan*, p. 64.
[3] Nicholas Withington in William Foster, ed., *Early Travels in India, 1583–1619* (Delhi: S. Chand, 1968), p. 225.

The Mughal Dynasty was a line of Muslim sultans who ruled India from 1526 to 1858. They were descendants of Turkish tribes who had for centuries tried to move south into India. The tribal leader called Babur finally did that in 1523. That achievement began the long succession of Mughal rulers. Once in power, they hoped to spread their faith across the subcontinent. At that time, the largest religious group in India was the Hindus. Converting Hindus, Jains, Buddhists, and other indigenous religious groups to Islam was a long, slow process that never included more than a third of the Indian population. Nevertheless, the Mughals ruled India successfully for centuries, until the empire was dismantled by the British in the eighteenth and nineteenth centuries.

Akbar the Great was the most successful of the Mughal sultans, ruling from 1556 to 1605, when he extended the empire south to the Deccan Plateau that includes most of southern India (Figure 5.1). At the end of Akbar's reign in 1605, the Mughal Empire was nearly at the peak of its wealth that so impressed European ambassadors, traders, and emissaries. Akbar had also forged an ethnic harmony between the many religious groups of the empire, the Hindus, Buddhists, Jains, Muslims, and even Jesuits who applied to his throne for special recognition. It was this great empire of economic strengths, military power, and ethnic peace that was inherited by Akbar's son Jahangir. Although Jahangir was deemed to be bright and curious about nature, painting, poetry, and many of the finer things in life, he was an exceedingly unstable person, prone to drink, drugs, violence, and dissolution. Fortunately for the kingdom, he was married to an exceptional woman, Nur Jahan, who in effect ruled the country as queen for much of Jahangir's reign. She was the daughter of a distinguished court official of Persian descent. In the vacuum created by Jahangir's intoxication and lack of attention to official duties, Nur Jahan acted on her own, appointing officials, undertaking public projects, and keeping the throne together.

In was into this scene that the officials of the British East India Company intruded when they arrived with their gifts. Sir Thomas Roe arrived in 1615, representing the East India Company, hoping to establish legal accords that would guarantee British freedom of trade and movement along with easy passage of goods throughout the empire. This was not an easy thing to accomplish because Mughal rulers had established themselves as absolute rulers. They felt no need for legal agreements with others and chaffed against restrictive covenants. Furthermore, European law was very different from Islamic law, especially because the latter had no concept of a legally autonomous corporation. Legally autonomous

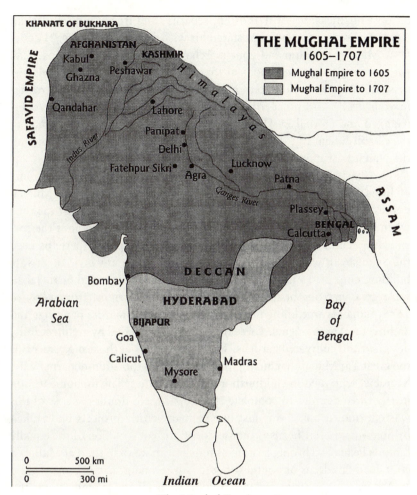

FIGURE 5.1. The Mughal Empire, 1605–1707.

entities such as cities and towns, charitable organizations, commercial enterprises, and professional guilds of surgeons and physicians had no existence in Islamic law. Nevertheless, after two years of persistence, Ambassador Roe maneuvered Jahangir into sending a letter of friendship to King James of England. This was after Roe transmitted a letter from King James to Jahangir. "The letter of love and friendship," the sultan wrote in reply, "which you sent and the presents, tokens of your good affections toward me, I have received by the hands of your ambassador, Sir Thomas Roe (who well deserves to be your trusted servant), delivered to me in an acceptable and happy hour." After assuring the king that all possible measures of friendship, security, and freedom of movement were to be respected, Jahangir stated his wish for full commerce: "I desire your Majesty to command your merchants to bring in their ships... all sorts of rarities and rich goods fit for my palace; and that you be pleased to send me your royal letters by every opportunity, that I may rejoice in your health and prosperous affairs; that our friendship may be interchanged and eternal."[4] These communications were but the prelude to the establishment of the British East India Company in India, along with the Dutch East India Company, the Portuguese, and the French.

The existence of great Mughal wealth rooted in the vast economy of India, along with outstanding military power, suggests that Mughal domestic policy was not constrained by monetary exigencies at the time when the telescope began to appear in the empire. Throughout the seventeenth century, the Mughals far surpassed Europeans (perhaps even the Spanish) in wealth and the ability to undertake any civic or private building enterprises that they desired. Sultans Jahangir and Shah Jahan were great patrons of the arts and, during their reigns, from 1605 to 1658, India became known for its extraordinary miniature paintings and architectural monuments. These were often surrounded by architecturally designed and cultivated gardens. The most famous of those monuments is the great Taj Mahal that Shah Jahan built for his too-soon-departed wife, Mumtaz, who died in childbirth in her late thirties. The monument took nearly three decades to complete (Figure 5.2). No doubt the cost of this was enormous, but it was just one of many costly projects undertaken by Shah Jahan. For he also commissioned the construction of the equally famous Peacock Throne, costing as much or more than the Taj Mahal. It contained hundreds of pounds of gold, silver, and precious gems.

A third outsized expense incurred by Shah Jahan (Jahangir's son and successor) was his military campaign to extend the western borders of his empire from Kabul into the Persian territories of Afghanistan and to capture Samarqand. The latter city had a special place in Muslim history because the Turkic ruler Ulug Beg had constructed a famous observatory there in the fifteenth century. For a time, he had supported scholars undertaking scientific studies, but that soon came to an end due to religious dissension, Ulug Beg's execution, and political instability. Scholars attached to his observatory and places of study had to emigrate to safer places.

---

[4] As cited in James Harvey Robinson, ed., *Readings in European History*, vol. 2, *From the Opening of the Protestant Revolt to the Present Day* (Boston: Ginn, 1904–6), pp. 333–35.

FIGURE 5.2. The Taj Mahal located northwest of Delhi. It took twenty-seven years to complete the construction of the monument, from 1632 to 1659. Photo courtesy of Daniel C. Waugh.

Unlike European universities established in the twelfth and thirteenth centuries that continued to function into modern times, these Islamic places of study such as Ulug Beg tried to support were very fragile and weakly supported by Islamic law and tradition. Nevertheless, Shah Jahan's so-called Balk campaign, hoping to reach Samarqand, went on for decades to no effect, except for draining the state coffers.

Spectacles and Telescopes among the Mughals

Ambassador Sir Thomas Roe appears to have been the first European to bring a telescope to India. However, a Jesuit missionary named Anthony Rubino, who was posted to India, had actually requested a telescope from his superiors in Rome in 1612. At that time, he did not know exactly what the device was, only that with it distant objects could be seen clearly.[5] The question of whether he received replicas of the device in India remains unsolved.

---

[5] Pasquale M. D'Elia, *Galileo in China* (Cambridge, MA: Harvard University Press, 1960), p. 17. Father Rubino's request was sent to Grienberger, the successor to Christoph Clavius at the Roman College.

But we know that Ambassador Roe presented his device to the Mughal court and that the next year, 1616, the chief minister to Sultan Jahangir, Asaf Khan, bought "the whole store" of spectacles, eyeglasses, and telescopes of a Venetian merchant. From Italian sources, we know that thousands of pairs of spectacles were being shipped from Venice to the Muslim world – often via Istanbul – from the fifteenth century onward.[6] In 1482, 1,100 pairs of spectacles were given to Giovanni d'Antonio Tornaquinci for sale in the Levant.[7] Hundreds of others were sent to Adrianople (Edirne) and Constantinople.

In the early seventeenth century, these items were popular in India, and British officials indicate that they were quickly bought at the fairs. In 1652, English officials known as factors who were running the great warehouses of goods in India requested shipments of telescopes.[8] A decade or so later, Jesuits brought 6- and 12-foot French-manufactured telescopes to India, which attracted public attention. With them, some minor discoveries were made by European missionaries.[9]

Moreover, eyeglasses had come to northern India with the Portuguese a century earlier. Records have recently been retrieved revealing the consignment of numerous spectacles in the period 1444–81 by the Cambini Company to agents in Portugal, Flanders, and England. In 1472, a consignment containing lenses and spectacles made in Florence was aboard a Portuguese ship bound for Lisbon. It included cargo for Maestro Latone, a Jewish merchant in Lisbon, including "a box with eighteen pairs of spectacles for all ages."[10] So it is not surprising that the Portuguese, the first Western colonizers in India, were the ones to introduce eyeglasses there.

By the middle of the sixteenth century, pictures and other references to Indian scholars with spectacles begin to appear. A scholar by the name of Vyasarya is described (ca. 1520) as reading a book with the help

---

[6] Vincent Ilardi, *Renaissance Vision from Spectacles to Telescopes* (Philadelphia: American Philosophical Society, 2006), pp. 118–19.

[7] Ibid.

[8] Ahsan Jan Qaisar, *The Indian Response to European Technology and Culture (A.D. 1498–1707)* (Delhi: Oxford University Press, 1982), pp. 35, 163, and William Foster, ed., *A Supplementary Calendar of Documents in the India Office Relating to India or the Home Affairs of the East India Company, 1600–1640* (British Home Office, 1928), p. 83.

[9] Kameswara N. Rao, A. Vagiswari, and Christina Lois, "Father J. Richaud and Early Telescope Observations in India," *Bulletin of the Astronomical Society of India* 12 (1984): 82–85; and Virendra Nath Sharma, *Sawai Jai Singh and His Astronomy* (Delhi: Motilal Banarsidarss, 1995), pp. 287ff.

[10] Sharma, *Sawai Jai Singh*, p. 126.

FIGURE 5.3. This portrait is of the highly accomplished Persian artist, Mir Musavvir who specialized in book illuminations. The portrait was painted by his son Mir Sayyid 'Ali between 1565–75. Both Musavvir and his son emigrated to India sometime before 1555 during the rule of Sultan Humayan. The son, Mir Sayyid 'Ali, achieved considerable fame during the rule of Akbar the Great (d. 1605) and his son Jahangir (d. 1627). That was a period when European style portraiture was spreading to Southeast Asia and the Muslim world.

of spectacles. This event led to the coinage of a new term, *uplochang-olak* (spectacles), because they were unknown in India prior to this time with no word in Sanskrit for them.[11] By the mid-sixteenth century, painters commonly used spectacles. Several of them had their own portraits executed with these aids to vision clearly depicted as in Figure 5.3.[12] In this portrait, the "spectacles are set in a wooden frame and there are arms that go towards the ears, the nose clip seems to be made of

[11] P. K. Gode, "Some Notes on the Invention of Spectacles," in *Studies in Indian Cultural History*, vol. 3, pt. 2, pp. 106–7, as cited in Iqbal Ghani Khan, "Medieval Theories of Vision and the Introduction of Spectacles in India, ca. 1200–1750," in *Disease and Medicine in India: An Overview*, ed. Deepak Kumar (New Delhi: Tulika Books, 2001), p. 35.
[12] J. M. Rogers, *Mughal Miniatures* (New York: Thames and Hudson, 1993), p. 33.

wire."[13] Other intellectuals of this period also referred to spectacles in their correspondence.

Likewise, Jesuits at the court of Abkar in the 1580s used them for reading, and this would have led to more gift-giving of spectacles to members of the court.[14] Other references to these aids to vision at the Mughal court appear in 1608–9 and 1625.[15] The brother of Queen Nur Jahan, Asaf Khan, received "a box containing two pairs [of spectacles] forwarded for him" by President Kerridge of the East India Company in September of that year. Both Dutch and English factory records contain references to the lively trade in these devices.[16]

As we have seen, by 1616, "Venetians" (it could have been other Europeans) were successfully selling lenses, spectacles, and other magnifiers in India, much to the discomfort of the British, who felt cut out of the optics trade.[17] These items could have been produced in Florence and then shipped outside Italy through Venice. Whether they were Florentine or Venetian in design, or whether spectacles from Flanders also arrived here, these optic items traveled around the world. It is fair to say that when the telescope arrived in northern India, scholars associated with the court, and probably other affluent nobles, had a century of experience under their belt with lens technology developed by Europeans in the Renaissance and earlier.

When the telescope arrived in India, the rulers and their officials were preoccupied with other things such as building monument gardens and conducting military campaigns. Although it is true that India under the Mughals had assimilated most of the Middle Eastern Islamic astronomical tradition, especially the tradition of compiling planetary tables called *zijs*, the Islamic presence in India did not have a deep scientific grounding. The golden age of Islamic science had occurred much earlier in the Middle East of the tenth and eleventh centuries. That was the age of such outstanding scholar-scientists as al-Biruni, Ibn Sina, and Ibn al-Haytham. Natural philosophers on this order were not to appear again in either the

---

[13] Iqbal Ghani Khan, "Medieval Theories of Vision and the Introduction of Spectacles in India, ca. 1200–1750," in *Disease and Medicine in India: An Overview*, ed. Deepak Kumar (New Delhi: Tulika Books, 2001), p. 35.

[14] Qaisar, *Indian Response to European Technology*, p. 75, who refers to the commentary of Father Monsarate at the Akbar court.

[15] Khan, "Medieval Theories of Vision," pp. 35–36.

[16] Ibid., p. 36.

[17] Qaisar, *Indian Response to European Technology*, esp. pp. 34–35, 71; and Foster, *Supplementary Calendar of Documents*, p. 83.

Middle East or Mughal India. On the other hand, very important Muslim astronomers continued to emerge in the thirteenth and fourteenth centuries in the Middle East. But no outstanding astronomers of the stature of such scholars as al-'Urdi, al-Tusi, and Ibn al-Shatir appeared in the Indian subcontinent in the time of the Mughals. By the time the Mughals established their suzerainty in the Indian subcontinent, the great intellectual achievements of the Arabic-Islamic past were only nostalgic memories.

Under Akbar, the tradition of Islamic educational institutions known as *khuttabs* and *madrasas* were established. These schools for village boys and older aspiring students were designed to preserve and pass on the Islamic religious tradition, not to undertake scientific studies such as al-Biruni, Ibn Sina, or Ibn al-Haytham undertook. They did not teach the natural books of Aristotle that discussed physics, plants and animals, mechanical motion, meteorology, or metaphysics and the logic of scientific investigation. Neither is there any evidence that India-based Muslim physicians achieved the level of medical insight of Ibn al-Nafis, the thirteenth-century physician and surgeon who has been credited with discovery of the lesser pulmonary circulation of blood. Likewise, the high level of optical studies pioneered by Ibn al-Haytham's eleventh-century achievements has no counterpart in the Indian subcontinent, either before or after the emergence of the Mughal Empire.

Nevertheless, Indian astronomers (Muslim and Hindu) maintained an interest in astronomy based on the Islamic models and the planetary tables compiled by Ulug Beg in Samarqand in 1428. Use of those tables came into play once again at the beginning of Shah Jahan's reign, when the tradition of creating a new calendar for the ascending monarch was followed.

Using the Samarqand data of Ulug Beg and the Ptolemaic system, a new calendar was presented to Shah Jahan in 1628 by the scholar Farid al-din Ibrahim Dihlawi. Asaf Khan, the scholarly Persian prime minister to Shah Jahan, was so impressed by the *Zij Shahjahani* that he asked another scholar, Nityananda, to translate it into Sanskrit.[18] That suggests that astronomy, time keeping, and calendar reform were important for the Muslims as well as the Hindus of India. Consequently, we can assume

---

[18] David Pingree, "The Sarvasiddhantaraja of Nityanda," in *The Enterprise of Science in Islam*, eds. Jan P. Hogendijk and A. I. Sabra (Cambridge, MA: MIT Press, 2003), pp. 269–84; and Pingree, "Indian Reception of Muslim Versions of Ptolemaic Astronomy," in *Tradition, Transmission and Transformation*, eds. F. J. Ragep and Sally Ragep (New York: Brill, 1996), pp. 471–85; as well as Pingree, "Islamic Astronomy in Sanskrit," *Journal of the History of Arabic Science* 2 (1978): 315–80.

that a new astronomical device, such as the telescopic discovery machine, would be of intrinsic interest to the Indian astronomers.

A still more intriguing connection concerns a certain Mulla Mahmud Jawnpuri (d. 1652), the leading scholar at Shah Jahan's court. He had spent time in Iran on scholarly pursuits and was very highly regarded. He drew up plans for an observatory and marked out a plot on the ground where an earlier observatory had once existed.[19] These events presumably transpired in the 1640s, but nothing came of the effort. Asaf Khan's knowledge of the telescope apparently did not trigger the thought that with the device, new astronomical discoveries might be made and that more precise observations would result in improved planetary tables, making calendars more reliable. Consequently, he did not champion Mulla Jawnpuri's proposal. Whether Mulla Mahmud Jawnpuri ever saw a telescope is difficult to say. He had studied in Persia, and telescopes began appearing there in the last several decades of the seventeenth century, probably after he returned to India, but they were already available in the Mughal capital five years after the appearance of Galileo's *Starry Messenger*. So it seems likely that conversations between Asaf Khan and Mahmud Jawnpuri would have acknowledged the existence of this new device, which had been in Asaf Khan's possession since 1615, when he served Jahangir.

In any event, his project was turned down, largely on the grounds that Shah Jahan hoped, through his Balk campaign, to conquer Samarqand where Ulug Beg's famous (pretelescopic) observatory was located. That plan never came very close to realization because all the Mughal military efforts failed to expand the empire into Persian territory. Of course, such an observatory as once existed in Samarqand was far from up to date in mid-seventeenth century.

In the meantime, the English astronomer Jeremiah Shakerley (d. 1653) arrived in Surat (on the west coast of India) with his telescope in 1651, the year before Mulla Mahmud Jawnpuri died. In October of that year, Shakerley observed the transit of Mercury across the face of the sun. This had been observed once before by European astronomers. Although it was an exciting event, it could by itself provide only modest evidence of the heliocentric hypothesis. Apparently, Shakerley's telescope was a

[19] Francis Robinson, *The 'Ulama of Farangi Mahall* (London: Hurst, 2001); and Zahir Ahmad Azhar, "Mulla Mahmud Jawnpuri," *Encyclopedia of Islam*, Urdu ed. (1984): 20:27–29. I am grateful to Professor S. M. Razaullah Ansari for this reference and the translation.

rather ordinary one called a "two-crown" telescope, which was widely available in England.[20]

Long after the telescope arrived, a Lahore family (the Allahdad family) was famous for making elegant astronomical instruments such as celestial globes, armillary spheres, and astrolabes. They continued making those instruments in the seventeenth century but made no telescopes. The engravings on their celestial spheres included 1,018 stars, the same as the number in Ulug Beg's tables but 10 fewer than in Ptolemy's compilation.[21]

With the ascension of Sultan Aurangzeb to the Mughal throne in 1666, concern for modern scientific developments faded even more into the background. His rulership became known for its vigorous imposition of Islamic law. Some historians suggest that he paid lip service to advancing modernized education, but no evidence has been brought forth indicating that such a reform was achieved. One can find individual scholars here and there who appreciated and avidly sought out the latest developments in Western science and philosophy brought to India by European travelers. For example, the French traveler and physician Francis Bernier, during his travels in India, spent twelve years (ca. 1658–70) as personal physician to Sultan Aurangzeb. During that time, he became fast friends with an official named Danishmand Khan (d. 1670), a former merchant. In 1660, he was appointed governor of Delhi. Bernier said of him:

He can no more dispense with philosophical studies in the afternoon than devoting the morning to his weighty duties as Secretary of State for Foreign Affairs and Grand Master of the Horse. Astronomy, geography, and anatomy are his favorite pursuits, and he reads with avidity the works of Gassendi and Descartes.[22]

Bernier reports that he personally translated same works of Gassendi into Persian. We cannot stress too much, however, the fact that science is a *social* activity embedded in a larger social context and that an isolated individual here and there is not enough to support the activity of science that leads to scientific progress.

[20] Allan Chapman, "Jeremy Shakerley (1626–1655?) Astronomy, Astrology and Patronage in Civil War Lancashire," chapter 7 in *Astronomical Instruments and Their Users* (Aldershot, UK: Variorum, 1996), p. 7.

[21] S. R. Sarma, *Astronomical Instruments in the Rampur Raza Library* (Rampur, India: Rampur Raza Library, 2003).

[22] As cited in S. A. A. Rizvi, *A Socio-intellectual History of the Isha Shias in India,* New Delhi (Munshivan Manoharlal Publishers, 1986), p. 225; and Francois Bernier, *Travels in India (1656–1668),* trans. Irving Brock, ed. and rev. A. Constable (New Delhi: Asia Educational Service, 1996), pp. 352–53.

In the end, no Mughal scholars undertook to use the telescope for astronomical purposes in the seventeenth century. Even the famous Jaipur observatory built by the Hindu scholar and ruler Jai Singh in the 1720s and 1730s near Delhi, long after the new-style observatories were built in Europe, made no use of telescopes. At the time, Europeans were using reflector telescopes, and modern-style observatories had been erected in Paris, London, and other places across Europe from the 1660s onward. Jai Singh's observatory, however, retained the old stone and masonry instruments that could not improve on the accuracy of telescopic observations, above all telescopes equipped with micrometers. Jai Singh's major observatory had only one scientifically useful instrument, a *Sasthansha yantra*, or pinhole-enabled sextant, located inside an observation room containing a scaled arch on which the sun's light would fall. It was a quadrant (or sextant) modeled after Ulug Beg's sextant in Samarqand but on a much smaller scale and with less versatility. It had no moving parts and thus could not be adjusted for measuring distances between adjacent stars or planets (Figures 5.4 and 5.5).[23]

Less than two decades after these monuments were constructed, the site ceased to function as an observation research site.[24] Reputedly, Jai Singh himself owned a telescope and had looked at Jupiter with it. But that experience did not jolt him into building observatories equipped with telescopes.

The Ottoman Middle East

Back in the greater Middle East, stretching from Morocco to Anatolia as well as to the northern part of Hungary, the Ottomans ruled an equally large empire. The Ottomans, more than the Mughals, were direct inheritors of the great Islamic traditions in astronomy, optics, medicine, and the physical sciences. Ottoman scholars had a long tradition of reading Arabic sources, and sixteenth- and seventeenth-century Ottoman

---

[23] For a picture of the Samarqand sextant, see http://en.wikipedia.org/wiki/Sextant_ (astronomical).

[24] V. N. Sharma, *Sawai Jai Singh and His Astronomy* (Delhi: Motilal Banarsidass, 1995), esp. pp. 58–59. Also Raymond Mercer, "The Astronomical Tables of Rajah Jai Singh Sawai," *Indian Journal of History of Science* 19, no. 2 (1984): 143–71; and Eric C. Forbes, "The European Astronomical Tradition: Its Reception into India, and Its Reception by Sawai Jai Singh II," *Indian Journal of History of Science* 17, no. 2 (1982): 234–43. For an insightful analysis of the instruments and their construction, see Andreas Volwahsen, *Cosmic Architecture in India* (New York: Prestel, 2001).

FIGURE 5.4 The Samrat *yantra*, the "Royal Instrument," at the Jaipur observatory outside Delhi functioned as a gnomon-based sundial. In the smaller towers in the western and eastern quadrants, there were dark chambers constituting a Sasthamsha *yantra*,[25] sextant, or aperture gnomon.

astronomy was based on the long-standing Arabic-Islamic tradition. Its earliest foundations were Greek, but by the thirteenth and fourteenth centuries, Islamic astronomers, especially al-'Urdi, al-Tusi, and Ibn al-Shatir, had made significant technical alterations of that tradition. But they did not generate a revolutionary break with the past. The Islamic astronomical tradition was the same Ptolemaic tradition that paved the way for the European revolution in astronomy of the sixteenth and seventeenth centuries.

By the sixteenth century, the Islamic astronomical tradition had lost all progressive momentum. There were scholars with very high mathematical skills, such as Shams al-din al-Khafri in the generation of Copernicus, but they were not able to make the revolutionary breakthrough to the heliocentric system. The last significant observatory was commissioned to be built in Istanbul in 1575 under the direction of the Syrian-born

[25] There are many variant spellings of this term, e.g., *Shastansha*, *Sasthamsa* (with diacritical marks), and even *Shashthamsa*.

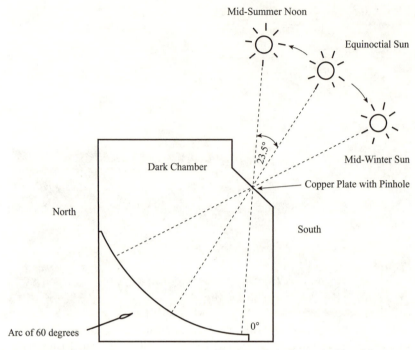

FIGURE 5.5 Aperture gnomon or sextant (Sasthamsha). On each side of the great Samrat *yantra* were large twin structures with enclosed dark chambers. Inside each chamber was a 60-degree arc marked off in degrees and minutes of a circle. Light entered the chamber through a pinhole and registered on the gradated arc. This aperture gnomon was the main precision instrument installed at Jaipur and could only be used to measure the height of the sun over the course of the year. As the sun's position changed, the cone of light entering the chamber would be registered on the scale at a moving point. This device was modeled after the stone and masonry sextant installed by Ulug Beg in Samarqand in the fifteenth century.

mathematician-astronomer Taki al-Din during the reign of the Ottoman sultan Murad III. The sultan was known for his very large collection of women in his harem, wars with Iran and Austria, and political decline. The observatory was immediately torn down before it was finished in the early days of 1580 because of religious opposition and political strife.[26] Time keepers (*muwaqqits*) with high levels of mathematical competence had been officially brought into the mosques in the Muslim world by

---

[26] Aydin Sayili, *The Observatory in Islam* (Ankara: Turk Tarih Kurumu Besimevi, 1960), pp. 289–95, 297–300, 302–5.

Mamluk times and as early as the thirteenth century. They could determine time using the old methods based on gnomons, sundials, and solar and lunar sightings. Even in the sixteenth and seventeenth centuries, Muslim use of mechanical clocks developed by Europeans was rare, and no clock towers were to be seen in public squares in the Middle East. In Europe, such devices were widely placed from the fourteenth century onward, with pocket watches appearing in the early sixteenth century.[27] Mechanical clocks and watches became highly prized gifts for Ottoman sultans, but adopting mechanical clocks for official time keeping was too daring and would have provoked a strong reaction by the religious scholars.[28]

The conservative Muslim tradition of opposing new technologies at that time prevented the use of the printing press, both in Ottoman lands and in Mughal India. Indeed, in the latter, regular use of the printing press did not occur until the nineteenth century, as we shall see.[29]

The arrival of the telescope early in the seventeenth century had the potential to level the playing field in astronomy so that Muslim astronomers could undertake the same investigations as Europeans. Insofar as it was also a revolutionary instrument capable of revealing a different cosmology, a heaven populated by many new stars, and planets surrounded by revolving satellites, it could also be seen as a threat to traditional Islamic astronomy.

Telescopes and the Ottomans

The Ottomans shared a long border with Europeans from Greece up through the Balkans to Hungary and Austria. Like the Mughals, the Ottomans were completely free from European domination. Parallel to the Ming Dynasty in China, the Ottomans maintained a many-centuries-long empire, one that far outlasted the Ming. From that perspective, the

[27] David Landes, *The Revolution in Time*, rev. and enl. 2nd ed. with a new preface (Cambridge, MA: Harvard University Belknap Press, 2000), pp. 90ff. Bernard Lewis, *What Went Wrong?* (New York: HarperCollins/Perennial, 2003), pp. 117ff.

[28] Otto Kurz, *European Clocks and Watches in the Near East* (London: Brill, 1975).

[29] Francis Robinson, "Islam and the Impact of Print in South Asia," pp. 62–976, in N. Crook, ed., *The Transmission of Knowledge in South Asia* (Oxford: Oxford University Press, 1996). For opposition in Ottoman lands, see Stanford Shaw, *History of the Ottoman Empire and Modern Turkey*, vol. 1 (New York: Cambridge University Press, 1976); and Niyazi Berkes, *The Development of Secularism in Turkey* (New York: Routledge, 1964), pp. 34–41.

Ottomans continued to look down on Europeans and considered their own civilization far superior to that of the West.

Until now, historians of Ottoman science placed the first mention of the telescope in Turkish in the publications of Ibrahim Müteferrika in the 1730s.[30] He was the Christian convert to Islam from Hungary responsible for setting up the first Turkish printing press in the Ottoman Empire. It had a short life, from 1728 to 1745,[31] but communications between the Ottomans and Europeans, especially those coming through Venice, were very extensive in the late sixteenth and early seventeenth centuries. During that time, dozens, if not hundreds, of Europeans sailed from ports in Italy, especially Venice, as travelers to the Ottoman capital in Istanbul.[32] So it is quite possible that travelers from Venice brought the spyglass during the very first decade of its invention. For we know that Florentine-made spectacles and other items of glassware were routinely shipped via Venice to Ancona, and then overland to Sarajevo, down to Adrianople, with a final destination in Pera, a section of Istanbul.[33] In September 1540, 24,000 pairs of spectacles were sent overland through Ragusa [Dubrovnik] to Pera in Constantinople.[34] Mosque lamps destined for the Middle East were also part of such shipments. Apart from these commercial channels, Ottoman soldiers during encounters with Europeans might have gained knowledge of spyglasses used by European military leaders.

It also seems likely that European ambassadors, such as our Sir Thomas Roe, would have brought telescopes with them as gifts in the 1620s. Ambassador Roe was reassigned to Istanbul after returning home from India. He arrived in Istanbul in 1621, but references to the telescope in his journal have not been found.

Our earliest certain knowledge of the presence of spyglasses in the Ottoman Empire comes from the 1630s. The very first reference to "Galileo's glass" in Istanbul is the hanging of a Venetian merchant who apparently used his spyglass to gaze at the harem of the royal palace.

---

[30] Adnan Adivar, *La Science chez les Turcs Ottomans* (1939), and Adivar, *Osmali Türklerinde Ilm* (Istanbul: Remzi Kitabevi, 1943), p. 160, in a translation of Aristotle's *Physics*, by Esad Effendi for Müteferrika's press; see also Berkes, *Development of Secularism in Turkey*, p. 50.

[31] Shaw, *History of the Ottoman Empire*, p. 235; Berkes, *Development of Secularism in Turkey*, pp. 36–47.

[32] Eric Dursteler, *Venetians in Constantinople: Nation, Identity and Coexistence* (Baltimore: Johns Hopkins University Press, 2006).

[33] Ilardi, *Renaissance Vision*, pp. 122–23.

[34] Ibid., p. 122.

For that offense, he was hanged in 1630 or 1631 by Sultan Murad IV,[35] known as Murad the Cruel because of his violent and impulsive disposition. Despite his impulsiveness, Murad coveted the telescope for spying on the French embassy nearby. But there were serious repercussions from this incident. Goods of other European merchants that might have contained telescopes also were impounded and the merchants imprisoned.[36]

Credit for teaching the sultan how to use the telescope has been claimed by another Venetian, Pietro Venier, who arrived in Istanbul in 1626 and served the royal court in the 1630s. He also converted to Islam to better serve the Ottomans.[37] So we can be sure that the telescope arrived in the Ottoman Empire at least by 1630 but probably earlier. It was also used by Murad's brother and successor, Sultan Ibrahim, who ruled from 1640 to 1648, before he was strangled by court officials who wanted him removed. He, too, used the telescope to spy on his subjects in their backyards.[38] This was a common use of the spyglass then, indulged in by Europeans as well. But with the close involvement of religious scholars in the royal court and the constant presence of the person who would have been the sultan's *hoja* (spiritual master), we must assume that the religious scholars knew about the telescope and probably had occasion to see and use it. The incident of the Venetian merchant spying on the royal harem must have created a stir in legal circles – that is to say, among the religious scholars. It would have been a basic legal question to ask whether such a new instrument and its use, above all one brought from infidels of the West, would be permitted or forbidden to Muslims.

By the 1650s, we have good evidence that the telescope was used by the Ottoman navy. According to the important Ottoman scholar, Katib Çelebi, telescopes in the 1650s were on board the "admiral's ship." This tells us that by mid-seventeenth century, the spyglass had been adopted by the Ottoman navy and probably by other military services.[39]

35 Dursteler, *Venetians in Constantinople*.
36 M. de Hammer-Purgstall, *Histoire de L'Empire Ottoman: Depuis son Origine Jusqu'a Nos Jours* (Paris: Parent-Desbarres, 1841), 2:472.
37 Dursteler, *Venetians in Constantinople*, p. 138.
38 Balthasar Monconys, *Journal des Voyages* [ca. 1647–48] (Lyon: Chez Horace Boissat and George Remeus, 1665–66), pp. 389, 391. He also looked at Venus and the moon while in Constantinople. Also see Sonja Brentjes, "Western European Travelers in the Ottoman Empire and Their Scholarly Endeavors (Sixteenth–Eighteenth Centuries)," in *The Turks*, eds. Hasan Celal Güzel, C. Cem Oguz, and Osman Karatay (Ankara: Yeni Türkiye, 2002), 3:795–803.
39 This is reported in Katib Çelebi's history of Ottoman naval warfare, *Tuhfet ul-Kibar* (ca. 1656). I thank Gottfried Hagen for this reference and translation.

The next indication of telescopes being used in the Ottoman Empire in the seventeenth century is found in the huge travel logs of Evliya Çelebi. He had been born in 1611, two years after Katib Çelebi. Evliya had an extraordinary career, mainly as a traveler throughout the Ottoman lands, with occasional service in the government. He had been an entertainer to Sultan Murad, and it was probably through his connection with the Sultan in the 1630s that he acquired his telescope.[40]

In any event, Evliya Çelebi carried his telescope with him on his travels all over the empire. In 1657, he "observed enemy movements from the top of a minaret during a siege of Özü," a small town on the Black Sea coast belonging to Ukraine. In 1663, he espied the banners of approaching Hungarian troops and in 1667 observed an unfolding naval battle off the Greek coast of Glarentsa. He also used it to decipher inscriptions too small for the unaided eye in Bosnia (1660). He did the same at the citadel in the Syrian town of Betlis in 1672.[41] Whether Evliya Çelebi took an interest in skyward observations is not known.

These diverse reports of telescopes being used in the Ottoman Empire suggest that they were widely available from the 1630s onward. Officials had them, naval vessels had them, and a traveler such as Evliya took his instrument all over the empire, from the borders of Eastern Europe to Ukraine, Greece, and Syria. No doubt other references to the presence of the telescope will be found throughout the Middle East and North Africa when scholars turn their attention to this topic. We know that telescopes were to be had in Egypt at about this time. For example, the French scholar and traveler, Balthasar de Monconys, visited Istanbul with his telescope in the 1640s. For some reason, during his travels, his telescope was misplaced. When he arrived in Egypt, he was able to buy another from a Cairo merchant. This suggests that the telescope had reached Egypt and probably North Africa by the 1640s, if not earlier.[42]

Likewise, telescopes had reached Isfahan in Safavid, Iran, by the 1660s and probably earlier. The French Capuchin priest Raphaël du Mans lived in Isfahan from May 1647 until his death in 1669. He appears to have been an extraordinary person, highly skilled in diplomatic as well as scientific matters, especially astronomy. A seventeenth-century Persian scholar by the name of Mirza 'Abdullah Isfahàni wrote a book called

---

[40] For Evliya Çelebi's biography and travels, see Robert Dankoff, *An Ottoman Mentality: The World of Evliya Çelebi* (Boston: Brill, 2004).

[41] Ibid., 191–92.

[42] Monconys, *Journal des Voyages*, and Brentjes, "Western European Travelers," 3:795–803.

"The Garden of the Scholars" (*Riyas al-'Ulama*) in which he refers to Father du Mans and a telescope that du Mans possessed. The instrument was 2.08 meters long and shaped "like a bamboo pipe, somewhat like the Indian reed from which spears are made." With it, many strange things could be seen. Apparently, Isfahani had the opportunity to view the sky through the telescope. He reported that many stars not seen before were revealed, including a vastly expanded Pleiades (Seven Sisters) that clearly could not be taken in with one view, given the narrow telescopic field of vision. Looking at the moon's rough and uneven surface suggested to viewers in Iran that there were "lands, jungles, and cities on the moon."[43]

In short, telescopes had been widely disseminated across the Ottoman Middle East, among the Safavids, and in the Mughal Empire within a decade or two of their invention in Europe in 1608–9 and the appearance of Galileo's revolutionary revelations in *The Starry Messenger* of 1610. Yet, the telescope's arrival in Muslim lands – in Mughal India, the Ottoman Empire, and elsewhere – hardly created a stir. No new observatories were built, no improved telescopes were manufactured, and no cosmological debate about what the telescope revealed in the heavens have been reported.[44]

Toward a Preliminary Assessment: The Muslim Context

Until the fourteenth century, students of optics in the Middle East did make progress and achieved a high level of development. These advances, associated with the name of Ibn al-Haytham and Kamal al-din al-Farisi, were partially transmitted to Europe, where the study of optics continued to develop. In the Muslim world, however, the study of optics went into abeyance. Although Western-manufactured lenses soon appeared in the Middle East, probably in the early fourteenth century, no interest emerged there leading to the development of a parallel industry in the optics invented by Europeans.

When the telescope arrived in the early seventeenth century, Muslim astronomers paid no attention to it. There is no indication that Middle Eastern scholars used the telescope for astronomical exploration, nor did

---

[43] I am grateful to Dr. Ali Paya, Visiting Professor, Center for the Study of Democracy, University of Westminster, United Kingdom, for this reference and translated passages. Personal communication, December 14, 2007.
[44] See Ekmeleddin Ihsanoglu, "The Introduction of Western Science to the Ottoman World: A Case Study of Modern Astronomy," chapter 2 in *Science, Technology, and Learning in the Ottoman Empire* (Aldershot, UK: Ashgate, Variorum, 2004).

they make any replicas of the Western-invented telescopes that arrived in the Muslim world in the second and third decades of the seventeenth century. Neither do we have any indication that Muslim scholars were in search of or open to new theoretical ideas in astronomy. Indeed, the shift to a Copernican and Newtonian worldview was greatly delayed in both the Middle East and China. In that regard, the curiosity deficit seen in China also prevailed in the Muslim world.

In a global context, there were significant cultural and intellectual differences between China and the Muslim world, yet the result was the same: stasis. The Christian missionaries were allowed into China, albeit in a controlled fashion, yet the Jesuits worked out a plan for converting the Chinese to Christianity by importing a rich body of Western scientific and philosophical writings. Unlike the Muslim world, there was an eager audience of Chinese scholars who wanted Western scientific assistance in correcting and perfecting their defective astronomical system. But success with establishing the new sciences of the seventeenth century in China – the new telescopically enabled astronomy as well as microscopy – was not to be forthcoming.

In the Muslim world, almost all the scientific corpus based on Greek natural thought was already shared by the two civilizations, Islamic and European. There were Jesuits traveling throughout the Muslim world in the seventeenth century, in the Ottoman lands and in Mughal India, but a master plan for converting Muslims to Christianity by teaching modern science in the Muslim world, such as Matteo Ricci conceived in China, could not possibly materialize there because conversion to Christianity, apostasy, was a capital offense. That injunction against proselytizing and conversion still officially prevails throughout the Muslim world today.

By way of summary, let us return to the three key areas on which we focused in the case of China to assess progress toward a scientific revolution on the order of the Newtonian synthesis. The first indicator would be use and modifications of the telescope such as the Europeans carried out during the seventeenth century. The second concerns the observational improvement of star charts and catalogues. Third, what can be said of progress in developing a unified physics of celestial and terrestrial motion? This latter subject also involves progress in mechanics and the science of motion.

As we have seen, no research up to the present has given us any indication that the Ottomans, Mughals, or Persians embraced the telescope, mastered its use, or improved on its design and construction.

All seventeenth- and eighteenth-century advances in telescopy came from within Europe.

It is agreed with regard to star charts that the last significant set of observations in the Muslim world were those done under the direction of Ulug Beg, who died in 1449. His *zij*, planetary tables, remained the most important and popular within the Muslim world long after major improvements over Ptolemy had been made in Europe. Ulug Beg's tables contained 1,018 stars, 10 fewer than Ptolemy's, but his improved coordinates for 992 stars and planets were the first major improvement over Ptolemy's compilation of the second century A.D. He also made a more precise calculation of the length of the solar year and some improvements on the orbits of the five major planets: Saturn, Jupiter, Mars, Venus, and Mercury. Because the planetary tables were not published in Europe until after Tycho Brahe's more exact observations, Ulug Beg's tables had no influence on European astronomy. But, as we saw earlier, eighteenth-century Indian astronomers continued to work with Ulug Beg's tables, seeking to improve them, rather than working with the more precise European tables.[45]

Furthermore, because the construction of daily and monthly tables (ephemerides) for the celestial bodies was perceived as an astrological exercise designed to predict the future, a domain of knowledge reserved to God, no new ephemerides were constructed in the Muslim world. Neither did the arrival of the telescope lead to the production of new star charts in the seventeenth century. Some Indian astronomers, under the direction of Jai Singh, sought to improve on the planetary tables of Ulug Beg in the early eighteenth century; but, without mastery of the new telescopic instrumentation, no significant improvements could be made. What they did produce appears to be a patchwork based on Ulug Beg's tables and values borrowed from late Europeans.[46]

Muslim and Indian astronomers, even those trained in the tradition of Greek spherical geometry, basically abandoned the field from the time of

---

[45] Allan Chapman, Introduction to *The Preface to John Flamsteed's Historia Coelestis Britannica or British Catalogue of the Heavens* (London: Trustees of the National Maritime Museum, 1982); and Joseph Ashbrook, *The Astronomical Scrapbook* (Cambridge, MA: Harvard University Press, 1984), p. 424.

[46] William A. Blanpied, "Raja Sawai Singh II: An 18th Century Medieval Astronomy," *American Journal of Physics* 43, no. 12 (1975): 1025–35; see also Raymond Mercer, "The Astronomical Tables of Rajah Jai Singh Sawai," *Indian Journal of History of Science* 19, no. 2 (1984): 143–71, and Eric C. Forbes, "The European Astronomical Tradition: Its Reception into India, and Its Reception by Sawai Jai Singh II," *Indian Journal of History of Science* 17, no. 2 (1982): 234–43.

the death of Ulug Beg to the early nineteenth century and the incursions of European scientists bearing the new scientific ethos. The first modern observatory in Turkey was built in the 1830s, but no systematic astronomical observations with a telescope were made until the early twentieth century.[47] Indian astronomers, collaborating with English scholars, launched their modern astronomical program in the early nineteenth century, but still two centuries after Galileo and Kepler.

On a more theoretical level, there was no interest in translating the major works of Copernicus, Galileo, or Kepler in the seventeenth century. Only toward the end of that century did Ottoman scholars translate the work of a minor European astronomer, Noel Durret, mainly because of interest in his planetary tables, not new astronomical theory. The Turkish historian of science, Ekmeleddin Ihsanoglu, has summed it up: "The basic European works by scientists such as Copernicus, Tycho Brahe, Kepler and Newton were not translated, indeed no attempt was made to understand or interpret them."[48] The arrival of Galileo's new discovery machine in Ottoman and Muslim lands provoked no new astronomical investigations.

Last, we must note that the science of mechanics – that is, physics – was an equally indispensable science on the path to the Newtonian synthesis. It was discussions in this field that eventually laid the foundations for a unified terrestrial and celestial physics in the thought of Newton. These were areas in which the combined work of Galileo, Kepler, and Newton, aided by many others, finally yielded Newton's grand synthesis – the New Mechanics. In the Western world, getting to that point involved a very long dialogue with the Aristotelian conception of physics (natural philosophy) and the science of motion. For practicing scientists, this concern for the fundamental properties of nature, their change and transformation, and above all the nature of motion, had always been the intellectual core of physics. This intellectual struggle lies at the very heart of physics and the science of mechanics as conceived by Aristotle. In one important sense, the Arab-Muslim world, unlike China and Asia more broadly, shared this Aristotelian heritage. The problem was that Aristotle's *Physics*, the text, and related sciences were not taught in the madrasas.

Nevertheless, there was one important Muslim scholar, living in Spain, who made an important contribution to the medieval science of mechanics

[47] Mustafar Kacar, personal communication, January 24, 2006.
[48] Ihsanoglu, "Introduction of Western Science," pp. 2ff, 32, 42.

that filtered through to Galileo. That man was Ibn Bajja, who was the first Aristotelian Muslim philosopher in Spain but who died, probably from poisoning by his coreligionists, in Fez, Morocco, in 1139. He had been born in Saragossa but traveled around the country, often advising the Almohad rulers in Seville and Granada.

Ibn Bajja's commentary on Aristotle regarding the dynamics of motion had elements of originality and, though it was far from a solution to the problem of explaining projectile motion and free fall, is in a direct path leading to Galileo's theory of free fall. It does not detract from Ibn Bajja to say that the theory was probably first invented by the Christian Neo-platonist Johannes Philoponus in sixth-century Alexandria. No Arabic manuscript of it seems to exist, though Ibn Bajja may have heard of it through various Arab traditions.[49]

Ibn Bajja's innovation was to suggest that the true motion of a projectile is revealed when we think of it as freed from other natural forces resisting its motion. Unlike other thinkers all the way back to Aristotle, who believed that if all resistance in the air surrounding an object were removed, the object would accelerate instantaneously to an infinite speed, Ibn Bajja dissented. He thought that if we imagine all resistance removed from an object, then that would reveal its true or "natural state" of motion. Although he was unable fully to work out the mathematics, his work suggested that the acceleration of an object freed from all resistance would be exponential, not instantaneously infinite.[50]

Ibn Bajja's ideas were acknowledged by the great twelfth-century Muslim philosopher, Ibn Rushd (Averröes), but he rejected them. Nevertheless, Ibn Bajja's ideas filtered into the thought of dozens of European scholars working on the science of mechanics at Merton College, Oxford, and others at Paris. The latter included Jean Buridan, Albert of Saxony, and Nicole d'Oresme.

In the Islamic world, the science of motion died out with Ibn Bajja. It was unable to make further contributions to the landmarks that marked the path to Galileo's law of free fall and inertia. The latter idea implied that an object unimpeded by external resistance would either remain in

---

[49] Ernest Moody, "Galileo and Avempace [Ibn Bajja]: Dynamics of the Leaning Tower Experiments," in *Roots of Scientific Thought*, eds. Philip P. Wiener and Aaron Noland (New York: Basic Books, 1957), p. 200.

[50] Ernest Moody, "Laws of Motion in Medieval Physics," *The Scientific Monthly* (January 1951): 18–23. For an easily accessible overview of the idea of impetus and the science of motion, see Herbert Butterfield, *The Origins of Modern Science, 1300–1800*, rev. ed. (New York: Free Press, 1954), pp. 13–28.

a steady state of rest or continue in its state of motion indefinitely. These were indispensable ideas for the new science of mechanics that was to be fashioned by Newton in 1687.

Samarqand: The Last Outpost

The last significant observational advances in the Muslim world, as we have noted, occurred under Ulug Beg in Samarqand in the mid-fifteenth century. Ulug Beg was the grandson of Timur (Tamerlane), who, in the fourteenth century, ruled a large territory in central Asia, extending south to Afghanistan and for a time into India. Ulug Beg's father, Shah Ruhk, inherited the empire, whose capital he moved to Herat in Afghanistan. With the latter's death in 1447, Ulug Beg became the titular ruler, reigning from Samarqand. However, he was a highly gifted mathematician, and his interests were in science and the arts, not political power. He had established a madrasa in Samarqand in 1420, to which he invited respected mathematician-astronomers to teach. These included the highly gifted Jamshid al-Kashi and a mathematical scholar known as Qadizade. Traditional legal scholars held forth at the madrasa, while students supported by Ulug Beg's generosity also learned the "mixed sciences" such as mathematics and astronomy. It is unlikely, however, that physics and the other natural books of Aristotle were taught there. Jamshid Kashi's letters to his father indicate that many people were versed in mathematics and some in astronomy, who had studied Euclid's *Elements*. He mentions "twenty-four calculators," some who knew "other sciences," probably experts in the construction of astrolabes and armillary spheres, and some with knowledge of Avicenna's medical canon. Ulug Beg's madrasa caused considerable controversy among legal scholars, with the result that Ulug Beg was executed by his son in 1449.

About eight years after the founding of Ulug Beg's madrasa in 1420, he built an observatory in another location because of its large size. The instrument was a sort of quadrant, a so-called Fakhri sextant, with a radius of 118 feet. A considerable portion of the instrument was located below ground, and only because of that fact were the ruins of the building later discovered through excavations in 1908.[51] Shortly after Ulug Beg's death, the observatory was abandoned and its location obscured by its destruction. With Ulug Beg's death, the scholars associated with this observatory had to flee for safer places. Nevertheless, Ulug Beg's new star catalogue was passed on to the Muslim world.

[51] For a photo image of the remains, refer to note 23.

One of the mathematician-astronomers attached to Ulug Beg was a scholar by the name of 'Ali al-Qushji, who died in 1474, one year after Copernicus was born. By some accounts, he was the son of Ulug Beg's falconer and the protégé of the Sultan Ulug Beg. In any case, he was a gifted mathematician and worked out a transformation routine for planetary orbits involving a parallelogram that appeared later in the work of the fifteenth-century German astronomer, Regiomontanus. He was the most important European astronomer of the fifteenth century and was virtually an exact contemporary of al-Qushji. Regiomontanus's rendition of an *Epitome* of Ptolemy's *Almagest* was a major tool for Copernicus as he attempted to work out the mathematical details of his new heliocentric system.

With the death of Ulug Beg, al-Qushji and others associated with the Samarqand observatory had to seek refuge elsewhere. Al-Qushji ended up in Istanbul during the last years of his life, where he taught mathematics in several madrasas. But al-Qushji belonged to an entirely different astronomical tradition as concerns human understanding of the nature of our cosmos. For one of the central issues in the march to the modern scientific revolution concerns the assumption that natural philosophy in Aristotle's view was queen of the sciences. This meant that only natural philosophers could decide what the shape of the world really is, and that search involved finding the *causes* of the changes and alterations of the natural world. That deep philosophical assumption underlay Galileo's wish to be named "philosopher and mathematician" to the grand duke in 1610 because he wanted to talk about cosmology, the real shape of the world, not arbitrary mathematical models.

Second, the long-standing Aristotelian tradition maintained that natural philosophers carried out their work by finding the real structure and causes of things. To explain something was precisely to explain the causes of it in a logically demonstrable manner. This had been understood by the great "philosopher of the Arabs," al-Kindi, and all Muslim philosophers who followed that path. When we arrive in the time of Galileo, the same assumptions prevail. In his first *Letter on Sunspots* (1613), Galileo insists that he was investigating "the true constitution of the universe – the most important and most admirable problem that there is." And then he claims, "For such a constitution exists; it is unique, true, real, and could not possibly be otherwise."[52]

---

[52] Galileo, *Letter on Sunspots*, in Stillman Drake, *Discoveries and Opinions of Galileo* (New York: Anchor Books, 1957), p. 97.

Likewise, more than a decade earlier, Kepler's *New Astronomy* begins by adopting Copernicus's heliocentric assumption and proceeds to find the foundations of modern astronomy, "Based Upon Causes or Celestial Physics." In other words, Kepler was bringing the principles of physics to bear on astronomy, and this marriage would yield a causal explanation of the movements of celestial bodies.

But in al-Qushji's writings, a very different attitude appears. Instead of founding the science of astronomy on the assumptions of natural philosophy, which had been taken to be a universal principle since the time of Aristotle, al-Qushji proposes to found it on the will of God. He wants to separate astronomy from the control of philosophy and thereby abandon the search for the real structure of the universe and the causes of it. He writes:

What is stated in the science of astronomy does not depend upon physical . . . and metaphysical . . . premises. . . . For of what is stated in this science some things are geometrical premises, which are not open to doubt; others are premises determined by reason in accordance with the apprehension of what is most suitable and appropriate. [On the contrary] it is sufficient for them [astronomers] to conceive, from among the possible models, the one by which the circumstances of the planets with their manifold irregularities may be put in order in such a way as to facilitate their determination of the positions and conjunctions of these planets for any time they might wish and so as to conform with perception and sight, this in a way that the intellect and the mind find wondrous.[53]

In this manner, al-Qushji departs entirely from those Arab and Muslim astronomers who were realists and sought an actual explanation of the heavens. Al-Qushji had taken a page directly out of the anticausal thought and language of al-Ghazali (d. 1111), the great Islamic religious philosopher who set out the most influential position denying all necessary connections. Al-Qushji asserts that, granted the existence of the all-powerful will of God,

it is conceivable that the volitional Omnipotent could by His will darken the face of the Moon during a lunar eclipse without the interposition of the Earth and likewise during a solar eclipse the face of the Sun (would darken) without the interposition of the Moon; likewise he could darken and lighten the face of the Moon according to the observed full and crescent shapes.[54]

---

[53] 'Ali al-Qushji, "Concerning the Supposed Dependence of Astronomy upon Philosophy," trans. F. Jamil Ragep, *Osiris* 16 (2001): p. 68. I have used Gerhard Endress's translation in his "Mathematics and Philosophy in Medieval Islam," in *The Enterprise of Science in Islam*, eds. Jan P. Hogendik and Abdelhamid I. Sabra (Cambridge, MA: MIT Press, 2003), p. 160.

[54] Al-Qushji, "Concerning the Supposed Dependence of Astronomy," p. 68.

Al-Qushji had adopted Islamic atomism, according to which all the motions and changes in the world are but the mere working of God's will, God's "habit" of arranging things. Al-Ghazali's famous denial of causality that reverberated throughout the Muslim world all the way to the twentieth century reads:

According to us the connection between what is usually believed to be a cause and what is believed to be an effect is not a necessary connection; each of the two things has its own individuality and is not the other, and neither the affirmation nor the negation, neither the existence nor the non-existence of the one is implied in the affirmation, negation, existence, and non-existence of the other – e.g., the satisfaction of thirst does not imply drinking, nor satiety eating, nor burning contact with fire, nor light sunrise, nor decapitation death, nor recovery the drinking of medicine, nor evacuation the taking of purgative, and so on for all the empirical connections existing in medicine, astronomy, the sciences, and the crafts. For the connection of these things is based on a prior power of God to create them in successive order, though not because this connection is necessary in itself and cannot be disjointed – on the contrary, it is in God's power to create satiety without eating, and death without decapitation, and to let life persist notwithstanding the decapitation, and so on with respect to all connections.[55]

Moreover, al-Qushji's position is fully in line with the orthodox religious position, expressed a century earlier by one of the most important Islamic religious philosophers of that age, al-Iji (d. 1355). His monumental summing up of the Sunni and atomist theology (*kalam*) was taught all the way into the early twentieth century at the great madrasa in Cairo, al-Azhar. According to al-Iji, "these [hypotheses of mathematical astronomy] are imaginary things that have no external existence . . . mere imaginings which are more tenuous than a spider's web."[56] In adopting the same metaphysical view, al-Qushji throws astronomy off the track that the European astronomers such as Galileo and Kepler pursued.

Within about thirty-five years of al-Qushji's death – that is, in 1510 – Copernicus would come up with his new heliocentric cosmology. In the meantime, Muslim astronomy in Ottoman lands languished. The arrival of the telescope in the late 1620s had no impact on Muslim astronomy.

Still another indication of the neglect of the new telescopic astronomy in the Middle East comes from the reports of the eighteenth-century Egyptian historian Abd al-Rahman al-Jabarti. In his multiple-volume work, he wrote a number of obituaries for Egyptian scholars. In a number of cases,

55 Al-Ghazali, as translated in Seyyed Hossein Nasr, *Science and Civilization in Islam* (New York: New American Library, 1968), p. 318.

56 As cited in A. I. Sabra, "Science and Philosophy in Medieval Theology: The Evidence of the Fourteenth Century," *Zeitschrift für Geschichte der Arabisch-Islamischen Wissenschaften* 9 (1994): p. 37.

he mentions their collections of scientific instruments but never mentions any telescopes.[57] Indeed, Jabarti himself was quite stunned by the level of scientific thinking that the French invaders brought to Egypt when Napoleon conquered it in 1798. He has been quoted as saying that the science of the French was "beyond minds the likes of ours."[58] This surprising delay in adopting the telescope in Muslim regions, given their long preoccupation with astronomical inquiries, especially for time keeping and for computing directions to Mecca, left the Middle East in an underdeveloped state of scientific inquiry. These contrasting paths of scientific development in the Muslim world and in Europe reveal the very different levels of scientific curiosity that prevailed in the West and in Muslim countries for centuries to come.

[57] Thomas Philipp and Moshe Perlmann, eds., *Abd al-Rahman al-Jabarti's History of Egypt* (Stuttgart, Germany: Franz Steiner, 1994).

[58] John Livingston, "Shaykhs Jabarti and 'Attar: Islamic Reaction and Response to Western Science in Egypt," *Der Islam* 74, no. 1 (1997): 92–106; Livingston, "Western Science and Educational Reform in the Thought of Shaykh Rifa'a al-Tahtawi," *International Journal of Middle Eastern Studies* 28 (1996): 543–64; and Albert Hourani, *A History of the Arab People* (New York: Warner Books, 1992).

PART II

PATTERNS OF EDUCATION

6

Three Ideals of Higher Education

Islamic, Chinese, and Western

Until the nineteenth and twentieth centuries, with the rise of globalization, societies and civilizations of the past were deeply rooted in their local cultures and traditions. This was especially so with regard to their practices of socialization and education. The educational traditions of Europe stood far from those of the Muslim Middle East and from those of China and Mughal India. Educational practices are always deeply embedded in religious and philosophical traditions, and those traditions in China, India, and Europe were considerably different.

Although Islam spread in many areas that had once been Christian, Islamic philosophy and institutional practices stand in contrast to Christian conceptions. Christianity from the outset had been deeply influenced by Greek philosophy and Hellenic culture that still survived at the time of Christ. On the other hand, when Islam arose, Hellenic culture had virtually disappeared. Furthermore, the Arabian peninsula had never been significantly penetrated by either Greek or Roman culture. Consequently, the metaphysical and philosophical foundations of the two civilizations were markedly different.[1] Even though there was an impressive translation movement of the eighth and ninth centuries that brought a huge

---

[1] I have discussed these differences in "Science and Metaphysics in the Three Religions of the Book," in *Intellectual Discourse* 8, no. 2 (2000): 173–98. A slightly revised version appears in "The Open Society, Metaphysical Beliefs and Platonic Sources of Reason and Rationality," in *Karl Popper: A Centenary Assessment, Metaphysics, and Epistemology*, eds. Ian Jarvie, Karl Milford, and David Miller (Aldershot, UK: Ashgate, 2006), 2:19–44.

stock of Greek philosophy into Arab areas,[2] differences in attitudes to the natural philosophy of Plato and Aristotle remained.

In addition to those dimensions of cultural embeddedness, there is always a legal side to cultural matters that serves to protect, preserve, and specify legitimate cultural practices. In the case of Islamic education, from the eleventh century onward, the madrasas became the center of the Muslim educational experience. These institutions, designed to pass on the transmitted sciences of Islamic life, were "pious endowments" (*waqf*, pl. *awaqaf*). That legal designation meant that they had to conform to the spirit and letter of Islamic law, which viewed the ancient sciences as foreign and alien to Islamic thought. In contrast to the Islamic tradition, the Western legal system had built within it a deeply grounded conception of legal autonomy, including formal procedures for institutional change based on election by consent. Lacking these ideas of legal autonomy represented by the fictive idea of a corporate legal personality, a corporation, the madrasas had no institutional procedures designed to effect change. Likewise, without the recognition of a legitimate sphere of secular studies distinct from Islamic studies in the eyes of dominant religious scholars, the madrasas had no way to allow or encourage the study of secular subjects within their midst.

In China, the cultural and legal setting was entirely different, though it too lacked the vital idea of legal autonomy. By the seventeenth century, the cultures of Confucianism, Buddhism, and Daoism were deeply entrenched and represented entirely different moral and philosophical orientations from either Greek or Christian modes of thought. And from a legal point of view, the idea of legal autonomy for any section of the Chinese population – professionals, scholars, or businessmen – was absent. The educational apparatus was firmly embedded in the official bureaucracy.

In Mughal India, Hinduism represented still another constellation of religious and philosophical thought distinct from Christianity, Greek philosophy, and even Islamic orientations that were superimposed over the greater part of India by the Mughals from the fifteenth century onward. Because of that Islamic dominance during the period of our concern, I focus on the Islamic institutions.

Let us turn first to the reconstruction of European thought and institutions that took place in the Middle Ages. These institutions created a

---

[2] Among others, Dimitri Gutas, *Greek Thought, Arabic Culture* (New York: Routledge, 1998); F. E. Peters, *Aristotle and the Arabs* (New York: Simon and Schuster, 1968); and Richard Walzer, *Greek into Arabic* (Columbia: University of South Carolina Press, 1962).

set of *breakthrough conditions* that set the stage for scientific, legal, and economic development ever after.[3]

Europe and Its Reconstruction

Although some scholars prefer to trace Europe's defining moments back to the so-called Axial Age between 800 and 300 B.C., the really defining transformative period took place during the Renaissance of the twelfth and thirteenth centuries. That is when the extraordinary fusion of Greek philosophy, Roman law, and Christian theology gave Europe a new and powerful civilizational coherence. What is also noticeable – if we compare the three other civilizations (China, India, and Islam) with Europe – is that the others did not go through the same kind of profound reconstruction that Europe experienced in the medieval period. Consequently, those civilizations lagged with respect to legal innovation, philosophical and theological innovation, scientific development, and ultimately economic development. Neither should we overlook that non-Western civilizations lagged for centuries in the development and promotion of representative democratic government and institutions.

An examination of the great revolutionary reconstruction of Western Europe in the twelfth and thirteenth centuries shows that it witnessed sweeping legal reforms, indeed, a revolutionary reconstruction, of all the realms and divisions of law – feudal, manorial, urban, commercial, and royal – and therewith the reconstitution of medieval European society. It is this great legal transformation that laid the foundations for the rise and autonomous development of modern science but also the rise of parliamentary government, the very idea of elective representation in all forms of corporate bodies, the legal autonomy of cities and towns, and a vast array of additional legal forms unique to Western law.[4]

---

[3] For an economic historian's analysis of this long path to the Industrial Revolution, see J. L. van Zanden, *The Long Road to the Industrial Revolution: The European Economy from 1000–1800* (Leiden, Netherlands: Brill, 2009), esp. chap. 2. Stressing the importance of the many institutional breakthroughs of the period 950–1350 that paved the way for European economic development, he refers to the many "institutional gadgets" that emerged while solving a variety of market, exchange, and collective-action problems. See also Tine de Moor, "The Silent Revolution: A New Perspective on the Emergence of Commons, Guilds, and Other Forms of Corporate Collective Action in Western Europe," *International Review of Social History* 53 (2008): 179–212.

[4] Throughout this chapter, I have adapted material from my book, *The Rise of Early Modern Science: Islam, China, and the West*, 2nd ed. (New York: Cambridge University Press, 2003), esp. pp. 118–19, 181–90.

At the center of this development, one finds the legal and political principle of treating collective actors as a single entity – a *corporation*. The emergence of corporate actors was unquestionably revolutionary in that the legal theory that made them possible created a variety of new forms and powers of association that were unique to the West because they were wholly absent in Islamic as well as Chinese law. Furthermore, the legal theory of corporations brings in its train organizational principles establishing such political ideas as constitutional government, consent in political decision making, the right of political and legal representation, the powers of adjudication and jurisdiction, and even the power of autonomous legislation. Aside from the scientific revolution itself, and perhaps even the Reformation, no other revolution has been as pregnant with new social and political implications as the legal revolution of the European Middle Ages. By laying the conceptual foundations for new institutional forms in legal thought, it prepared the way for the two other revolutions.

With this summary, I am suggesting that the unparalleled legal revolution of Europe in the twelfth and thirteenth centuries implied the creation of numerous innovative conditions in the three major spheres of societal development: scientific development, the emergence of representative democratic politics, and economic development. The papal revolution of the later twelfth and early thirteenth centuries, setting off the legal revolution, radically altered the foundations of virtually all social, political, and economic relationships. In thinking about this, the comparison should not be made with contemporary science and its ideals or with current democratic institutions and their practice but rather with the contemporaneous institutions and practices of the early modern era of the three other civilizations.

For example, the theory of corporate existence, as understood by Roman civil law and refashioned by the Canonists and Romanists of the twelfth and thirteenth centuries, granted legal autonomy to a variety of corporate entities such as cities and towns, charitable organizations, and merchant guilds as well as professional groups represented by surgeons and physicians. Not least of all, it granted legal autonomy to universities. All of these entities were thus enabled to create their own rules and regulations and, in the case of cities and towns, to mint their own currency and establish their own courts of law. Nothing like this kind of legal autonomy existed in Islamic law or Chinese law or Hindu law of an earlier era.

Furthermore, as concerns both the economic and political spheres, the theory of corporate existence, as understood by Roman civil law, distinguished between the property, goods, debts, liabilities, and assets of the corporation and those of individual members. A debt owed by the corporation was not owed by the members individually. Likewise, *ownership* of property by the corporation was not equivalent to *jurisdiction* by the head of the corporation because those empowered to adjudicate within or for the corporation were distinguished from the owners of the property. Most important, the allegiance of the individual members was said to be to the corporation, not to other members of the corporation personally. These ideas served to create a foundation for a *public* versus a *private* sphere of action and commitment – a clear distinction still lacking in many parts of the world.[5]

In short, the theory of corporate existence that was worked out uniquely by twelfth- and thirteenth-century Western legalists (but neglected by the Byzantines) created a whole new bundle of rights. These included the right to own property, to have representation in court, to sue and be sued, to make contracts, and to be consulted when one's interests were affected by actions taken by others, especially kings and princes, following the Roman legal maxim, "What touches all should be considered and approved by all."[6] Of course, it was a slow process putting all these new ideas into action, but the door had been opened, a new legal framework had been gestated. In the centuries to come, those ideas were to be transported across Europe and around the world.

But notice that this whole constellation of ideas and institutions can be called *civilizational complexes* – that is, segments of cultural life – and they give a civilization a distinctive identity when they are shared by multiple societies or peoples encompassed by that civilization.[7] Most if not all of these components of the European civilizational domain were disseminated over vast expanses of time and space without the force of

---

[5] Ibid., p. 144. For the earlier work on the idea of a public sphere, see Jürgen Habermas, *The Structural Transformation of the Public Sphere*, trans. T. McCarthy (Boston: Beacon Press, 1989); Craig Calhoun, ed., *Habermas and the Public Sphere* (Cambridge, MA: MIT Press, 1992); David Zaret, "Religion, Science, and Printing in the Public Sphere in Seventeenth Century England," *American Sociological Review* 54 (1989): 163–79.

[6] This was thoroughly studied by Gaines Post, *Studies in Medieval Legal Thought: Public Law and the State, 1150–1322* (Princeton, NJ: Princeton University Press, 1964); and see Harold Berman, *Law and Revolution* (Cambridge, MA: Harvard University Press, 1983).

[7] Here I have used some of the distinctions regarding "civilizational analysis" pioneered by Benjamin Nelson and, before him, Emile Durkheim and Marcel Mauss; see Nelson, "Civilization Complexes and Intercivilizational Encounters," chapter 2 in *On the Roads to Modernity: Conscience, Science and Civilizations: Selected Writings by Benjamin Nelson*, ed. Toby E. Huff (Totowa, NJ: Rowman and Littlefield, 1981).

empire. They were largely spread voluntarily as scholars – especially in the universities – met, discussed, and debated all the fundamental issues of law, governance, philosophy, logic, religious doctrine, and medical education. Unlike other parts of the world, physicians benefited from this legal revolution that enabled them to legally protect and preserve their ideas of proper medical treatment against outsiders, imposters, and all others not certified in their jurisdiction.[8]

Of course, there were religious controversies, some met by the use of force and violence such as the crusades in southern France against the Albigensians, Waldensians, and Catherai. Yet, the central intellectual foundations of European civilization, including theological and doctrinal issues, were hammered out in the disputatious but nonviolent setting of colleges and universities scattered across Europe. Converting so-called pagans to Christianity, especially in Scandinavia and elsewhere, was a long and drawn-out affair, with some acts of force, yet the successes were mainly achieved through intellectual persuasion. All the other components of European civilization, many rooted in Greek philosophy and Roman law, were freely adopted and transmitted throughout Europe.

Universities and the Scientific Agenda

How, then, can we explain the exuberant astronomical and scientific curiosity of the Europeans in comparison with the three other situations? The answer lies in the educational experience of Europeans in contrast to that of the Chinese, Muslims, and Hindus.

As noted earlier, the legal revolution of the twelfth century and later, along with the communal movement that acknowledged the legitimate collective character of human associations, created the framework for legally autonomous entities – namely, universities. As these new educational institutions were founded all over Europe from this era onward, the leaders of the movement imported the great scientific and philosophical heritage of the Greeks, along with supplements by Arab commentators.

The Europeans refocused the curriculum of the universities on the three philosophies: natural philosophy, moral philosophy, and metaphysics. Then they placed at the center of this new curriculum the natural books of Aristotle. These included his *Physics, On the Heavens, On Generation*

[8] See Huff, *Rise of Early Modern Science*, pp. 189–90.

and Corruption, On the Soul, Meteorology, and *The Small Works on Natural Things,* as well as biological works such as *The History of Animals, The Parts of Animals,* and *The Generation of Animals.* It is with these books, Ed Grant observed, that we find "the treatises that formed the comprehensive foundation for the medieval conception of the physical world and its operation."[9] This was indeed a core experience that was essentially scientific. Put differently, the Europeans institutionalized the study of the natural world by making it the central core of the university curriculum.

Moreover, the universities developed a method for taking up and disputing all sorts of naturalistic questions. It was a product of the lectures that the masters gave. It was based on compilations of the summaries of major physical questions as well as original treatises.[10] This literature and its use resulted in a concerted form of skeptical probing of a large set of questions in the natural sciences – physics, astronomy, cosmology, mechanics, and so forth. These probings included questions about the ultimate constitution of nature and the conditions of its transformation. Questions were asked about whether the world is singular or plural, whether the earth turns on its axis or is stationary, "whether every effecting thing is the cause of that which it is effecting," whether things can happen by chance, whether a vacuum is possible, whether the natural state of an object is stationary or in motion, whether luminous celestial bodies are hot, whether the sea has tides, and so on, for virtually every known field of inquiry.[11] It is hard to imagine a more concentrated diet

[9] Edward Grant, "Science and the Medieval University," in *Rebirth, Reform, and Resilience,* eds. James M. Kittelson and Pamela J. Transue (Columbus: Ohio State University Press, 1984), p. 78.

[10] The *questio* literature is illustrated in Edward Grant, ed., *A Source Book in Medieval Science* (Cambridge, MA: Harvard University Press, 1974), pp. 199–204, and has been discussed by many recent historians of medieval science, e.g., James Weisheipl, "The Curriculum of the Faculty of Arts at Oxford in the Early Fourteenth Century," *Medieval Studies* 26 (1975): 143–85; John Murdoch, "From Social to Intellectual Factors: An Aspect of the Unitary Character of Late Medieval Learning," in *The Cultural Context of Medieval Learning,* eds. John Murdoch and Edith Sylla (Boston: Reidel, 1975), pp. 271–338; Grant, "Cosmology," pp. 265–302 in *Science in the Middle Ages,* ed. David C. Lindberg (Chicago: University of Chicago Press, 1978); Grant, "The Condemnation of 1277, God's Absolute Power, and Physical Thought in the Late Middle Ages," *Viator* 10 (1979): 211–44; Grant, "Science and the Medieval University," esp. pp. 80ff; and M. McLaughlin, *Intellectual Freedom and Its Limits in the Twelfth and Thirteenth Centuries* (New York: Arno Press, 1977), among others.

[11] See Grant, *A Source Book,* pp. 199–200; and Grant, "Science and the Medieval University," pp. 82ff.

of scientific questions about the natural world and how it works. On the other hand, there is no evidence that Muslim scholars in the madrasas, or Chinese scholars prepping for the Civil Examinations, indulged in such skeptical probing of the natural world (as we shall see later).[12]

In a word, by importing and digesting the corpus of the "new Aristotle" and its methods of argumentation and inquiry, the intellectual elite of medieval Europe established an impersonal intellectual agenda whose ultimate purpose was to describe and explain the world in its entirety in terms of causal processes and mechanisms. This disinterested agenda was no longer a private, personal, or idiosyncratic preoccupation but rather a publicly shared set of texts, questions, commentaries, and, in some cases, centuries-old expositions of unsolved physical and metaphysical questions. These inquiries set the highest standards of intellectual investigation. By incorporating the natural books of Aristotle in the curriculum of the medieval universities, a disinterested agenda of naturalistic inquiry had been institutionalized. It was institutionalized as a curriculum, a course of study, and it was this curriculum that remained in place for the next 400 years in the European universities. It thereby laid the foundation for the breakthrough to modern science. It did so by instilling a profound skepticism about the sources of knowledge and a deep curiosity about the workings of nature. It inculcated a spirit of scientific curiosity that was unmatched anywhere in the world.[13]

The prime piece of evidence of this profound interest in scientific questions is the radically different reactions to the telescope and other experimental procedures in the four different civilizations that we saw in earlier chapters. Furthermore, this deliberately cultivated curiosity that flooded all over Western Europe in the seventeenth century flowed into a multiplicity of new experimental studies: in medicine and microscopy, in hydraulics and pneumatics, and in electrical studies (see Chapters 7–9.)

---

[12] For the Chinese, see Benjamin Elmans's masterful study, *A Cultural History of Civil Examinations in Late Imperial China* (Berkeley: University of California Press, 2000), and his recent study *On Their Own Terms: Science in China, 1550–1900* (Cambridge, MA: Harvard University Press, 2005). With regard to the madrasas, no one has yet shown that scholars there read and taught Aristotle's natural books such as his *Physics*, *Metaphysics*, or any of the other naturalistic books with their questioning epistemological posture – more on which follows.

[13] Also see Toby E. Huff, "Some Historical Roots of the Ethos of Science," *Journal of Classical Sociology* 7, no. 2 (2000): 193–210.

The Madrasas and the Transmitted Science

If we look next at the madrasas and their founding purpose, the context is radically different. From their emergence in eleventh-century Baghdad, their purpose was to preserve and pass on the religious or "transmitted sciences."[14] From a Muslim point of view, the Word had been delivered from on high: the Quran was both the speech and command of God. Coupled with the Sunna of the Prophet – that is, the reported sayings (hadiths) and actions attributed to the prophet Mohammad collected nearly a century after his death – the Holy Book and the example of the Prophet were seen by the Believers as a blueprint for a new civilization. Consequently, they quickly set about creating a new civilizational entity across the Middle East and North Africa.

When the time came to establish a system of education, much later in the eleventh century, these documents – the Quran and the hadith collections – were seen as the essential heart and core of what Islamic education ought to include. That is why Islamic education was always seen as comprising the transmitted sciences. They include these sacred documents along with the intellectual disciplines needed to understand them. The Quran, the hadith collections, Arabic grammar, genealogy and history, Quranic commentary, and Islamic law were the foremost subjects of study in the madrasas from that time to the modern age. Medical studies, which were generally highly valued in Middle Eastern culture, were almost never allowed into the madrasas.[15] Postmortem examinations were forbidden by Islamic law and tradition,[16] thereby stalling the advances in anatomical knowledge that were so dramatically revealed in Vesalius's *Fabric of the Human Body* in 1543.

At the outset, Islamic madrasas prohibited the inclusion of any of the so-called foreign or ancient sciences. A revealing insight into how the madrasas narrowed the subjects of study down to the religious sciences

---

[14] George Makdisi, *The Rise of the College: Institutions of Learning in Islam and the West* (Edinburgh: Edinburgh University Press, 1981); Jonathan Berkey, *The Transmission of Knowledge in Medieval Cairo* (Princeton, NJ: Princeton University Press, 1992); Michael Chamberlain, *Knowledge and Social Practice in Medieval Damascus, 1190–1350* (New York: Cambridge University Press, 1994).

[15] Makdisi, *Rise of the Colleges*; cases of strict prohibition are reported by Said Amir Arjomand, "The Law, Agency, and Policy in Medieval Islamic Society: Development of the Institutions of Learning from the Tenth to the Fifteenth Century," *Comparative Studies in Society and History* 41, no. 2 (1999): 263–93.

[16] Huff, *Rise of Early Modern Science*, 193–204.

is seen by looking at the curriculum of study in Baghdad in the early ninth century. For example, Job of Edessa, a Syriac Christian living in the Syriac community near Baghdad in the 800s, published a *Book of Treasures* containing an outline of the philosophical and natural sciences taught in Baghdad in about 817. It was arranged in a series of discourses. For example, the chapters discussed

> the cause of the coming together of the elements;...the fact that the elements were created from nothing, that they had no beginning, and that they are not infinite...the cause of the coming into existence of the genera and species of sea, land, and air, from the compound elements...the reasons why the bones are below, while the veins and the muscles are above...; the reasons why the five fingers came into existence on every hand, and five toes on every foot, and the thumb and big toe are thicker than the rest...; the reasons why the first composition of the genera and species of man, horse, ox, et cetera came into existence from the elements, and did not vary from the beginning up till now; while other species come into existence in our days, such as flies, midges, tapeworms, et cetera.[17]

These inquiries are clearly based on the naturalistic assumptions of Aristotle and Greek philosophy. They run directly counter to the anticausal Islamic metaphysics put in place by the intellectual elite of Islam, especially al-Ash'ari (d. 925) and later al-Ghazali (d. 1111). Job of Edessa's course of study shared the same underlying philosophy of Aristotle's natural books that were imported into the European universities but rejected by the madrasas. Instead of centering inquiry on naturalistic inquiries, the madrasas were created to preserve and protect the Islamic religious tradition. This meant that the Islamic sciences, the study of Islamic law, Quranic studies, and hadith collections and their interpretation were the central subjects. It was the scholars versed in religious law (*fuqaha*), scholars specializing in one of the four major schools of Islamic law, who were at the center of each of the madrasas. Over time, and even under the influence of al-Ghazali's trenchant defense of the faith, Greek logic and its various modes were adopted among the religious scholars.[18] But the study of logic is not the same as the study of the natural sciences such as physics, the science of motion, meteorology, or biology. Logic and mathematics are tools used to test or demonstrate propositions, given the

---

[17] Job of Edessa, *Book of Treasures*, Syriac text ed. and trans. with a critical apparatus by A. Mingana (Cambridge: W. Heffer, 1935). Each of the subjects cited is a chapter in the book.

[18] A. I. Sabra, "The Appropriation and Subsequent Naturalization of Greek Science in Medieval Islam: A Preliminary Statement," *History of Science* 25 (1987): 223–43.

initial premise. Natural science seeks to explore the actual world, adding to our knowledge of it, using logic and math in the process. But without a focus on the *natural world*, logical tools are of little avail.

By the thirteenth century, the study of the so-called rational sciences, such as logic, mathematics, and astronomy, were accepted by some religious scholars associated with the madrasas.[19] But teaching within the madrasas never included the study of Aristotle's natural books. It did not include the study of his *Physics*, *Metaphysics*, and *Meteorology*, nor *On Generation and Corruption*, *Plants*, *Animals*, and so on. Neither did it include the study of optics.

Moreover, according to Aristotle's classification of the sciences, which was followed by both Arab scholars and Europeans, astronomy was included among the mathematical sciences, not the natural sciences. This meant that only natural philosophers could decide what the shape of the world really is, and that search involved finding the causes of the changes and alterations of the natural world. As discussed earlier, that deep philosophical assumption undergirded Galileo's wish to be named "philosopher and mathematician" to the Grand Duke in 1610 because he wanted to talk about cosmology, the real shape of the world, not arbitrary mathematical models.

Some scholars have suggested that there were modifications of the texts studied in the madrasas in the seventeenth and eighteenth centuries, yet the materials in that revised study list had been in use for at least a century in Mughal India, in the Ottoman Empire, and among the Safavids of Persia.[20] There is very little evidence in these lists of books that the natural sciences, physics, the science of motion, meteorology, biology, and plants and animals had any place within them. Even astronomy was often absent, as was optics. Students of India in particular have

---

[19] Sonja Brentjes, "On the Location of the Ancient or 'Rational' Sciences in Muslim Educational Landscapes (AH 500–1100)," *Bulletin of the Royal Institute for Inter-faith Studies* 4, no. 1 (2002): 47–71; Ekmeleddin Ihsanoglu, "The Introduction of Western Science to the Ottoman World: A Case Study of Modern Astronomy," chap. 2 in *Science, Technology, and Learning in the Ottoman Empire* (Aldershot, UK: Ashgate, Variorum, 2004); and Ihsanoglu, "Institutionalisation of Science in the *Medreses* of Pre-Ottoman and Ottoman Turkey," in *Turkish Studies in the History and Philosophy of Science*, eds. G. Irzik and G. Güzeldere (Netherlands: Springer, 2005), 265–84.

[20] Francis Robinson, "Ottomans-Safavids-Mughals: Shared Knowledge and Connective Systems," *Journal of Islamic Studies* 8, no. 2 (1997): 151–84. Reprinted in Robinson, *The 'Ulama of Farangi Mahall and Islamic Culture in South Asia* (London: Hurst and Company, 2001). Both works include three appendixes listing the books. Also see Muhammad Qasim Zaman, *The Ulama in Contemporary Islam* (Princeton, NJ: Princeton University Press, 2002), pp. 68ff.

claimed a revised madrasa curriculum in the seventeenth century associated with either Akbar's sixteenth-century reforms[21] or, later, with the name of Mulla Nizam al-din Muhammad, who died in 1748.[22] Sultan Akbar's great interest in comparative religions and interreligious dialogue, although commendable from a modern point of view, seems to have generated the opposite response among the ulama surrounding his court. The result was a greater emphasis placed on the transmitted sciences – that is, Quranic and hadith studies.[23] Even Sultan Aurangzeb (d. 1707), Abkar's great grandson with strong Islamist convictions, criticized what the religious scholars taught him. Instead of teaching him useful things, such as geography and statecraft,

you taught me to read Arabic. Forgetting how many important subjects ought to be embraced in the education of a prince, you acted as if it were chiefly necessary that he should possess great skill in grammar, and such knowledge as belongs to a doctor of law; and thus did you waste the precious hours of my youth in the dry unprofitable, and never-ending task of learning words.[24]

A greater interest in such practical things as crafts, engraving, military ware, goldworking, sewing, dying, and pottery hardly qualifies as a new interest in science or technological innovation in the seventeenth century.

Consequently, a review of the contents of the reading list of the Mughal madrasas as well as the Safavid and Ottoman lists reveals no books from Aristotle's natural sciences. There are lots of books on grammar; some on rhetoric, logic, and theology (*kalam*); and many on law and Islamic jurisprudence but nothing representing the natural sciences. The basic division of subjects was between the transmitted and the rational sciences, where the latter represent the nontransmitted disciplines, not natural philosophy. They could include logic, philosophy (*hikmat*), mathematics, and sometimes medicine (which generally was not taught in madrasas). Yet, even Islamic jurisprudence (*fiqk*) is sometimes classified among the rational sciences. The philosophy taught (*hikmat*) was not Greek natural philosophy but rather a more mystical wisdom-of-the-East orientation

[21] Iqbal Ghani Khan, "Technology and the Question of Elite Intervention in Eighteenth-Century North India," in *Rethinking Early Modern India*, ed. Richard B. Barnett (New Delhi: Manohar, 2002), 255–88.
[22] Robinson, "Ottomans-Safavids-Mughals," and Francis Robinson, "Nizam al-Din, Mulla Muhammad," *EI²* 8 (2009): 68.
[23] Robinson, "Ottomans-Safavids-Mughals," p. 173.
[24] Francois Bernier, *Travels in the Mughal Empire (1656–1668)*, trans. A. Constable (Delhi: Asian Educational Services, 1996), p. 156.

that had something in common with Sufism.²⁵ It did not include epistemology and metaphysics, as one finds in Aristotle's writings. In some
periods, and some regions of the Muslim world, the founding documents
of the madrasas explicitly excluded Greek philosophy. The *transmitted
sciences* or Islamic sciences remained the center of educational experience. Those curricular materials of the three Muslim empires focused on
the transmitted sciences continued to be used all the way to the twentieth
century.²⁶

Still, this curriculum was not imposed in any uniform manner but
depended on the wisdom of the individual religious scholar to interpret
it. The scholar's task was to give the student an *ijaza*, a "permission to
transmit" the works studied once the scholar was convinced that the student had memorized a particular manuscript. But the timing and selection
of the materials to be taught depended entirely on the religious scholar.

All this absence of training in the natural sciences helps explain why
reaction to the arrival of the telescope in Muslim lands was so muted.
The idea of using the telescope to gather new observations or to test
the assumptions of existing astronomical knowledge did not occur. The
discipline of astronomy had become moribund in Muslim lands. The late
Indian scholar M. Athar Ali was bemused that when Jai Singh, the Indian
ruler who built the huge stone and masonry observatory in Jai Pur in the
early eighteenth century, ordered up a new set of planetary tables and
a star catalogue, it was a verbatim copy of the work of Ulug Beg, who
had died nearly 300 years earlier.²⁷ Completed between 1733 and 1738,
more than 100 years after the invention of the telescope, the observatory
relied on unaided human vision. Only the numbers in some of the tables
were changed, but these were copied from a set of tables produced by

²⁵ For a tabulation of the books in various categories, see Robinson, "Ottomans-Safavids-
Mughals," p. 153.
²⁶ In addition to ibid., see Ekmeleddin Ihansanôglu, "The Emergence of the Ottoman
Medrese Tradition" (unpublished paper, September 2006); Ihansanôglu, "Institutional-
ization of Science in the Medreses of Pre-Ottoman and Ottoman Turkey," in *Turkish
Studies in the History and Philosophy of Science*, eds. G. Irzik and G. Güzeldere (New
York: Springer, 2005), pp. 265–83; Makdisi, *Rise of the College*; Jonathan Berkey, *The
Transmission of Knowledge in Medieval Cairo: A Social History of Islamic Education*
(Princeton, NJ: Princeton University Press, 1994); Michael Chamberlain, *Knowledge and
Social Practice in Medieval Damascus, 1190–1350* (New York: Cambridge University
Press, 1994); and Gary Leiser, "The Madrasa and the Islamization of Anatolia before
the Ottomans," in *Law and Education in Islam: Studies in Honor of Professor George
Makdisi* (London: E. W. J. Gibb Memorial, 2004), pp. 174–91.
²⁷ M. Athar Ali, "The Passing of Empire: The Mughal Case," *Modern Asian Studies* 9,
no. 3 (1975): 391.

the French astronomer La Hire.[28] Jai Singh's observatories, built at great cost, were inadequate for the purposes of modern astronomy.

In the Middle East and outside the madrasas, there had been scholars who took it on themselves to pursue privately some of the great questions in astronomy and optics until the end of the fourteenth century, but thereafter creative advances in astronomy, optics, and physics are hardly ever found. None of the scholars toward the end of the Mamluk period has yet been shown to have used the telescope in the seventeenth century to advance astronomical thinking. Even sources reporting on scientific curiosity in late-*eighteenth*-century Cairo – that is, al-Jabarti's famous *History of Egypt*, with its extraordinary collection of biographies of scholars – fails to reflect any use of the telescope among them.[29]

In the meantime, the rote learning and memorization of a limited set of the transmitted sciences, vouchsafed by a scholar's *ijaza*, formed the core of Islamic learning. Well into the twentieth century, the system of memorizing and preserving the religious tradition permeated the famous madrasas such as the oldest one in Cairo, al-Azhar. Here, for example, is the report of one of the most famous graduates of that institution, Taha Husayn (1889–1973), describing his daily routine in the madrasa in the opening decades of the twentieth century. He was a blind student, yet later wrote a daring commentary on the Quran, questioning its objectivity as an historical source, and was severely criticized and prosecuted for doing so. About his experience at al-Azhar, he wrote:

The four years I spent [at Al-Azhar] seemed to me like forty, so utterly drawn out they were. . . . It was life of unrelieved repetition, with never a new thing, from the time the study began until it was over. After the dawn prayer came the study of Tawhid, the doctrine of unity; then fiqh, or jurisprudence, after sunrise; then the study of Arabic grammar during the forenoon, following a dull meal; then more grammar in the wake of the noon prayer. After this came a grudging bit of leisure and then, again, another snatch of wearisome food until, the evening prayer performed, I proceeded to the logic class which some shaikh or other conducted. Throughout these studies it was all merely a case of hearing re-iterated words and traditional talk which aroused no chord in my heart, nor taste in my appetite. There was no food for one's intelligence, no new knowledge adding to one's store.[30]

[28] Raymond Mercier, "The Astronomical Tables of Rajah Jai Singh Sawai," *Indian Journal of History of Science* 17, no. 2 (1984): 143–71.

[29] Thomas Philipp and Moshe Perlmann, eds., *Abd al-Rahman al-Jabarti's History of Egypt* (Stuttgart, Germany: Franz Steiner, 1994).

[30] As cited in Donald Malcolm Reed, *Cairo University and the Making of Modern Egypt* (Cambridge: Cambridge University Press, 1990), p. 13.

Higher education following European models of colleges and universities arrived in the Middle East only in the 1860s, when Americans founded Robert College in Istanbul in 1863 and the precursor to the American University in Beirut (Presbyterian Men's College) in 1867. The Indian subcontinent reflects a similar pattern with the founding of the Anglo-Oriental College in Aligarh by Ahmad Khan in 1875. In 1920, it became Aligarh Muslim University.[31]

China and the Examinations

A still different model of education sharply diverging from European universities developed in China. The ideals of this system were embedded in a different metaphysical worldview. It was not based on the Islamic anticausal worldview of a supremely governing deity, yet it too postulated a world of constant change and transformation that lacked the Greek notions of push–pull causality or atoms buzzing in predicable patterns. Instead, it was based on the idea of a flow of energy, a complimentarity of Yang and Yin forces that in the end followed recurrent, albeit long-term cyclical patterns. By the end of the Sung Dynasty (960–1279) and the triumph of Neo-Confucian thought, it was assumed that the great ancestors of the past had actually achieved the perfection of moral and ethical knowledge so that the ideal of thought and education was to regain that lost wisdom, not to pioneer new thoughts. So strong was that belief in China at the end of the seventeenth century that critics of the Jesuits and the new science proclaimed that although the ancestors of the past perhaps did not have an astronomy that could predict planetary movements as well as the new European system, yet those great wise men of the past had grasped the real essence and meaning of living, so it was far better to return to their ways than to adopt the new ways of foreigners.[32]

From the time of the Sung Dynasty, Chinese rulers thought it necessary to have a universal system of education. That system was designed, controlled, and administered by the scholar-bureaucrats headquartered in the capital city. The unchallenged assumption was that wisdom exists vouchsafed from the past, and the educational process centered on examinations was meant to guarantee that the moral and ethical wisdom of

---

[31] Among others, see "Aligarh," in *Oxford Encyclopedia of Modern Islam* (New York: Oxford University Press, 1995), pp. 1:73–74; and Christian Troll, *Sayyid Ahmad Khan: A Reinterpretation of Muslim Theology* (Atlantic Highlands, NJ: Humanities Press, 1978).

[32] This follows my analysis in Huff, *Rise of Early Modern Science*, pp. 280–84.

the past was mastered by young scholars before they would be qualified to serve in the imperial service. The examination system started in
the villages, progressed to the provinces, and finally to the metropolitan
examination, each held every three years. Later, a Palace Examination
conducted by the emperor himself was added.

The subject matter of the examinations was the Confucian classics,
poetry, and official histories. From an early age, young boys were taught
to memorize the Confucian classics: the "Four Books" and the "Five
Classics." The four included *The Greater Learning, Analects, Book of
Mencius (Meng-Tzu)*, and *Doctrine of the Mean.*[33] The Five Classics
included *Book of Changes (I ching), Book of Poetry (Shih ching), Book
of History (Shu ching), Book of Rites (Li-chi)*, and *Spring and Autumn
Annals (Ch'un-ch'iu)*. In the Sung Dynasty, the examinations were based
on Zhu Xi's Four Books and his commentary on them but later expanded
to include the Five Classics. This diet of literary, moral, and historical
recollections was clearly intended to emphasize what we would call the
humanities and remained distant from natural science.

Initially, young pupils learned these works without knowing the meaning of what they were memorizing. They also learned Chinese calligra-
phy and to write classical poetry.[34] The examinations largely asked students to recite passages from the classics, to comment on selections from
them, or, at more advanced levels, to write about the appropriate conduct
for the wise and virtuous ruler. The so-called eight-legged essay, which
emerged in early Ming times, was a composition based on a quotation
from the classics that was to be presented in rigid stylized form. It has
been described as an exercise similar to composing a fugue based on a
few introductory notes.[35] Jonathan Spence gives a good example:

To cite an example of this style – the Ming Examination of 1487 had set as a topic
a six-character quotation from the Mencius: *Lo t'ien che, pao t'ien-hsia*. The standard translation . . . reads: "He who delights in Heaven, will affect with his love
and protection the whole empire." (A literal translation would be "Love Heaven
person, protect Heaven-below.") In his eight-legged essay the candidate would be
expected to proceed as follows – make a preliminary statement (three sentences),

---

[33] Among others, see Daniel K. Gardner, trans., *Learning to Be a Sage: Selections from the
Conversations of Master Chu* (Berkeley: University of California Press, 1990).

[34] Thomas H. C. Lee, *Government Education and Examinations in Sung China* (Hong
Kong: Chinese University Press, 1985), chap. 1.

[35] Wolfgang Franke, *The Reform and Abolition of the Traditional Chinese Examination System*, East Asian Monographs, 10 (Cambridge, MA: Harvard University, 1963),
pp. 19–20.

treat the first half ("love Heaven person") in four "legs" or sections, make a transition (four sentences), treat the second half ("protect Heaven-below") in four "legs," make a recapitulation (four sentences), and reach a grand conclusion. Within each four-legged section, his expressions should be in antithetic pairs, such as con and pro, false and true, shallow and profound, each half of each antithesis balancing the other in length, diction, imagery, and rhythm.[36]

The examinations were held first at the district level, and then successful candidates went to the prefectural examinations. Those who passed became *shengyuan* (cultivated talents). Following that, successful students every three years could take the provincial exams, and if they passed, they were awarded the *juren* (recommended man) title. During some periods of time, this qualified the candidate for a position in the government such as district magistrate or possibly a lower-level position in the capital under the supervision of a higher official. The *juren* certificate is sometimes equated to the Western bachelor's degree. The successful candidates for a provincial *juren* examination in 1669 performed the following feats:

[They] had pondered three passages chosen that year by the Shantung examiners; they had placed them in their correct context and explicated them. From the Confucian Analects there was the phrase "They who know the truth" from Book VI, chapters 17 and 18: "The Master said, 'Man is born for uprightness. If a man lose his uprightness and yet live, his escape from death is the effect of mere good fortune.' The Master said, 'They who know the truth are not equal to those who love it, and they who love it are not equal to those who delight in it.'" From the Doctrine of the Mean came the phrase "Call him Heaven, how vast he is!" From the closing sentences of Book XXXII, on the man of true sincerity: "Shall this individual have any being or anything beyond himself on which he depends? Call him man in his ideal, how earnest he is! Call him an abyss, how deep he is! Call him Heaven, how vast is he!" And from the Book of Mencius there was "By viewing ceremonial ordinances" from Book II, Part I, where Mencius quotes Confucius's disciple Tzu-Kung in his absolute praise of his teacher (and of the historian's power): "Tzu-Kung said, 'By viewing the ceremonial ordinances of a prince, we know the character of his government. By hearing his music, we know the character of his virtue. After the lapse of a hundred ages I can arrange, according to their merits, the kings of a hundred ages – not one of them can escape me. From the birth of mankind till now, there has never been another like our master.'"[37]

---

[36] J. K. Fairbank, Edwin O. Reischauer, and A. M. Craig, eds., *East Asia: The Modern Transformation* (Boston: Houghton-Mifflin, 1965), p. 122. For an analysis of the eight-legged essay and additional examples of it, see Elman, *A Cultural History* (Berkeley: University of California Press, 2000), pp. 380–99.

[37] Jonathan Spence, *The Death of Woman Wang* (New York: Viking, 1978), p. 16.

Next, the candidate could take the metropolitan examination in the capital. Although there were quotas allotted for each province, those who passed this examination had reached the top, except for the final Palace Examination administered by the emperor himself. The Palace Examination was pro forma, but emperors could and did fail candidates as well as assign them final rankings among each other.[38] Successful candidates were called "presented scholars" (*jinshi* or, literally, "advanced scholars"),[39] and this was the most prestigious award available, sometimes equated to the doctorate.

It is obvious that such a system of universal examinations, based on examination questions created by a board of senior bureaucrats, established an extraordinary uniformity of attitude and opinion. Because it standardized the very texts that were to be studied in addition to the exams themselves, this educational system, particularly from Ming times (1368–1644) on, created a virtual state dogma, "an unparalleled uniformity of thought [that] was enforced not only among the officials but throughout the whole leading class.... There remained almost no opportunity for the development of original ideas, for any deviation from the orthodox interpretation led certainly to failure."[40]

But this is not to say that the system was incapable of producing and selecting experts with technical knowledge. Robert Hartwell, for example, has shown that the Northern Sung Dynasty (970–1127) experienced unprecedented economic growth and development and that the services of financial experts were necessary to accomplish this result. Accordingly, during the eleventh century, "nearly ninety percent of the chief financial officials were brought into the administration through examination,"[41] a portion of which involved problems of policy analysis. It seems that financial experts played a role not only in setting economic policy but

---

[38] Franke, *Reform and Abolition*, p. 6; Edward Kracke Jr., *Civil Service in Early Sung China, 960–1067* (Cambridge, MA: Harvard University Press, 1953), pp. 60–70; and Robert M. Hartwell, "Financial Experience, Examinations, and the Formulation of Economic Policy in Northern Sung," *Journal of Asian Studies* 30 (1971): 300–2; John W. Chafee, *The Thorny Gates of Learning in Sung China: A Social History of Examinations* (New York: Cambridge University Press, 1985), pp. 223, 49.

[39] See Charles O. Hucker, *A Dictionary of Official Titles in Imperial China* (Stanford, CA: Stanford University Press, 1985), "chin-shih," no. 1148; Adam Yuen-chung Liu, *The Hanlin Academy, 1644–1850* (Hamden, CT: Archon Books, 1981), chap. 1; and Ho Ping-ti, *The Ladder of Success*, rev. ed. (New York: Columbia University Press, 1967), pp. 12–14.

[40] Franke, *Reform and Abolition*, p. 13; and Derk Bodde, *Chinese Thought, Science, and Society: The Intellectual and Social Background of Science and Technology in Premodern China* (Honolulu: University of Hawaii Press, 1991), pp. 185, 193.

[41] Hartwell, "Financial Experience," p. 300.

also in formulating questions for the Civil Service Examination at both provincial and imperial (metropolitan) levels. To make the system yield the experts it needed, a set of complex administrative problems was added to the examination at the palace. One might also note that in other areas, such as history, law, ritual, and the classics, no original compositions were required because the exam depended entirely on the recall of memory and the elucidation of passages of text. Benjamin Elman estimates that after 1787, "over 500,000 characters of textual material had to be memorized to master the examination curriculum of the Four Books and the Five Classics," and that was not all. Other material related to dynastic histories had to be committed to memory.[42] In short, even though one might say that practical problems in the domain of state-craft were part of the examinations during certain periods of time, it cannot be said that the system encouraged scientific interests in general – it did not. This was not a system that instilled or encouraged scientific curiosity.

The most significant qualification that needs to be added to this assess-ment stems from the recent discovery that the late Ming examiners actu-ally introduced questions on the examinations that required knowledge, some of it quite technical, of astronomy, calendrical calculations, and mathematics as applied to musical harmonics. This was never a very large portion of the examinations, but it does indicate that for a period of time during the Ming Dynasty, some candidates were required to answer questions about the Chinese astronomical system, whether there was a method of explaining the celestial movements, and why there were errors in calendars and how they were rectified. Likewise, questions were asked on Chinese musical harmonics and on the mathematically stated basis of the relationship between pitch and the length of an instrument.

In all these policy questions in the domain of natural studies, the main concern was preserving the present political harmony and explaining why things were done as they were. As Elman points out, the minority of can-didates who qualified on these examinations were not licensed to become scientists. Although some technical material was involved, the material presented came from the relevant passages of the classical texts that had been memorized. At best, it might be suggested, as Elman does, that these scholars functioned as historians of science who understood how the stud-ies in question had evolved and perhaps how they functioned currently.[43]

---

[42] Elman, *A Cultural History*, p. 373; and Ichisada Miyazaki, *China's Examination Hell* (New Haven, CT: Yale University Press, 1981), pp. 16–17.
[43] Elman, *A Cultural History*, pp. 482, 483n63.

Moreover, the candidates were required both to place their responses in the format of the eight-legged essay and to relate the issues at hand to larger issues of governance, for which they were being recruited. During this period, and prior to the reaction against this trend that set in with the arrival of the Qing Dynasty (in 1644), it appears appropriate to say "that the ability to deal with astronomical, medical, mathematical, and other technical questions was an essential tool of the new classical studies emerging in late Ming and early Qing China."[44] What seems remarkable is not that the Ming examinations included such questions (dropped soon thereafter in the Qing) but rather that the Ming literati successfully encapsulated natural studies within a system of political, social, and cultural reproduction that guaranteed the long-term dominance of the dynasty, its literati, and the Neo-Confucian orthodoxy.[45]

Because this system was tightly controlled by the government hierarchy (the Directorate of Education), no tradition of independent learning emerged, and no agency was given autonomous control of a curriculum of education. Everything centered around passing the Civil Service Examinations, and consequently, students were interested only in mastering the material required for the state examinations. One learned official wrote in 1042, "When the examination year comes, the Directorate School is flooded with more than a thousand students . . . and then when the examination is over, they all disappear, and teachers find nothing to do except sit in their chairs."[46]

The official Civil Service Examination system created a structure of rewards and incentives that over time diverted almost all attention away from disinterested learning into the narrow mastery of the Confucian classics. Astute scholars recognized this but were powerless to change it. In the thirteenth century, another official wrote that "schools are considered to be the business of officials and examinations are considered to be the vocation of scholars, alas!"[47]

Entrenched Worldviews

It was into this rigid world of educational training that the Christian missionaries, discussed in Chapter 4, stumbled in the seventeenth century.

---

[44] Ibid., p. 468.
[45] Ibid.
[46] Lee, *Government Education*, p. 76.
[47] Yeh Shih as cited in Chafee, *Thorny Gates*, p. 88.

It is apparent that both Chinese and Islamic mentors of educational virtue were fearful of foreign influences. Both began with the assumption that they already possessed all the wisdom necessary for life. There were many brave and curious Chinese scholars who found the new learning from the West fascinating in the seventeenth century and did what they could to absorb it. Xu Guangqi (Dr. Paul), the leader of astronomical reform in the Bureau of Astronomy, was a brilliant example of that. He was the major architect of the plan to translate as much of Western science and philosophy as possible into Chinese, looking ahead to the future when the new or Western system of astronomy might break down. Having access to that deeper background would then make it possible to understand its original foundations and make repairs.

Unfortunately, the Xu Guangqis of China did not carry the day. In the end, the Chinese civil service examination system remained intact all the way to the twentieth century. No room could be made within that system for any of the foreign scientific ideas that were found useful in other parts of the world. When Ferdinand Verbiest (1670s), then in charge of the Chinese Bureau of Astronomy, submitted a memo to the emperor suggesting that it would be good to incorporate a considerable part of Greek natural philosophy into the Chinese educational system, he was rebuffed; the emperor declined to let the report be published.[48] Even after three decades of running the Chinese Bureau of Astronomy according to Western scientific assumptions, translating thousands of books containing Western science and philosophy, and training scores of Chinese for the task, no progress was made toward instilling a commitment to advancing the new sciences and astronomy. The scholar bureaucrats had too tight a control over the system to allow reform. Consequently, scientific thought and activity stagnated in China, whereas Europe experienced one of its most revolutionary eras ever.

One cannot resist the thought that one consequence of that fear of the foreign was the economic decline that overcame China from the eighteenth to the twentieth centuries, even taking into consideration the heavy hand of Western imperialism in the early nineteenth century. In the seventeenth century and the time of the Jesuits, scientific exchanges were possible on an egalitarian basis, without militarist interference, but at

---

[48] Noël Golvers, "Verbiest's Introduction of *Aristoteles Latinus* (Coimbra) in China: New Western Evidence," in *The Christian Mission in China in the Verbiest Era: Some Aspects of the Missionary Approach*, ed. N. Golvers (Leuven, Netherlands: Leuven University Press, 1999), 33–51.

the end of that century, China's leaders took a different direction. No Chinese Galileos, Keplers, Leeuwenhoeks, or Newtons were to emerge during that long run. That is a loss for both the Chinese and the rest of the world.

In the Muslim world, the Quran and the hadith collections were good enough for building a complete system that would preserve and transmit the inherited Islamic message to all future generations. Of course, some scholars saw the need for adding Aristotle's logic as a balance for weighing arguments, and mathematics was needed for dividing inheritances prescribed by passages in the Quran. Because the Quran was a legal document in Muslim eyes, the study and teaching of the Sharia (Quran and hadiths) was the epitome of an Islamic science.

Scholars largely outside the madrasas thought it necessary to learn geometry and develop trigonometry so that the directions to Mecca, the *qibla*, could be ascertained. Others, such as al-Biruni, even worked out universal directions to Mecca from all parts of the world, though even he was criticized for doing so.

Still, there were limits within the madrasas to the importing of foreign subjects such as the study of physics, meteorology, metaphysics, plants and animals, and even medicine. A scholar might study those subjects on his own but, clearly in the long run, energy to support those inquiries flagged.

Europeans, on the other hand, managed to create public spaces (universities) within which the free flow of critical thought could flourish. Yet, these were not the only branches of *legally autonomous* entities that were found all over Europe. Other forms included cities and towns, professional guilds, merchant guilds, charitable organizations, and daily newspapers that flourished from the 1640s to the present.

Moreover, European educators – above all, Christians – knew very well that their university studies included the wisdom of ancient peoples who had lived in foreign places far away such as Greece and the Arab Middle East. They knew that many parts of their educational materials had passed through the hands of "our Arab masters," as Adelard of Bath put it, and they were grateful for it. Likewise, they revered Ibn Rushd, known as Averröes, and studied Avicenna's medical *Canon* in the universities for nearly 400 years, knowing full well that Ibn Sina had mastered and improved on the medical works of the Roman physician Galen.

Debates of all sorts took place within the strictly dogmatic domains when European religious scholars asked whether it was prudent to read the writings of "pagans" because this might irreparably damage one's mind and soul. St. Thomas Aquinas contributed to those debates with the result that the works of Aristotle, Plato, and many other pagan authors, even Averröes, gained approval. From this point of view, Europeans had a long tradition of learning from others. The European university experience from that perspective seems to have been more ecumenical than either the Chinese or the Islamic experience. By 1500, the universities composed a thick web of educational institutions across Europe. The scientific worldview that was instilled in those who attended the universities would reap amazing benefits in the early and mid-seventeenth century. At the same time, the new scientific worldview would increasingly spread across Europe. To those outcomes I now turn.

PART III

SCIENCE UNBOUND

7

Infectious Curiosity I

Anatomy and Microbiology

Our fascination with the revolutionary heliocentric hypothesis of Copernicus, carried forward by Galileo and Kepler, has led us to overlook the revolutionary discoveries tumbling out of other scientific investigations in the seventeenth century. The Copernican revolution has an additional fascination because it seems to pit a great scientific hero, Galileo, against an oppressive religious structure. But the Church outside of Italy controlled neither the press, the dissemination of telescopes, nor the exploration of nature. Neither could it suppress the anatomical or microscopic study of nature and the human body.

In this way, the workings of the omnipresent *European ethos of science* was operative in many fields in England and from Scandinavia to Italy on the Continent. It can be seen in medicine and in the broad range of microscopic studies that gave birth to microbial studies. This was made possible by the invention of the microscope, both single- and compound-lens versions. Likewise, significant empirical advances were made in the field of hydraulics, pneumatics, and electrical studies. All these came out of the ubiquitous scientific curiosity that we saw earlier in the Europe-wide fascination with the telescope. That curiosity had been bred in the universities and both preceded the scientific revolution and served to keep it going.

The Royal Society and Scientific Academies

Before turning to the many new scientific inquiries that were ongoing outside astronomy and the science of motion in the early seventeenth century, we should notice the following. The infectious scientific curiosity that had been gestating in universities since the establishment of the study of Aristotle's natural books in the twelfth and thirteenth centuries had two pronounced manifestations in the seventeenth century. The first was simply the fascination with studying natural phenomena epitomized by the range of scholars, ambassadors, merchants, and religious officials who took up telescopes to either confirm or reject Galileo's surprising discoveries. Yet, that curiosity had been equally evident in the study of human anatomy for centuries. It took graphic form in the publication of Vesalius's epic work on *The Fabric of the Human Body* in 1543. Furthermore, anatomical dissections of human bodies had been based in the universities for hundreds of years.

Second, this widespread scientific curiosity began to manifest itself in the desire among many scholars and educated nobles to establish new scientific institutions that could support hands-on experimental investigations better than the slowly changing universities of that era. This trend began to emerge in the early years of the seventeenth century, when we witness the formation of a bevy of new scientific societies. The earliest of these was the Accademia dei Lincei, founded in Rome in 1603 by Frederico Cesi and into which Galileo was inducted in 1611. Slowly, the idea percolated across Europe that such new organizations were needed.

In England, the idea of an invisible college devoted to investigations in natural philosophy appeared in the 1640s. According to various accounts, in 1645, "divers worthy persons" founded a group at Gresham College that met once a week to study "what hath been called the New Philosophy or Experimental Philosophy . . . under a certain penalty, and a weekly contribution for the charge of experiments, with certain rules amongst us, to treat and discourse on such affairs."[1] Part of this group, or at least some of its members, adjourned to Oxford, where they met with others either in Wadham College of the university or later in the quarters of Robert Boyle. Apparently, others met in a coffeehouse, where occasional experiments were mounted.[2] After a period of significant violence due to the English civil war, the two groups were reunited in London, where they founded the Royal Society of London. It was given an official charter in 1661 by King Charles II under the restored monarchy. Scholars have

---

[1] Reported in the diary of John Wallis, as cited in Henry Lyons, *The Royal Society, 1660–1940* (Cambridge: Cambridge University Press, 1944), pp. 8–9.

[2] Ibid.; also Adrian Johns, "Coffeehouses and Print Shops," in *The Cambridge History of Science* (New York: Cambridge University Press, 2006), pp. 3:320–22. Also see Steven Shapin and Simon Schaffer, *Leviathan and the Air-Pump* (Princeton, NJ: Princeton University Press, 1985).

often pondered the fact that the new society was indeed a legal entity – officially, the Royal Society of London for Improving Natural Knowledge – with a "constitutional structure." In other words, it became a legally autonomous corporation with all the rights and privileges of that station. It could, just like the medieval universities, have its own official public representations, buy and sell property, print its own books, and sue and be sued.[3] It also had privileged access to the bodies of executed criminals for scientific proposes.

Thanks to the perspicacity of several of its members, and especially its first secretary, Henry Oldenberg, it became the most famous and possibly the most successful of such seventeenth-century scientific institutions by reaching out to other natural philosophers across Europe.[4] The correspondence of Oldenberg then became a major source of the Royal Society's publications – that is, "Letters" published in the *Transactions of the Royal Society*. These contained reports on a very broad range of scientific subjects for centuries to come.

Similarly, the Paris Academy of Sciences, founded in 1666, was the inspiration of mathematicians, natural philosophers, and other correspondents of Marin Mersenne (1588–1648). Mersenne was a monk living in Paris who was also a brilliant student of mathematics and natural philosophy who gathered around him many of the brightest scientific lights of his generation. These included René Descartes, the great mathematician Pierre Fermat, Blaise Pascal, the astronomer and mathematician Pierre Gassendi, and among many others, Nicolas-Claude Fabri de Peiresc. The latter was among the first astronomers in southern France to confirm Galileo's discoveries of the satellites of Jupiter in 1610. Peiresc himself was the most prolific correspondent of his age, having written between 10,000 and 14,000 letters exchanged with scholars in more than 100 cities across Europe.[5] Building on this French network, Jean-Baptiste Colbert, the controller general of France, established the Paris Academy

---

[3] See Michael Hunter, "The Importance of Being Institutionalized," in *Establishing the New Science* (Woodbridge, UK: Boydell Press, 1989), pp. 1–43, esp. p. 3.

[4] The classic study of such societies is Martha Ornstein, *The Role of Scientific Societies in the Seventeenth Century* (Chicago: University of Chicago Press, 1928). For more on the function of such academies in Italy, see Mario Biagioli, *Galileo, Courtier: The Practice of Science in the Culture of Absolutism* (Chicago: University of Chicago Press, 1993), esp. pp. 357ff.

[5] Robert A. Hatch, "Between Erudition and Science: The Archive and Correspondence Network of Ismaël Boulliau," in *Archives of the Scientific Revolution: The Formation and Exchange of Ideas in Seventeenth-Century Europe*, ed. Michael Hunter (Woodbridge, UK: Boydell Press, 1998), p. 51. Also see the graphic plot of the letters, p. 52.

of Sciences in 1666. Like its counterpart in London, the Paris Academy invited distinguished foreign scholars to join its members, as did the Dutch experimentalist Christiaan Huygens. He had been a member of the Royal Society of London since 1663.[6]

Overshadowed by the greater success of the other two academies in London and Paris was the Accademia del Cimento, founded in 1667 in Florence. It was the brainchild of two of Galileo's students, Evangelista Torricelli and Vincenzo Viviani. Torricelli became famous for many things but especially the invention of the mercury barometer. Viviani wrote the classic biography of Galileo and established the definitive collection of Galileo's papers.

All three of these new scientific institutions reflected the long-standing inclination of scholars across Europe to communicate about their individual enterprises, to join together as a community of practitioners, and to compile a written record of their experimental results. Although there is reasonable evidence that in England, the rise of Puritanism strongly influenced the founders and many members of the Royal Society,[7] the whole spirit of modern science had already been implanted prior to the rise of these institutions as well as prior to the rise of English Puritanism. No doubt in England there was a strong influence flowing from the writings of Francis Bacon and his call for a "Great Instauration," a great new scientific beginning. Yet, Bacon was neither a scientist nor a particularly well informed student of the history of science of his day. He seems not to have been influenced either by Galileo's discoveries or the pioneering and clearly experimental work of William Gilbert on magnetism. On the other hand, Galileo's continental influence as a model of the new science is evident in the crest alternatives drawn up by John Evelyn for the Royal Society in 1661. One of these alternatives was a drawing containing two telescopes in repose against each other in the form of a cross and above them a rough sketch of Jupiter with its satellites, three on the right side and one of the left, imitating Galileo's own style of

---

[6] Among others, see Roger Hahn, *The Anatomy of a Scientific Institution: The Paris Academy of Sciences, 1666–1803* (Berkeley: University of California Press, 1971).

[7] Among others, see Robert K. Merton, *Science, Technology, and Society in Seventeenth Century England* (New York: Harpers, 1938/1970); Charles Webster, *The Great Instauration: Science, Medicine, and Reform, 1626–1660* (Cambridge: Cambridge University Press, 1976); Christopher Hill, *The Intellectual Origins of the English Revolution* (Oxford: Clarendon Press, 1965); and I. Bernard Cohen, ed., *Puritanism and the Rise of Modern Science: The Merton Thesis*, with the assistance of K. E. Duffin and Stuart Strickland (New Brunswick, NJ: Rutgers University Press, 1990).

notation.[8] Written beneath the crest were the Latin words "How much we don't know!"

It is fair to say that the scientific movement was aided, indeed significantly so by the Royal Society; nevertheless, the movement was rooted deeply in European culture and practice that antedated these new institutions. This pattern of founding new scientific institutions outside the universities has been a common practice from that day to the present. Various sorts of think tanks or experimental laboratories are now longstanding adjuncts to universities, especially in the United States. But in no case are they replacements for universities themselves. The universities of seventeenth-century Europe are very much with us today and remain the primary sites for training each new generation of students. Just as the Royal Society of London took on no separate pedagogic function (though its members considered doing that) but provided a place for "advanced" experimental inquiry by a mature group of scholars, so too "centers for advanced study" periodically emerged across Europe from the seventeenth century onward. The creation of such advanced centers attached to universities is a major trend across Europe in the early twenty-first century. Professional societies – physical, chemical, biological – have served, since the seventeenth century, as publishing and clearinghouses for scientific research produced elsewhere. In no sense have they replaced the functions of universities. Professional societies, however, were unique European institutions for several hundred years thereafter.

In contrast to that, there was no legal possibility within Islamic law for the establishment of legally autonomous entities such as the professional societies and scientific institutions that Europe enjoyed. Except for the period between 1728 and 1745, there was no Ottoman Turkish or Arabic press until the early nineteenth century. Consequently, a free press was unavailable. In addition, the experimental phase of modern science did not emerge in the Middle East or other Muslim countries for centuries more.[9]

---

[8] This sheet of alternative crests was published by Michael Hunter in *Establishing the New Science: The Experience of the Early Royal Society* (Woodbridge, UK: Boydell Press, 1989), p. xiv. However, the Galileo-inspired drawing is misidentified as "earth and planets."

[9] See Francis Robinson, "Islam and the Impact of Print in South Asia," in *The Transmission of Knowledge in South Asia*, edited by N. Crook (Oxford: Oxford University Press, 1996), pp. 62–97. Also see A. Demeerseman, "Un ètape décisive de la culture et de la psychologie sociale islamique: Les données de la controverse autour du problème de l'Imprimerie," *Institute des Belles Lettre Arabs* 16 (1953): 17, (1954): 1–46, and 113–40; and "Matba'a," *EI²* 6 (1960): 794–807.

The Chinese had invented block printing in the late seventh century with a cultural flowering in the Sung Dynasty (960–1127). In Ming times, but only after 1500, various forms of local printing did emerge as scholars, monasteries, and private individuals printed their own pamphlets and tracts on various subjects, including travel, the arts and crafts, and novels. But even so, manuscript printing probably dominated all the way into the Ming Dynasty.[10] Even further away was the emergence of a free press. Not until the nineteenth century did this occur in China, if not later.[11] Scientists or natural philosophers as a group did not emerge in seventeenth- or eighteenth-century China. In the first place, there was no recognition of science as an intellectually separate enterprise,[12] neither could legally autonomous entities composed of private individuals emerge in China because the law of China had no such legal possibility. The legal concept of a "whole body," a legally autonomous entity (corporation), was absent.[13] Others have pointed out that "there was no occupational group sufficiently autonomous or coherent to be called a 'profession.'"[14]

Medicine and the Study of Anatomy

It is well to recall a strand of empirical research in Europe entirely separate from physics and astronomy that goes back to the Middle Ages: the study of anatomy. The most famous practitioner of this tradition at the

---

[10] Joseph McDermott, "The Ascendance of the Imprint in China," in *Printing and Book Culture in Late Imperial China*, eds. Cynthia J. Brokaw and Kai-wing Chow (Berkeley: University of California Press, 2005), pp. 56ff.

[11] Joseph Needham and Tsien Tseun-hsuin, *Science and Civilization in China: Paper and Printing* (New York: Cambridge University Press, 1983), 5/1:159–83.

[12] Nathan Sivin reminds us that there was no overall, coherent natural philosophy such as one finds among the Greeks, Arabs, or medieval Europeans. This follows from his reminder that the sciences in China were a heterogeneous mixture of inquiries far wider in scope than those of the Western tradition. China "had sciences but no science, no single conception or word for the overarching sum of them all"; in "Why the Scientific Revolution Did Not Take Place in China – or Didn't It?," in *Transformation and Tradition in the Sciences*, ed. Everett Mendelsohn (New York: Cambridge University Press, 1984), p. 533. What is more, "philosophers were in no position to define a common discipline among them, as Aristotle and his successors had done in Europe, and so philosophers had practically no influence on the development of these pursuits"; ibid., p. 535.

[13] Among others, see Derk Bodde and Clarence Morris, *Law in Imperial China* (Cambridge, MA: Harvard University Press, 1967); and for further discussion of these issues, see Toby E. Huff, *The Rise of Early Modern Science*, 2nd ed. (New York: Cambridge University Press, 2003), pp. 253–63, 271–76.

[14] Sivin, "Why the Scientific Revolution Did Not Take Place," p. 545.

beginning of the seventeenth century was William Harvey (1578–1657), who discovered the circulation of blood throughout the body. Using a variety of experimental procedures on man and animals, he proved the one-way flow from the heart through the arteries. At the extremities of the body, it flows back into the veins on its return to the heart. The connecting link between the arteries and veins, however, the capillary system, could not be seen until the invention of the microscope. Consequently, that quest kept the study of the circulatory system of the body in the forefront of anatomical research.

Harvey's research program, however, was the continuation of the great tradition of Andreas Vesalius, who had been born in Belgium in 1514 and who took a position teaching and performing public dissections in the amphitheater at the University of Padua. He was the author of the pathbreaking publication, *The Fabric of the Human Body*, published in 1543, the same year as Copernicus's *On the Revolutions of the Heavenly Spheres*. Vesalius's book laid the foundation for the new anatomical studies, both those of Harvey and the microscopists of the seventeenth century.

Yet, that tradition went all the way back to the thirteenth century, though its history has sometimes been neglected. Early in the twentieth century, the historian of science, Charles Singer, published a long extract from the work of the Italian physician Mondino de' Luzzi (1265–1326) titled "Anatomy Based on Human Dissection."[15] Since then, Mondino has been "universally given credit for the reintroduction of systematic human dissection into anatomy."[16] Recent scholarship reveals that Europeans had considerable knowledge of human anatomy, not just Galen's second-century A.D. knowledge based on animal dissections. For the Europeans had performed significant numbers of human dissections, especially postmortem autopsies throughout the medieval and early modern era. From the twelfth century onward, there was a pronounced growth of hospitals in Europe, and this coincided with the establishment of medical faculties and medical training in universities. By 1480, there were dozens of medical teaching faculties connected to major universities across Western Europe.[17]

---

[15] See Mondino de' Luzzi, "Anatomy Based on Dissection," in *A Source Book in Medieval Science*, ed. Edward Grant (Cambridge, MA: Harvard University Press, 1974), pp. 729–39.

[16] Ibid., p. 729n.

[17] Vivian Nutton, "Medicine in Medieval Western Europe, 1000–1500," in *The Western Medical Tradition, 800 BC to 1800*, ed. Lawrence I. Conrad, Michael Neve, Vivian

Already in the early twelfth century, Europeans in Salerno were performing dissections of pigs. In a document that has come down to us,
written before 1150, the author says, "Although some animals such as
monkeys, are found to resemble ourselves in external form, there is none
so like us internally as the pig, and for this reason we are about to conduct an anatomy upon this animal."[18] Clearly, this early-twelfth-century
document was meant to be accompanied by an actual dissection.

It goes without saying that a Muslim (or Jewish) physician of that
era would find operating on a pig highly repulsive. Consequently, it is
necessary to treat the pioneering work and influence of the Syrian-born
Muslim physician Ibn al-Nafis (1210–88) with caution. Al-Nafis is credited with the discovery of the lesser (pulmonary) circulation of the blood
between the heart and the lungs. In his *Commentary* on Ibn Sina, Ibn
al-Nafis writes:

This is the right cavity of the two cavities of the heart. When the blood in this cavity
has become thin, it must be transferred into the left cavity, where the pneuma is
generated. But there is no passage between these two cavities, the substance of
the heart there being impermeable. It neither contains a visible passage, as some
people have thought, nor does it contain an invisible passage which would permit
the passage of blood, as Galen thought. The pores of the heart there are compact
and the substance of the heart is thick. It must, therefore, be that when the blood
has become thin, it is passed into the arterial vein [pulmonary artery] to the lung,
in order to be dispersed inside the substance of the lung, and to mix with the
air. The finest parts of the blood are then strained, passing into the venous artery
[pulmonary vein] reaching the left of the two cavities of the heart, after mixing
with the air and becoming fit for the generation of pneuma.[19]

This passage seems remarkable and unprecedented for its time. It took
nearly four centuries more before Europeans were able to prove this possibility by performing the experiments that demonstrate the flow of blood
from the lungs to the heart. This was done by Realdo Columbo (1510–
59), the successor to Vesalius at Padua, who performed vivisections

Nutton, Roy Porter, and Andrew Wear (New York: Cambridge University Press, 1995),
pp. 153ff. A map showing the locations of medical faculties and universities is reprinted
in Huff, *Rise of Early Modern Science*, p. 2:194, figure 7.

[18] Anonymous, "Anatomical Demonstration at Salerno (the Anatomy of the Pig)," pp. 724–
26, in Grant, *A Source Book*, p. 725.

[19] Albert Iskandar, "Ibn al-Nafis," *Dictionary of Scientific Biography* 9 [1968]: 603. For
other versions of this passage, see Max Meyerhof, "Ibn Nafis (XIIIth cent.) and His
Theory of the Lesser Circulation," *Isis* 22 (1935): 100–20; and Emilie Savage-Smith,
"Dissection in Medieval Islam," *Journal of the History of Medicine* 50 (1995): 102.

on dogs.[20] There are some problems, however, with the passage from al-Nafis.[21]

What is especially problematic is that al-Nafis tells us that he avoided the practice of dissection because of the Sharia (the religious law) and his own "compassion" for the human body. He also says that "we will rely on the forms of the internal parts [of the human body] on the discussion of our predecessors among those who practiced this art [of dissection], especially the excellent Galen, since his books are the best of the books on this topic which have reached us."[22] Ibn al-Nafis, it should be noted, was also a specialist in Islamic jurisprudence so that his construal of the practice of dissection as un-Islamic carries special weight. This suggests that he did not perform dissections, leaving the reader to wonder how he arrived at his conclusion about pulmonary circulation and making his description of "a heart" mysterious.

Throughout Muslim history, there was a prohibition against post-mortem examinations that was parallel to the same prohibition in Judaic culture. In the Islamic case, one source of this prohibition came from a saying attributed to the Prophet Muhammad incorporated in the first rendition of Islamic law called the *Muwatta*.[23] That passage declares that "mutilating" the human body, especially in war, is forbidden. This argument was applied to the human body in general, and hence it prohibited postmortem examinations. That prohibition was reclaimed in the twentieth century when the Ayatollah Khomeini came to power with the Iranian revolution of 1979 and banned postmortems for the same reasons.[24]

In addition to that, hands-on study of anatomy was impeded because madrasas were not medical schools and, in some cases, founders of madrasas specifically forbade the teaching of medicine.[25] There was no progression in the Muslim world leading to the institutionalization of

---

[20] Andrew Wear, "Medicine in Early Modern Europe, 1500–1700," in Conrad, *Western Medical Tradition*, p. 328.

[21] I have discussed these broader issues in Toby E. Huff, "Attitudes towards Dissection in the History of European and Arabic Medicine," in *Science: Locality and Universality*, ed. Bennacer El Bouazzati (Rabat: Publications of the Faculty of Letters and Human Sciences, 2002), pp. 53–88.

[22] Savage-Smith, "Dissection in Medieval Islam," p. 100.

[23] *Al-Muwatta of Imam Malik ibn Anas: The First Formulation of Islamic law*, rev. in whole and trans. Aisha Abdurrahman Bewley (Granada: Madinah Press of Granada, 1989).

[24] See Huff, "Attitudes towards Dissection," p. 79.

[25] Said Amir Arjomand, "The Law, Agency, and Policy in Medieval Islamic Society: Development of the Institutions of Learning from the Tenth to the Fifteenth Century," *Comparative Studies in Society and History* 41, no. 2 (1999): 263–93.

medical education. Institutions of higher learning dedicated to medical education that would have been parallel to the European institutions did not emerge until the nineteenth and twentieth centuries. The study and teaching of medicine remained a private matter that relied on scholars privately tutoring aspiring students.

When we return to the European situation, we see that there was an entirely different attitude that gave permission for autopsies to be performed on both pigs and, later, human cadavers. Many of the autopsies in the thirteenth century were performed to determine whether the deceased had died of natural causes or whether there had been foul play, poisoning, or physical assault. Very early in the thirteenth century, the pope, Innocent III (1198–1216), ordered a postmortem autopsy of a person whose death was suspicious.[26] In 1286, an Italian cleric by the name of Salimbene reported that in response to the plague that had devastated several Italian cities, a physician opened the bodies of human victims of the plague as well as some chickens. He hoped to determine what was happening to the internal organs of the deceased, both animals and humans. Salimbene's remarks are so offhand as to suggest that this practice of postmortem autopsy had happened before. Likewise, in 1302, a scholar in Bologna died suddenly, raising the fear of poisoning. A postmortem was conducted with the conclusion that no poisoning was evident and that a large amount of blood had congealed around the heart, presumably causing death.[27]

From a modern point of view, one might be skeptical that medieval Europeans actually knew enough about the structure and function of the human body to draw sensible scientific conclusions from their autopsies. But, as a leading historian of the history of dissection put it, they had to have an idea of the normal anatomical condition to compare with the postmortem findings.[28]

A fifteenth-century autopsy of a young boy came to the following conclusion, expressed in modern medical terminology: "the autopsy seems to have revealed that the boy suffered from multifold metastatic abscesses of

---

[26] See Ynez Violé O'Neill, "Innocent III and the Evolution of Anatomy," *Medical History* 20 (1977): 429–33; Katherine Park, "The Criminal and the Saintly Body: Autopsy and Dissection in Renaissance Italy," *Renaissance Quarterly* 47, no. 1 (1994): 1–33; Roger French, *Dissection and Vivisection in the European Renaissance* (Aldershot, UK: Ashgate, 1999), p. 11; and C. D. O'Malley, "Pre-Vesalian Anatomy," in *Andreas Vesalius of Brussels, 1514–1564* (Berkeley: University of California Press, 1965), pp. 1–20.

[27] French, *Dissection and Vivisection*, p. 13.

[28] Ibid., chaps. 2 and 3.

the liver, the result of septicemia or pyleophlebitis."[29] In layman's terms, this was blood poisoning and a diseased liver.

By the end of the thirteenth century, postmortem examinations to determine the causes of death were well established. More important, one could say that the Europeans, unlike Muslims or the Chinese, had launched a program of empirical inquiry into the constitution of the human body. Part of that inquiry necessitated dissecting human bodies.

For very different reasons, postmortem dissections were not performed by any independent group in China. Instead, the process was controlled by the imperial bureaucracy. If a suspicious death occurred, then a state official was sent to the scene, where the examination was done "by the book" called *The Washing Away of Wrongs (Hsi Yüan chi lu)*. With this little book, published in the thirteenth century, Chinese authorities ensconced within the imperial bureaucracy all authority to inquire into wrongful deaths led by a district magistrate who had no medical training. Otherwise, autopsies were not performed because physicians had no freedom to do so. For this reason, the study of human anatomy in China remained underdeveloped.[30]

In contrast, by the end of the thirteenth century, European medical specialists, especially in Bologna, were using the practice of dissection to train students. These public dissections were carried out in a formal and solemn way with religious and public authorities in attendance, along with the presiding physician dressed in academic robes. Very soon thereafter, textbooks of human anatomy, based on dissection, were in circulation. The first of these was that of Luzzi de' Mondino, published in 1316. His book – appearing barely twenty-five years after the death of Ibn al-Nafis – later became something of a model for medical scholars training their students. It led to the publication of more texts of this sort, many with detailed anatomical illustrations. By the end of the fourteenth century, observation of a human dissection, usually taking place over four days, became part of medical training throughout Europe. And, by 1500, the universities composed a thick web of educational institutions across Europe.

Another impediment inhibiting anatomical knowledge in Muslim lands is the Islamic aversion to the artistic representation of the human body, though some individuals did violate the prohibition. Many hadiths

[29] Bernard Tornius, "A Fifteenth-Century Autopsy," in Grant, *A Source Book*, pp. 740–42, 740n1.
[30] See Huff, *Rise of Early Modern Science*, Appendix, pp. 205–8.

(sayings attributed to the Prophet Muhammad) refer to the opprobrium that attached to "pictures" and the "picture makers." For example, the hadith complier "Muslim" cites a tradition according to which the Prophet Mohammad "cursed the picture-makers."[31] Likewise, is it said in the same source that "angels do not enter a house in which there is a picture."[32] The same warning appears in the Islamic legal handbook called the *Muwatta*.[33]

Nevertheless, a number of Persian artists did produce renderings of human subjects in later centuries. But, as the art historian B. W. Robinson[34] has pointed out, Persian artists hardly ever drew from nature; they drew from their heads, using a set of traditional formulas. Their aim was to combine clear illustration with a pleasing decorative effect, and they felt no obligation to reproduce such optical accidents as shadow or perspective.[35] Likewise, the early Mughal miniatures often provide sharply rendered portraits of people and sometimes birds and animals, but the faces of people are very stylized. While there is a discernible move toward greater realism,[36] they do not approach the level of realism, emotional expression, and individuality of the late-sixteenth-century European painter Pieter Bruegel and his *Peasant Dance* of 1568. By the time of Emperor Aurangzeb's reign in the 1650s, the tradition faded so that it never reached the level of Dutch and Flemish realism of the seventeenth century.[37]

The Persians who did draw medical depictions of the human body were untutored, lacking in visual perspective and realistic detail as well as scale (Figure 7.1). The point is that realistic and detailed depiction of the natural world requires considerable training in a realist artistic tradition which

---

[31] *Muslim*, vol. 3, book 34, no. 299.

[32] *Muslim*, book 24, no. 5253.

[33] Imam Malik ibn Anas, *Al-Muwatta: The First Formulation of Islamic Law*, sect. 54.3.

[34] B. W. Robinson, Introduction to *The Metropolitan Museum of Art Miniatures: Persian Painting* (New York: Metropolitan Museum of Art, 1953).

[35] For an account of the gradual assimilation of the practice of dissection and anatomical depiction in the Ottoman Empire, see Gül Russell, "'The Owl and the Pussy Cat': The Process of Cultural Transmission in Anatomical Illustration," in *Transfer of Modern Science and Technology to the Muslim World*, ed. Ekmeleddin Ihsanoglu (Istanbul: Research Center for Islamic History, Art, and Culture, 1992), pp. 180–212, esp. pp. 195–208.

[36] Among others, Shelia Blair and Jonathan M. Bloom, *The Art and Architectures of Islam, 1250–1800* (New Haven, CT: Yale University Press, 1994), chap. 19.

[37] Ibid., p. 296. Bloom and Blair observe that "the naturalism that had been apparent in the borders of the Jahangir albums became increasingly stiff and formal, especially in the Late Shahjahan Album" of the 1650s.

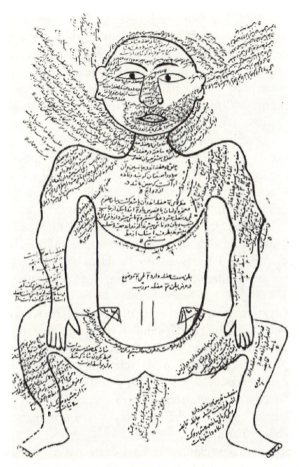

FIGURE 7.1. Mansurian illustration of the body's muscle system. From *The Anatomy of the Human Body* (a Persian text, *Tashrih-i badan-i insan*, ca. 1396). As reproduced on the National Library of Medicine Web site: http://www.nlm .nih.gov/exhibition/islamic_medical/islamic_10.html.

was absent in the Muslim world. Transfer of that tradition from the West to the Muslim world began to occur in the sixteenth and seventeenth centuries. This was true also for early Mughal miniatures in which traces of European themes are easily visible.[38]

In the meantime, scholarly consensus asserts that the first illustrated medical treatise in the Muslim world was that of the Persian physician

[38] Ibid., chap. 19; and J. M. Rogers, *Mughal Miniatures* (New York: Thames and Hudson, 1993).

Mansur ibn Ilyas (fl. late fourteenth century). His *Mansurian Anatomy* of 1396 contains a number of illustrations of the various anatomical systems of the body such as the bones, veins, nerves, and internal organs.[39] These illustrations, however, appear to be modeled after the pre-Islamic Alexandrian originals[40] (which had been known in Europe for some time). The same untutored illustrations, lacking realistic detail, continued to be reproduced into the nineteenth century. It seems evident that the attitudes toward the study of human anatomy, including its depiction, and above all the practice of dissection, were very different in the two civilizations, and the consequences were likewise very different for the progress of science. The contrast between these technically very primitive Middle Eastern sketches and the richly detailed anatomical drawings of Vesalius is pronounced (Figure 7.2). That contrast makes graphic the nature of the medical revolution that occurred in the sixteenth and seventeenth centuries in Europe, which was enabled by the institutional breakthrough that permitted the detailed study, dissection, and depiction of the human body.

In this context, the anatomical work of Vesalius represents an extraordinary leap forward in our understanding of human anatomy. Its depiction of the five anatomical systems of the body remains unsurpassed and graphically illustrates the nature of the scientific revolution as it emerged in the medical sciences. The printed copy of *The Fabric of the Human Body* pioneered the use of illustrations for scientific works. It labeled each anatomical part with a letter that was keyed to a text that provided its technical Greek or Latin name.

Furthermore, Vesalius claimed to have corrected more than 200 mistakes in the classic work of Galen that had been the bible of anatomical studies for more than a millennium. For these reasons, the University of Padua became a magnet for medical students throughout the sixteenth century. That is where William Harvey was recruited to his lifelong study of human anatomy. This was the same university at which Galileo taught in 1609, when he made his famous telescopic discoveries.

Soon after his graduation with a medical degree in 1602, Harvey traveled back to England, where he proceeded to challenge the centuries-old

---

[39] Lawrence Conrad, "The Arab-Islamic Medical Tradition," in *Western Medical Tradition*, pp. 120–21.

[40] For possible sources of the various "Five Figure" and "Nine Figure" models, see Roger French, "An Origin for the Bone Text of the Five-Figure Series," *Sudhoff's Archive* 68, no. 2 (1984): 143–58; now reprinted in French, *Ancients and Moderns* (Aldershot, UK: Ashgate, Variorum, 2000).

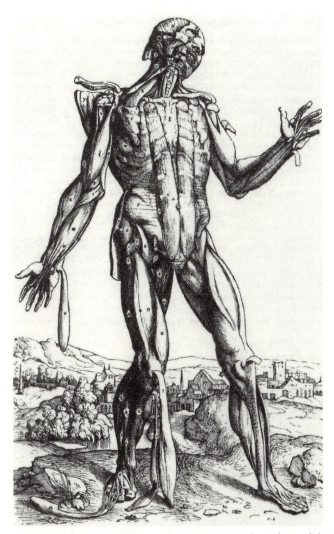

FIGURE 7.2. The body's muscle system, from Vesalius, *The Fabric of the Human Body* (1555 ed.). Each of the parts is labeled with a letter that was keyed to an anatomical description.

theory of Galen regarding the circulation of the blood. By 1628, and after years of experimentation, Harvey could demonstrate that there is indeed a one-way flow of blood throughout the body, not just between the heart and lungs, as demonstrated by Realdo Columbo.

Another example of the great gulf of anatomical learning that existed between European physicians and those in the Muslim world comes from the French physician Francois Bernier. He personally attended Mughal Sultan Aurangzeb between 1656 and 1668. He reported that

> it is not surprising that the Gentile [nobles] understand nothing of anatomy. They never open the body either of man or beast, and those in our household always ran away, with amazement and horror, whenever I opened a living goat or sheep for the purpose of explaining to my Agah [assistant] the circulation of the blood, and showing him the vessels, discovered by Pecquet, through which the chyle[41] is conveyed to the right ventricle of the heart. Yet notwithstanding their profound ignorance of the subject, they affirm that the number of veins in the human body is five thousand, neither more nor less, just as if they carefully reckoned them.[42]

In short, the European practice of experimental medicine, the detailed examination of the human body through the use of manipulative techniques, was far in advance of China, the Middle East, or other parts of the Muslim world. As a result, the Europeans developed a considerable stock of empirical knowledge about human anatomy that was not available outside Europe. Inspired by the pursuit of scientific knowledge, European physicians engaged in a variety of practices that would have been forbidden in a Muslim context and restricted by Chinese imperial authorities. These practices included (1) the dissection of human bodies, (2) the dissection of pigs, (3) the performance of the operation in a public forum, and (4) the publication of richly detailed drawings of the human anatomy in all its minute, and some would say offensive, detail.

Microscopy

Microbes, microorganisms, and microbiology have the ring of contemporary science, but they are rooted in the seventeenth century. They were discovered, and could only have been discovered, by the use of the microscope – that is, either a single- or compound-lens system that could magnify tiny objects 250 to 500 times normal size. In this case, they were discovered by an implausible investigator, a cloth merchant without university training, in Holland in the 1670s.

According to some historical reports, the compound microscope consisting of two lenses was invented in the same small Netherlands town of

---

[41] This is not the chyli of Galen but rather a milky fluid of lymph and emulsified fat that is a by-product of digestion passed into the bloodstream.

[42] F. Bernier, *Travels in the Mughal Empire, 1656–1680*, trans. Irving Brock, rev. A. Constable (New York: Asian Educational Service, 1996), p. 339.

Middelburg (where the telescope was invented) by John and Zacharias Janssen in 1590. However, that line of invention disappeared when a man named Cornelius Drebbel (1572–1623) gained possession of the Janssen microscope but failed to either preserve it or make it available for histori-cal examination.[43] Later, as a Dutch instrument maker living in England, Drebbel made and sold microscopes across Europe. These instruments seem to have been Keplerian in the sense that they used two convex lenses. One or more of these devices came into the hands of the queen of France, Marie de' Medici. The astronomer Nicolas-Claude Fabri de Peiresc, who attended that French court, is said to have procured other copies of the Drebbel microscope. Through Drebbel's contacts (and pos-sibly others) with notables in Rome, the microscope arrived there by the early 1620s. But even before this, there are good indications that Galileo had transformed his telescope into a microscope as early as 1610. This was reported by the Scotsman John Wodderborn in the same year.[44] So when Galileo was consulted in Rome in 1624 about one of the telescopes sent to a cardinal who did not know what the instrument was for, he clearly understood how the device was meant to work.[45] At the same time, Galileo had constructed new models of his own microscopes, send-ing one to the Duke of Bavaria and presenting another to the head of the Lincean Accademia, Frederico Cesi. Another member of that academy, Johannes Faber, in a letter dated April 13, 1625, gave the name "micro-scope" to the device.[46] In this circuitous way, microscopic technology spread across Europe, with Italian, Dutch, German, and English inven-tors developing their own designs.

Thus, long before 1623, Galileo had used this new magnifying device to examine flies, fleas, and other tiny creatures. In 1624, he reported to Cesi that he had observed "many tiny animals with great admiration, among

---

[43] Alfred N. Disney, C. F. Hill, and W. E. W. Baker, eds., *Origin and Development of the Microscope* (London: Royal Microscopical Society, 1928).

[44] See Gilberto Govi, "The Compound Microscope Invented by Galileo," *Journal of the Royal Microscopical Society* 9 (1889): 575.

[45] Ibid., p. 576.

[46] Marian Fournier, *The Fabric of Life* (Baltimore: Johns Hopkins University Press, 1995), pp. 10–11. Also see Alfred Disney et al., *Origin and Development of the Microscope* (London: Royal Microscopical Society, 1928), esp. pp. 94–99; G. L'E. Turner, *Essays on the History of the Microscope* (Oxford: Senecio, 1980), pp. 3–5; and Edward Ruestow, *The Microscope in the Dutch Republic* (Cambridge: Cambridge University Press, 1996), pp. 6–8. A reproduction of the earliest sketch of the compound microscope, by Isaac Beeckman, is in David Bardel, "The Invention of the Microscope," *Bios* 75, no. 2 (2004): 82.

which the flea is quite horrible, the gnat and the moth very beautiful; and with great satisfaction I have seen how flies and other little animals can walk attached to mirrors, upside down... the greatness of nature can be infinitely contemplated, and [we may now see] how subtly and with what unspeakable care she works."[47] In this letter to Cesi, Galileo provided more instruction on how to use his microscope:

the object must be placed on the movable circle which is at the base, and moved to see it all; for that which one sees at one look is a small part. And because the distance between the lens and the object must be most exact, in looking at objects which have relief one must be able to move the glass nearer or further, according as one is looking at this or that part.[48]

But it was his fellow members of the Lincean Academy, Cesi and Francesco Stelluti, who published the first microscopic study with impressively detailed drawings of the bee in 1625.

Exploratory use of the microscope, however, did not take off with the same enthusiasm as did the use of the telescope. Perhaps this was because it was so difficult to describe and discuss the minute and startling exotic structures of the microscopic world that were newly revealed. For it was one thing to discover the hidden structure of flies, fleas, and plants; it was quite another to comprehend the existence of hundreds, thousands, or millions of organisms living invisibly in a nearby freshwater pond or in table water infused with grains of pepper.

Nevertheless, microscopy, the study of the hidden world of nature using the microscope, flourished throughout the latter half of the seventeenth century. A study of publications in this field between 1625 and 1750 reveals that fifty books by twenty-nine authors were published on the subject. During the same period, more than 140 letters reporting on microscopic studies were published in the *Philosophical Transactions* of the Royal Society of London that had been founded in 1661. Another seventy-six publications on microscopy appeared in the German journal *Miscellanea Curiosia Medio-physica* (1670–1750).[49] By the end of the century, microscope kits, complete with instructions and mounting tools, were commercially available to an eager public.

[47] Stillman Drake, *Galileo at Work* (Chicago: University of Chicago Press, 1978), p. 286. This is from the letter sent to Frederico Cesi on September 23, 1624.
[48] From the translation of Govi's article, "The Compound Microscope," p. 577.
[49] Fournier, *Fabric of Life*, appendix.

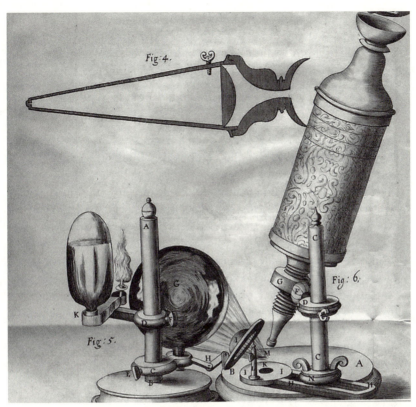

FIGURE 7.3. Hooke's microscope from his *Micrographia* (1665). The vertical height of the apparatus can be adjusted with the collar and thumbscrew. By rotating the barrel that is mounted in a threaded collar (G), finer adjustments can be made.

The combined work of the most active microscopists of this period revolutionized our understanding of the basic forms of life, their hidden structure, and our comprehension of the very nature of human and animal reproduction. The most influential of these early uses of the microscope were reported by the peripatetic Robert Hooke working in London. Using the microscope he had probably bought from the instrument maker Christopher Cock, which he modified to get sharper images (Figure 7.3), he looked at a great miscellany of small objects such as fleas, gnats, the spider, plant life, minerals, household mold, and many other things. He made sketches of some of these objects that presented extraordinary images. His study of a cross section of the cork tree led to the first use of the term *cell* to describe the interlocking units of the invisible structure of nature. His sketches of magnified poppy seeds also displayed an interlocking cell structure, as did the elder pith and the shaft of quill pens.[50] The idea of the cellular structure of biological phenomena became a cornerstone of microbiology to the present.

Marcello Malpighi

Still earlier in the 1660s, the Italian physician Marcello Malpighi used his microscope to reveal the hitherto invisible structure of animal organs. Born in 1628, he took up medical studies at the university in Bologna, where he joined a select group of nine students who were invited by a professor named Massari to witness dissections and vivisection in his home – something we have noted that would have been forbidden to Muslim and Jewish scholars and unavailable to Chinese students.

Consequently, Malpighi's experience in Bologna prepared him well to undertake a systematic study of the hidden structures of human and animal organs. The traditional anatomical view was that the organs of humans and animals were coagulated clusters of blood. But Malpighi's work, accompanied by vivid sketches, began to reveal a whole new level of fine anatomical structure.[51] Using the frog as the subject of study, and aided by his microscopic devices, Malpighi was able to identify the extraordinary fine venous network of the lungs. He could discern the arteries and the connecting capillary threads that joined up with the veins. He saw the expanding and contracting sacks of the lungs and the passage of blood through this network. These discoveries of 1661 reported in his work, *On the Lungs*, were a landmark event that confirmed the revolutionary work of William Harvey. Back in 1628, Harvey had predicted the existence of capillaries connecting arteries and veins in the human body. Malpighi's work soon became very influential among the Dutch anatomists and microscopists. Within a decade or so, several other leading microscopists would visually confirm Malpighi's observations of the capillary system. Nevertheless, Malpighi's most famous drawings of his

---

[50] See Lisa Jardine, *Ingenious Inventions: Building the Scientific Revolution* (New York: Doubleday, 1999), pp. 44ff; and Brian J. Ford, "The van Leeuwenhoek Specimens," *Notes of the Royal Society of London* 36, no. 1 (1981): 37–59.

[51] See Catherine Wilson, *The Invisible World* (Princeton, NJ: Princeton University Press, 1995), pp. 94–95; and Fournier, *Fabric of Life*, pp. 55–62, 112–21.

pioneering anatomical work are those of the gestation and early life of the chicken from 1673.[52]

The Dutch, Leiden, and Microscopy

Although a great deal of attention has focused on the Royal Society of London and its part in publishing the letters and reports of leading microscopists in the seventeenth century, it is important to note that the Dutch Republic was experiencing its greatest intellectual flowering during this century, especially between 1620 and 1672. A group of brilliant anatomists had gathered around the medical faculty of the University of Leiden. Student enrollment at the University of Leiden, in particular, grew very significantly during the first half of the century, attracting the attention of many foreign students, especially in medicine.[53]

At the time when the anatomists and microscopists were working their revolution in our understanding of the human body, as well as the fine structures of biological creatures, the mechanical view that even animals and humans were just mechanical devices held sway among the natural philosophers that we would today call physicists. This was a by-product of the great successes of the Newtonian worldview that saw all matter as just particles in motion and human beings just a collection of so many mechanical devices. Descartes, for example, among others, tried to explain muscle movements using mechanical principles. Muscles could move bones by expanding or contracting, according to him and others, thereby seeming to replicate the mechanical movements of levers. Descartes believed that some kind of *spiritus animalis* flowed into the muscles, causing their movement. No doubt there was circulation of particles and chemicals in the body, but chemical analysis had hardly begun.

The anatomists, physiologists, and emerging protobiologists, however, saw something entirely different, just as Vesalius, Harvey, and Malpighi had. The fine anatomical structures they saw were *not* very like mechanical devices. Malpighi's early work had already shown that the organs of the body were not just coagulated masses of fluids but rather had their

---

[52] For a replica of this extraordinary illustration, see Malpighi under Google Images.
[53] See the table on student enrollments at Leiden in Jonathan I. Israel, *The Dutch Republic: Its Rise, Greatness, and Fall 1447–1806* (New York: Oxford University Press, 1995), p. 901.

own discernible structure. By 1664, Jan Swammerdam demonstrated that the muscles of the hind legs of frogs could be stimulated to action either by touching with a wire or simply contact by a human finger.[54]

Likewise, the research of the microscopists focused on the organs and processes of reproduction in insects, animals, and humans. These were more than mechanical motions. Those studies argued against the idea of spontaneous generation of new life. Going all the way back to Aristotle, it was thought that many organisms spontaneously regenerated themselves. The idea was that plants and animals could re-create themselves without the contribution of another life partner or biological input. It was also thought that decaying flesh generated new organisms and new life. This view received a powerful setback when, in 1668, the Italian physician Francisco Redi experimentally demonstrated that the maggots commonly found on decaying flesh were actually caused by flies who laid their eggs in the flesh, which then hatched into the larvae (maggots). He did this by observing jars of discarded meat and fish, some covered completely and some left open. In the tightly sealed jars, no maggots emerged.[55] Redi's demonstration, for those who believed in it, created further puzzles when experiments were done to find out whether Leeuwenhoek's "little animals" could reproduce themselves when held tightly in a sealed vial. They could, but the question was how. This was the larger background against which the early microscopists carried out their research, discovering all sorts of new biological structures and finding even more microorganisms than had been imagined.

Pioneering Advances and Conflicts of Priority

Among the graduates of the University of Leiden medical faculty during this early phase, we find pioneers such as Jan Swammerdam (1637–80), Reiner de Graff (1641–73), and the Englishman Nehemiah Grew (1641–1712). Grew received a medical degree from Leiden before returning to London, where he published his landmark works on the anatomy of plants (1661–82).

Jan Swammerdam earned his MD from the University of Leiden in 1667. Although he had qualified to study medicine in 1665, he went to Paris, where he studied with other important physicians before returning to Leiden for his terminal degree. After graduating, he turned his attention

---

[54] Fournier, *Fabric of Life*, pp. 106–7.
[55] E. G. Ruestow, *The Microscope*, p. 203; Wilson, *Invisible World*, pp. 199–201.

to entomology and became the first scientist to identify the life cycle of insects, growing from egg, larva, and pupa to adult insect. Later, in 1668, while studying human anatomy, he was the first to identify red blood cells using his microscope.

Swammerdam, however, was a gloomy person, obsessed both with his scientific work and his commitment to a godly life. He believed deeply that the hidden structures of nature reveal the hand of God. He often repeated a story attributed to a friend, marveling at the unity of God and nature: "Oh wondrous God! Who would not know you from this and, knowing you love you!"[56] At the same time, he experienced severe ambivalence over the call of God and the call of science, which drove him into a spiritual crisis.

After seeing Malpighi's pioneering microscopic work reveal the invisible structure of the silkworm and other creatures, Swammerdam was challenged to hone his microscopic skills that eventually produced extraordinary images of the intricate miniature structure of the ovaries of the bee (Figure 7.4). This was quite a remarkable achievement. This and other anatomical discoveries revealing the structures of reproduction in animals and insects spurred Swammerdam to great ambitions and contentiousness over issues of scientific priority. These wrangles over priority of discovery – so common in science and which the sociologist Robert Merton made famous[57] – occurred both with Malpighi and, tragically, with the young Reiner de Graaf.

Along with the idea that some plants and animals could spontaneously regenerate themselves without any input from another life-creating substance, there was the belief that the parts of the body were already preformed in the semen of the creatures involved. Consequently, the question began to focus more intensely on describing the microscopic parts, especially the ovaries and the eggs and the process by which they developed into new life-forms. No one in the seventeenth century did more to oppose the idea of spontaneous generation than the untutored Antoni van Leeuwenhoek, who was born in 1632. Of course, there are forms of biological life that do indeed regenerate or replicate by subdivision into new units of life, but it was very difficult to determine which forms replicated without interaction with others, going through a form of asexual

---

[56] As cited in Ruestow, *The Microscope*, p. 116.
[57] See chap. 14, "Priorities in Scientific Discovery," in Robert K. Merton, *The Sociology of Science: Theoretical and Empirical Investigations* (Chicago: University of Chicago Press, 1973), 286–324.

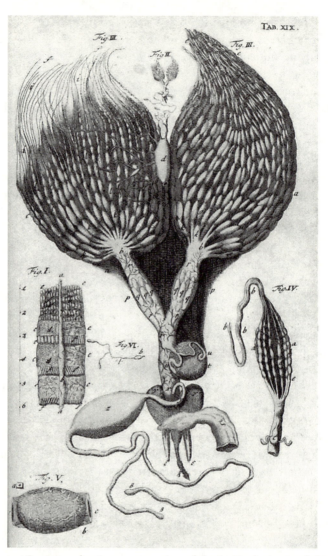

FIGURE 7.4. Swammerdam's representation of the internal organs of the immature queen bee. II shows the ovaries as they appear to the naked eye. III shows the
ovaries of a mature (on the right) and immature queen bee. From Jan Swammerdam, *Bybel der nature*, 1737–38.

reproduction, and which were actually the product of sexual fertilization.
The idea of regeneration through putrefaction, though opposed by many
leading researchers, was still believed by some physicians, despite Redi's
experimental work.

The bridge between the self-taught Leeuwenhoek and the broader scientific community was provided by Reiner de Graaf (1541–1673), a gifted young anatomist with whom Swammerdam had a fateful priority clash. Reiner de Graaf was a young, colorful investigator and Catholic whose religious background precluded him from a university career at Leiden. After graduating from Leiden, he traveled around France, where he had access to numerous cadavers. He reports that he "took great delight in opening them and particularly when I set my knife in the Pancreas or the Reproductive Parts, since I daily observed things in and about these Organs, which had never before been brought to light by any dissector."[58]

Moreover, de Graaf had submitted a dissertation on the pancreas for his medical degree to the University of Leiden. Even granting that the process of examining medical candidates at that time was not as rigorous as twentieth-century standards, this was something that could only happen in Western Europe. For as we have seen, no postmortem access to bodies in the Muslim world was allowed, as by law and tradition each cadaver had to be interred within a day of the person's death, and otherwise, postmortem examinations were forbidden. Even if surreptitious access had been gained, there were no institutions of higher learning in the Muslim world, where the practice of submitting dissertations on anatomical topics existed. There was no faculty of medical experts to which such a work of scholarship could be submitted because madrasas were not medical schools. In the madrasas, receiving an *ijaza*, "permission to transit," was a personal authorization given by a senior scholar to a student that authorized the student to transmit certified knowledge, not a collective act of a faculty approving an independent piece of new scholarship.[59]

Likewise in China, physicians did not have the independence to conduct postmortem exams, and even if someone did it surreptitiously, there was no institutional training structure for submitting such a scholarly study. The Chinese educational system was entirely geared toward mastering traditional subjects for the triennial examinations.[60]

Reiner de Graaf's attention after graduating, however, was focused on the male and female reproductive systems. He sought to explain the female reproductive process by focusing on the structure of female ovaries, the fallopian tubes, and the emergence of an ovum (egg). At the

---

[58] As cited in Wilson, *Invisible World*, p. 25n41.

[59] Among others, see George Makdisi, *The Rise of Colleges: Institutions of Learning in Islam and the West* (Edinburgh: Edinburgh University Press, 1981).

[60] See Benjamin Elman, *A Cultural History of Civil Service Examinations in Late Imperial China* (Berkeley: University of California Press, 1994).

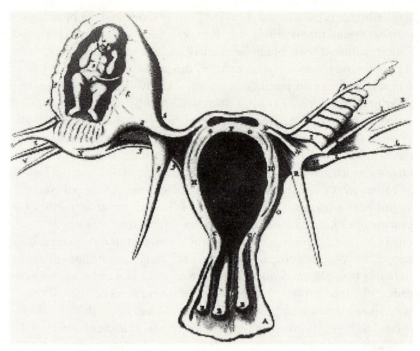

FIGURE 7.5. Illustration of an ectopic pregnancy that de Graaf adapted from a letter by Benoit Vassal that was published by the Royal Society of London in January 1669. De Graaf surmised that the fallopian tube had a special function of carrying the ovum forth in the reproductive cycle. Reiner de Graaf, *De Mulierum Organis Generationi Inservientibus* (1672).

time, anatomists believed that male and female reproductive organs were analogous and consequently referred to "testicles" and "semen" in both males and females. Neither eggs (ova) nor spermatozoa in human sperm had been identified by medical scholars before the 1670s.

De Graaf's extraordinary investigations took him increasingly away from the accepted view, though he continued to use the term *testicle* when describing some parts of female reproductive anatomy. Fallopian tubes had been identified by Gabriello Fallopio, an assistant to Andreas Vesalius, back in 1561. De Graaf and others, especially the French anatomist Benoit Vassal, refined the details of that anatomy. But it was de Graaf who correctly identified the function of the fallopian tube, pointing out in particular that its terminus in a healthy female was open, not closed, as some anatomists thought. A closed terminus of the tube meant that it was abnormal and would lead to a dangerous ectopic pregnancy (Figure 7.5). In this diagram borrowed from Benoit Vassal, and first published in the *Philosophical Transactions* of the Royal Society of 1669, we see the same labeling of all the anatomical parts, A through M, as was established by Vesalius a century earlier in his classic study.

Separately, de Graaf correctly sketched the healthy and unhealthy forms of the fallopian tubes while pointing out that their "very elegant shape" could be seen "with the aid of a microscope or some other device."[61] From his investigations of rabbits, he asserted that "not only do the expansions at the ends of the tubes embrace the 'testicles' from all sides but, in rabbits, the eggs themselves on the third day after coitus can be seen to pass through the tubes."[62] This idea that an egg was actually released from the human ovary was a radical idea. It was becoming increasingly recognized that this was so in animals, but not until 1827 was this fact actually observed in the human body. At the same time, de Graaf was unaware that the fertilization of the egg only took place with the union of a spermatozoon (later discovered by Leeuwenhoek) and the female egg. Nevertheless, de Graaf understood how the female reproductive system worked: "All of these cases [cows, ewes, rabbits]," he wrote in 1672,

prove that the eggs from which fetuses are to be generated pass from the "testicle" through the tubes to the uterus and that a fetus is generated in a tube from no other cause than that an already fertilized egg gets caught for some reason or other in its transit. As such a fetus grows it prepares death for its mother.[63]

Here, de Graaf signals that ectopic pregnancies often led to death for pregnant mothers in the seventeenth century. In the case of the young mother whose autopsy was reported by Vassal and whose diagram de Graaf adopts, the deceased had had eleven successful deliveries, yet the twelfth resulted in the fatal pregnancy. "Having duly weighed all these considerations," de Graaf went on, "we judge said Fallopian tubes in women and every kind of female animal are the real 'delivering vessels' or, if you prefer, the oviducts. It is through them that the eggs of the 'testicles' are transferred to the uterus."[64]

Unfortunately, de Graaf's work, suggesting that the so-called testicles of the female are really ovaries, provoked bitterness and then a priority dispute initiated by Swammerdam. Swammerdam had worked with

---

[61] *De Mulierum Organis*, as cited in W. M. Ankum, H. L. Houtzager, and O. P. Bleker, "Reinier de Graaf (1641–1673) and the Fallopian Tube," *Human Reproductive Update* 2 (1996): 366.

[62] Ibid., p. 367.

[63] Ibid., p. 368.

[64] Ibid.

another anatomist, Johannes van Horne, isolating the ovaries and reproductive process in insects, but this work had not been published before van Horne died and de Graaf published his report. Consequently, Swammerdam submitted an appeal to the Royal Society of London asking it to arbitrate his claims. Sadly, de Graaf died at the young age of thirty-two from depression-related circumstances before the dispute could be resolved.

Leeuwenhoek's Little Animals

The next step in this process of discovery was either the identification of the actual human ovum or the identification of the male fertilizing component – that is, spermatozoa, the sperm cells that had neither been seen nor imagined hitherto. Semen was well known, and both males and females were thought to have it, but the spermatozoon had yet to be discovered.

At this critical juncture, and just a few months before he died, de Graaf sent a letter to the Royal Society of London on behalf of Leeuwenhoek and his unusual work. De Graaf had met Leeuwenhoek because they both lived in Delft. In the letter sent to the Royal Society of London (April 28, 1673), de Graaf spoke of "a certain most ingenious person here, named *Leewenhoeck* [one of many spellings of his name] [who] has devised microscopes which far surpass those which we have hitherto seen."[65] This was an entirely surprising recommendation because Leeuwenhoek had no university training, had been apprenticed as a young boy to a cloth draper, and otherwise knew little about anatomy or microscopes, or so people thought. Most unexpected was that he had crafted his own variety of microscope made by using strands of glass drawn out over a flame and further heated to make a tiny ball. These were polished and then installed in a tiny hole between two brass plates with a mounting spike on the opposite side, adjustable with several screws (Figure 7.6). Peering through the little lens and facing a source of light, organic materials could be seen magnified hundreds of times. This innovation seemed to run counter to best practice of the compound microscope, but in Leeuwenhoeks's hands, these new single-lens instruments were put to spectacular use.

On the way to Leeuwenhoek's important work on reproductive systems in nature, he hit on revolutionary discoveries that opened up the

---

[65] As cited in Brian J. Ford, "The Van Leeuwenhoek Specimens," *Notes and Records of the Royal Society of London* 36, no. 1 (1981): 45.

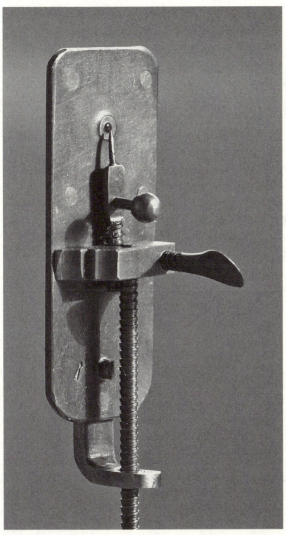

FIGURE 7.6. A replica of a Leeuwenhoek's microscope by Jacapo Werther. The small opening toward the top of the instrument houses the small glass bead lens with which a specimen could be viewed mounted on the needle. Adjusting screws could reposition the specimen up or down and closer or farther from the lens. From http://upload.wikimedia.org/wikipedia/commons/d/de/Leeuwenhoek_Microscope.png.

whole world of microorganisms and microbiology. His very careful work preparing and preserving specimens, his attempts to devise a scale for comparing microorganisms, along with his broad and indefatigable collection of new life specimens, gave coherence to the new science of microscopy.[66]

Leeuwenhoek's interest in using a microscope for biological purposes was in part inspired by his peculiar interest in what it is that causes spices to taste the way they do. In that quest, he tried putting salt, sugar, and peppercorns in snow water, which was the purist form of water available at the time, and observing the solutions day after day. The pepper infusion revealed an extraordinary sight, something never before seen: tiny creatures, some smaller than a human hair, swimming about. But even that outcome was made possible by Leeuwenhoek's ingenuity in fashioning his own glass-bead single-lens microscope that was more powerful than ones being used by leading members of the Royal Society of London in 1674.

However, the really decisive event leading to Leeuwenhoek's microscopic career probably occurred in the summer of 1674, when he visited a local pond. In it, he saw a variety of odd biological forms that prompted him to take samples of the water. After carefully observing the water at home over several days, he wrote a short report that he sent to the secretary of the Royal Society of London, Henry Oldenburg. In the letter of September 7, 1674, he reported seeing "floating therein divers earthly particles and some green streaks, spirally wound serpent-wise, and orderly arranged, after the manner of the copper or tin worms which distillers use to cool their liquors." In addition, he saw many "green globules joined together" and among these "very many little animalcules." These creatures seemed to move with astonishing speed in many directions.[67]

For microbiologists, this discovery is the birth of microbiology and perhaps bacteriology. For Leeuwenhoek had indeed discovered the world of protozoa and algae, of single-celled microbes that exist in great

---

[66] There have been many disputes about Leeuwenhoek's methods and techniques, but the reexamination of Leeuwenhoek's actual specimens and his methods of work by Brian J. Ford seem to put most of them to rest. See Ford, *Single Lens* (New York: Harper and Row, 1985); Ford, "The Leeuwenhoek Specimens," *Notes and Records of the Royal Society of London* 36, no. 1 (1981): 37–59; and Ford, "Dilettante to Diligent Experiment: A Reappraisal of Leeuwenhoek and Microscopist and Investigator," *Biology History* 5, no. 3 (December 1992), http://www.brianjford.com/a-av101.htm.

[67] As reprinted, Clifford Dobell, *Antony van Leeuwenhoek and His "Little Animals"* (London: Cavendish Press Limited, 1932), p. 109.

abundance in the purest water, in rainwater, snow water, and of course in seawater. Although Leeuwenhoek thought all these creatures were of one kind, that the smaller ones were babies soon to grow into adults, they fell into a variety of biological categories or phyla. Technically, some would be called flagellates, amoeba, ciliates, voticella, rotifers, and such. Some had long miniature tails, others had hundreds of the tiniest feet (cilia) that propelled them forward, and some were just round globules or immobile heliozoans.

It is certain that Leeuwenhoek observed several forms of bacteria, not only in the water samples but also in the decaying matter of the human mouth, though neither he nor anyone else at the time understood the nature of bacteria, its superabundance in nature, its typical mode of reproduction by cell division, or its potential to cause lethal diseases. Still, his careful descriptions of what he saw, and the aid of an engraver, provided later scientists with specimens that were identified as forms of bacteria (Figure 7.7).

Only in the early nineteenth century would European physicians identify the anthrax bacilli, invent the Petri dish for cultivating such organisms, and then develop a vaccine, as was accomplished by Robert Koch in the 1860s. Leeuwenhoek's little animals represented a discovery that paved the way for developments that would only emerge nearly 200 years after his work.

In the meantime, Leeuwenhoek went back to his microscope, studying more samples of water and describing in further detail many microbial animals never before seen. When his observations were recorded and sent to the Royal Society, they created quite a stir; even the king of England wanted to see the Dutchman's "little animals."

In October 1676, Leeuwenhoek composed a long, seventeen-and-a-half-page letter (in his native Dutch) reporting a long series of observations made by using his microscope with its tiny glass bead. These included many more details about his animalcules in pond water, rainwater, snow water, seawater, and peppercorn-infused water.

This letter of October 1676 likewise was sent to Secretary Oldenburg of the Royal Society of London. Because it was written in Dutch (not Latin or English), there was a delay in reporting this landmark set of biological observations. After its translation, Oldenburg read extracts of the letter to the members of the Royal Society at their meeting on February 2, 1677. Oldenburg called this the "third part" of the report. He went on to mention Leeuwenhoek's "living creatures" found in water, and especially in pepper-infused water.

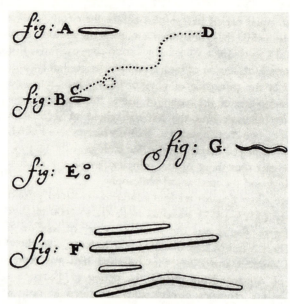

FIGURE 7.7. Leeuwenhoek's bacteria in tartar. They were described in a letter dated September 17, 1683, as follows: *A*, The big sort that "had a very strong and swift motion, and shot through the water or spittle like a pike through the water"; *B*, "The little sort [that] often spun round like a top and every now and then took a course like that shown here between C and D"; *E*, The third sort are organisms that "seem to be long and round while at another time they appeared to be round . . . and therewithal they went forward so rapidly and whirled about among one another so densely that one might imagine to see a big swarm of gnats or flies flying about together"; *G*, "The big sort which were very plentiful, bent their body into curves while going forward" as in this figure.[68] From *Ondervindingen En Beschouwinger* (1694).

Finally, on April 5, Nehemiah Grew, the English microscopist who had been trained at Leiden, was requested to determine whether he could replicate Leeuwenhoek's observations. At first he could not, which prompted further inquiries into Leeuwenhoek's methods of work. Earlier in March, Leeuwenhoek had supplied some additional comments on his little animalcules, noting at the outset, "Nor do I wonder, they [members of the Royal Society] could not well apprehend, how I had been able to observe so vast a number of living creatures in one drop of water."[69] But

[68] Leeuwenhoek's letter translated in P. Smit and J. Heniger, "Antoni van Leeuwenhoek (1632–1723)," in *Antonie van Leeuwenhoek* 41 (1975): 220.
[69] *Philosophical Transactions* 12 (1677–78): 844.

the failure of the experimenters in London was bound to cause alarm for the Dutch pioneer. The situation was similar to Johannes Kepler's alarm when others could not verify Galileo's sighting of the satellites of Jupiter, when skeptics began to call Galileo's reputation into question, and when he then assembled witnesses to record secretly what they saw when looking at Jupiter through the borrowed telescope that finally came into Kepler's hands in August 1610.

In this case, Leeuwenhoek began to think that he should gather witnesses to his discoveries. Having done that in the fall, he sent another letter to the Royal Society containing the testimonials. Although they were mentioned at the October 15, 1677, meeting, they were put off until the next meeting of November 1 because Oldenburg had died in September. Because Grew had not succeeded with his observations of Leeuwenhoek's little creatures, Robert Hooke was asked to make a new microscope similar to Leeuwenhoek's to verify the Hollander's observations. At the November first meeting, as reported in Thomas Birch's *History of the Royal Society*, an extract of Leeuwenhoek's letter was read, noting that it contained a number of "testimonials," four of which came from civic professionals – that is, two ministers, a public notary, and "others of good credit to the number of eight" – testifying as to "the truth of his former assertions concerning the almost incredible number of small animals wriggling in pepper-water, some of whom estimated that they saw ten thousand, others thirty thousand, others forty-five thousand little animals in a single drop of water as big as a millet seed."[70]

By the middle of November, Hooke and others had succeeded in fashioning a series of very tiny glass vials in which to observe water samples suggested by Leeuwenhoek's work. These together with a good microscope would give the magnifying power needed to confirm the discoveries. Indeed, by November 15, it was determined "that there could be no fallacy in the appearance" of Leeuwenhoek's little animals. They had now been observed by Mr. Henshaw, Sir Christopher Wren, John Hoskyns, Sir John Moore, and Dr. Mapletoft, among others, "so that there was no longer any doubt of Mr. Leeuwenhoek's discovery."[71]

With this triumph of experimental discovery, Leeuwenhoek was launched on a new career that would make him the leading microscopist of his generation and, according to many, the founder of the science of microscopy. Subsequently, he was made a member of the Royal Society

---

[70] Thomas Birch, *History of the Royal Society* (New York: Johnson, 1968), 3:347.
[71] Ibid., p. 352.

of London in 1680.[72] The former cloth draper was now showered with honor and remained grateful for the rest of his life for the singularity of this membership. He continued to be an inexhaustible researcher who examined hundreds of biological specimens from his miniature animals, including spermatozoa, to bees and insects, numerous plants, ruminant animals, the eyes of oxen, whales, and humans, but especially the dragonfly.

His discovery of human spermatozoa was the lucky result of a young physician in training at Leiden, a Mr. Ham, who brought him semen collected from a patient suffering from gonorrhea. Under a microscope, the physician saw what he thought were new life-forms spontaneously generated by "putrefaction." Leeuwenhoek, however, recognized the microorganisms as spermatozoa, constituents of sperm that he considered normal. To prove that they are the agents of fertilization of the female-produced egg, he undertook a program of isolating spermatozoa in the semen of many other creatures, including the rooster, the cow, and other animals.[73]

Perhaps the most extraordinary example of Leeuwenhoek's persistence and ability to work with exceedingly small biological specimens is his work on the eye of the dragonfly. Leeuwenhoek had examined a number of different kinds of eyes, the compound eyes of insects and those of invertebrates. The latter included the cow, pig, hare, and rabbit. He also examined the eye of a whale and even fishes and birds.[74]

After undertaking preliminary inquiries, Leeuwenhoek often put his results aside, only to return later and repeat his work. In his work on the structure of the eye in nature, Leeuwenhoek mounted the cornea of the dragonfly – the outer crystal-clear portion on the surface of the eye – on his microscope and then arranged a candle so that the light came through the specimen. Then he saw

thus represented the flame of the candle upside down through the Cornea, so that I saw not one, but several hundreds of candle flames; nay, I even saw them so plainly, however small they were, that [I] could distinguish in each of them the movements of the candle flame.[75]

Unlike the human eye, the cornea of the dragonfly is a composite cluster of networks, rete, or hemispheres, each of which projects a ray of light

---

[72] Dobell, *Antony van Leeuwenhoek*, p. 46.
[73] A. Schierbeek, *Measuring the Invisible World: The Life and Works of Antoni van Leeuwenhoek, F.R.S.* (London: Abelard-Schulman, 1959), pp. 86–87.
[74] Fournier, *Fabric of Life*, pp. 161–62.
[75] Leeuwenhoek letter dated April 30, 1694, as cited in Fournier, *Fabric of Life*, p. 162.

to the retina. Later, in 1698, van Leeuwenhoek studied the substance adjacent to the cornea and saw what he called particles. He then reached more surprising conclusions:

Each of this large number of particles which I observed was an optic nerve and that the thick and round end of the optic nerves had been placed in the small cavity in the Cornea, briefly: there are as many optic nerves there as there are facets of the Cornea.[76]

The human eye is no doubt quite different, but nerve endings are attached to the cornea, and this is a reason why some individuals cannot wear contact lenses. But the nerves from the cornea are attached to the oph-thalmalic nerve, not the optic nerve, in humans.[77]

Conclusion

This sketch of the early history of microscopy should be sufficient to remind us of the extraordinary and revolutionary work that was being done in the life sciences in the seventeenth century in Europe. Our understanding of the human body, of blood circulation, and of the reproductive systems of plants, insects, and animals was radically changed during this period. Likewise, the discovery of invisible microorganisms living in all aspects of the human and natural environment brought another radical shift in scientific thinking. Antoni van Leeuwenhoek did not get everything right, nor should we expect any scientist to do so in the early stages of investigation. Although he was the most prolific of the microscopists, he had many rivals, and they too often claimed to have found that for which he got credit. This tells us how likely it is within modern science for there to be multiple discoveries. This was pointed out long ago by Robert Merton, and before him, by William F. Ogburn.[78] Yet, the remarkable fact in the broader canvas is the absence of scientists equivalent to William Harvey, Robert Hooke, Malpighi, Swammerdam, Reiner de Graaf, and Antoni van Leeuwenhoek, among others, in the three other civilizations of the world.

---

[76] Letter of May 9, 1698, as cited in ibid., p. 163.
[77] See University of Wisconsin Hospitals and Clinics Authority, "Cornea," http://apps
.uwhealth.org/health/adam/hie/2/9909.htm.
[78] Robert Merton, "Singletons and Multiples in Science," chap. 16 in *The Sociology of Science* (Chicago: University of Chicago Press, 1973); and W. F. Ogburn and Dorothy Thomas, "Are Inventions Inevitable?," *Political Science Quarterly* 37 (March 1922): 83–98.

In previous chapters, it was pointed out that by the late 1400s, European glass and spectacle technology had spread to the Ottoman Empire, to the Middle East, and to India. In the late 1400s, hundreds of pairs of spectacles were sent to the Levant, and by the early sixteenth century, the scale of exporting jumped into the thousands of pairs shipped to the Ottoman Empire. So it is fair to say that European lens technology, probably the highest that existed, centered in Florence, was available around the world.

Yet, no evidence has turned up indicating that microscopes were developed or used in India until the nineteenth century.[79] A poignant reflection of the state of knowledge of microscopy in India in the nineteenth century is found in a recent book by William Dalrymple. There one finds the report of the "arrival of Messrs Trood and Co.," who brought a traveling exhibition to Delhi that "included several microscopes, which were reported by the *Delhi Gazette* [dated February 10, 1847] to have caused 'great consternation' among native gentlemen [at] the curiosities revealed to their wondering eyes."[80] Yet, this was 200 years or more after Galileo's microscopic observations.

In July 1679, the Italian Luigi, Count de Marsigli, while accompanying the new Venetian envoy to the Ottomans, mentions using his microscope on his journey by sea to Istanbul. Soon thereafter, a ban on Venetian ships was pronounced because the envoy's ship was said to have "unaccustomed goods" aboard that were being smuggled ashore – but probably not microscopes.[81]

Likewise, by the 1620s, the Jesuits had brought European lenses to China. More than that, they published books in Chinese explaining the construction and use of telescopes and many other Western devices. Joseph Needham and Lu Gwei-Djen published a study of two seventeenth-century Chinese scholars famous for making spectacles and then a

---

[79] One looks in vain for such evidence in Deepak Kumar, ed., *Science and Empire: Essays in Indian Context, 1700–1947* (Delhi: Anamika Prakashan, 1991); Kumar, ed., *Disease and Medicine in India: A Historical Overview* (New Delhi: Tulika Books, 2001); Kumar, *Science and the Raj: A Study of British India*, 2nd ed. (Oxford: Oxford University Press, 2006); or Ahsan Jan Qaisar, *The Indian Response to European Technology and Culture (A.D. 1498–1707)* (Delhi: Oxford University Press, 1982). Likewise, David Arnold is unaware of microscopes in India before the nineteenth century; personal communication, January 15, 2009.

[80] William Dalrymple, *The Last Mughal: The Fall of a Dynasty, Delhi, 1857* (New York: Penguin Books, 2007), p. 83. I thank Professor S. R. Sarma for this reference.

[81] John Stoye, *Marsigli's Europe, 1680–1730* (New Haven, CT: Yale University Press, 1993).

telescope. However, both of these scholars, Po Yü and Sun Yün-qiu were born after 1628, when the Jesuit volumes explaining the telescope, lenses, and human vision had already been made available in Chinese. This casts doubt on Needham's claim that Po Yü was a "Chinese Lipperhey."[82] Chinese astronomers and the Jesuits together had been using European- and Chinese-made telescopes in connection with Xu Guangqi's impressive plan to reform Chinese astronomy in the 1630s, when these individuals were teenagers.

Even so, there have been no reports indicating that the Chinese then went on to invent or use microscopes in this period, as the Dutch spectacle makers did and as Galileo did back in 1610.[83] Beyond that, we have no reports of Chinese microscopic studies of plant, animal, or human physiology such as took place in seventeenth-century Europe.

A recent synthetic study of Chinese science during the seventeenth cen- tury notes that the Chinese knew little of the new medical advances in Europe and consequently focused their efforts on recovering and renew- ing the ancient Chinese medical classics. According to this account, the Chinese used "general assumptions about the application of yin-yang, the five phases, and the system of circulation tracts (*jingluo*) to under- stand the human body and its susceptibility to illness."[84] But this was not accompanied by the kind of anatomical or microscopical study widely practiced in Europe. The treatises that the Chinese possessed enabled doctors to map acupuncture and moxibustion points on the skin fol- lowing the belief that *qi* (energy) flows through various channels of the body. Illness was defined as an imbalance of such forces throughout the body. Yet, "they did not distinguish between the pulse and palpita- tions associated with the nervous system as Greek physicians had done since Herophilus."[85] This suggests that the Chinese lacked the anatomical sophistication that Europeans had acquired during centuries of empiri- cal study while performing human postmortems, extending from the late medieval to the early modern period. As we have seen, this reached a

---

[82] Joseph Needham and Lu Gwei-Djen, "The Optick Artists of Chinagu," in *Studies in the Social History of China and South-East Asia: Essays in Memory of Victory Purcell*, eds. Jerome Ch'en and Nicolas Tarling (Cambridge: Cambridge University Press, 1970), pp. 197–224.

[83] Minghui Hu reports that he has never come across a reference to "any term referring to something similar to a microscope before 1800"; personal communication, January 11, 2009.

[84] Benjamin Elman, *On Their Own Terms: Science in China, 1550–1900* (Cambridge, MA: Harvard University Press, 2005), p. 227.

[85] Ibid., pp. 227–28.

hiatus in the work of Vesalius in the sixteenth century and William Harvey's demonstration of the circulation of the blood throughout the body in 1628. These advances were just the prelude to what European anatomists and microscopists achieved in the seventeenth century, including the creation of microbiology.

The naturalistic curiosity of the seventeenth century was widely represented across the borders of Western Europe throughout that century. Accounts highlighting the role of the Royal Society of London in the seventeenth century are indispensable, but they fail to stress the fact that the Royal Society was not only the product of events in England itself, or its budding Puritanism, but also of the broader landscape within which the European ethos of science was working its wonders. One has only to consider the many foreign members of that society to see how much it was an international organization. Furthermore, other scientific societies existed across Europe, and many of their members were also members of the Royal Society. Furthermore, in Italy, the Accademia del Cimento, dedicated to experimental science, had been founded in 1657.[86]

The outstanding attention that historians of science have given to astronomy and the science of motion, though well deserved, has prevented us from seeing how truly widespread the modern spirit of scientific inquiry was in the seventeenth century. Consequently, the putative obstructive role of the Catholic Church with respect to the scientific revolution of that era has been exaggerated. Whatever unofficial qualms Church officials might have had regarding human or animal dissections, they never reached a significant level to impede any scientific work in medical-biological studies after the thirteenth century. This stands in contrast to the religious, legal, and cultural contexts of the Muslim world of the Middle East and India as well as China.

But there are still more fields of scientific innovation in the early seventeenth century that must be considered if we are to grasp the full impact of the scientific spirit on the social, economic, and industrial development of Europe after the seventeenth century.

---

[86] See Luciano Boschiero, *Experiment and Natural Philosophy in Seventeenth-Century Tuscany: The History of the Accademia del Cimento* (Dordrecht, Netherlands: Springer, 2007).

8

Infectious Curiosity II

Weighing the Air and Atmospheric Pressure

In the 1630s, when the official debate over Galileo's provocative defense of the Copernican system was starting to heat up again, physical inquiry began shifting its focus to another part of the natural world. It concerned hydraulics, the limitations of siphons and suction pumps to lift water, and the idea that the air of our atmosphere has weight. If true, that idea would have momentous implications for human life. Within seventy years, Europeans would be pioneering the effort to harness that principle of nature as a new source of energy. First steam power and soon thereafter electric power would follow.

Such technological advances could only be harvested by advances in basic science itself. Furthermore, each of these inquiries was rooted in ancient conceptions that had been studied continuously from the time of Aristotle. In the early 1600s, Italy was a leader in hydraulics and in the construction of mechanical devices for lifting water. Some of these mechanical devices were also used to power machines for the grinding and processing of other materials. Vittorio Zonca (b. ca. 1580) had published a book in 1607 with dozens of illustrations of such devices, some powered by water, some by beasts, and some by human agents. It went through many editions.[1] Consequently, Rome had a band of hydraulics experts in the 1630s experimenting with various hydraulic devices. They found the question of why water can be raised hydraulically only ten meters needing an explanation. This problem was mysteriously linked to the question of a vacuum.

---

[1] Vittorio Zonca, *Novo teatro di machine et edificii per varie et sicure operationi* (Padova, Italy: Appresso Pietro Bertelli, 1607).

Is a Vacuum Possible?

The Aristotelian view was that a vacuum, a space devoid of all air or substance, was impossible. This place where "nothing" existed was said to be prohibited by "nature's abhorrence" of a vacuum. Nevertheless, natural philosophers in the Middle Ages engaged in all sorts of thought experiments, trying to imagine what would happen if, contrary to Aristotle, a vacuum should exist, perhaps in outer space.[2] For medievals, the location of this metaphysical possibility was between the curved surface of the moon's orbit and the earth itself. It was concluded that such a vacuum could not exist naturally but that God could create such a "void," a place where "no body" existed, no air and no water, if he wished. Scholars tried to imagine a variety of counterfactual situations that might reveal further reasons why such a physical thing was not possible. For example, if such a remarkable thing as a vacuum were to exist, what would happen to sound or light? Would people in a vacuum be able to hear each other talking? And if, against all reasonable thought, a vacuum formed, would not objects fly through the air instantaneously? This was a claim that Galileo would demolish about 300 years later, just as others would want to test the idea of whether sound travels through a vacuum.

In the fourteenth century, European philosophers in Paris and elsewhere developed additional arguments and sought to bring experimental demonstrations to bear on the problem. Both Albert of Saxony (ca. 1316–90) and Jean Buridan (1300 to ca. 1358) used an argument based on the humble bellows. According to Albert, a void cannot exist either as a matter of metaphysical reasoning or as a matter of physical reality. In the first case, a vacuum, a place where there is nothing, no air or water, cannot exist because even if such a *place* should exist, there would still be something there because a place is still *something*. A more spiritually potent argument was that the void of extraterrestrial space was identified with God. This was something beyond the physical idea that space must have three dimensions.

But then Albert turns to a sort of experimental approach: if all the openings in a bellows were stopped up so that no air whatsoever could enter, then "no power could raise one handle from the other unless a break occurred somewhere through which air could enter. . . . This [experience]

---

[2] For a now classic study of this, see Thomas Kuhn, "A Function of Thought Experiments," in *The Essential Tension* (Chicago: University of Chicago Press, 1977), pp. 240–65.

seems to be a sign that nature abhors a vacuum."[3] Even here we find a premonition that something in the atmosphere must have great force.

Virtually simultaneously, Jean Buridan in Paris, who also studied the question, arrived at the same conclusion but foresaw an historic experimental test. He, too, thought it impossible for a vacuum to exist in which there would be nothing. Then he turned to the bellows argument for an "experimental induction." Using the stopped-bellows argument, he averred that "if all the holes of a bellows were perfectly stopped up so that no air could enter, we could not separate their surfaces." He meant that the two handles of the bellows could not be pulled apart. This was similar to the ancient case of two plane surfaces placed together that were said to be bonded because of the impossibility of a vacuum.[4] But here Buridan added this extraordinary claim: "Not even horses could do it if ten were to pull on one side and ten on the other; they would never separate the surfaces of the bellows unless something were forced or pierced through and another body could come between the surfaces."[5]

For us moderns, this prescient thought experiment cries out for an experimental try. What if someone fashioned a large sphere, separated into two parts with good seals, fitted with a petcock for drawing air out of the sphere, attached horses pulling in opposite directions: What would happen? Would a vacuum be created by evacuating the air, and would the mighty horses be unable to pull the hemispheres apart? This was a perfect experiment waiting for a trial. But technically, creating an air pump, a device that could remove all the air within a tightly sealed container, was beyond late medieval capacities. It would require the construction of a metal sphere, divided into two halves, sealed all around but fitted with an exhaust port; a cylinder pump with an intake port; a plunger or piston that fitted airtight within the cylinder for withdrawing air; another port that could be opened and closed by hand through which air could be expelled after being drawn in from the top. This was beyond medieval capabilities. Consequently, three centuries would pass before the mayor of Magdeburg, Germany, would craft such a sphere, invent just such an air pump, and perform this very experiment.

---

3 Albert of Saxony, "Questions on the Physics of Aristotle," in Edward Grant, ed., *A Source Book in Medieval Science* (Cambridge, MA: Harvard University Press, 1974), p. 325.
4 For these convoluted arguments, see Edward Grant, *Much Ado about Nothing* (New York: Cambridge University Press, 1981), pp. 86ff.
5 Buridan, "Questions on the Eight Books of the Physics of Aristotle," in Grant, *A Source Book*, p. 326.

On the other hand, the bellows experiment could have been performed had a genius like the young Blaise Pascal turned his attention to it. For Pascal did indeed devise a means for testing the weight needed to counterbalance that of air in a vacuum using the bellows in 1647. But that is to get ahead of the story.

In the meantime, there were other approaches to the problem of the vacuum or the weight of air, but not until the early 1600s.

Weighing the Air

For centuries, it had been known that water could not be raised above a certain limit by a siphon or a pump. By Galileo's time, that distance was determined to be ten meters (about thirty-four feet). This value of thirty-four feet, it will turn out, varies because it is based on the weight of air, which itself varies, and is what European experimenters were trying to determine. At the same time, some locations are located several hundred or perhaps a thousand feet above sea level, thereby resetting the base line. As we shall see later, Blaise Pascal's very careful experimenting arrived at a value of thirty-two feet for the location where he lived and experimented in northwest France. This suggests that the air in his part of France tends to be lighter than, for example, parts of the Mediterranean coast.

The Greek mathematician and polymath Archimedes, who lived in Syracuse on the island of Sicily (287–212 B.C.), is credited with inventing the basic machinery of water-lifting technology. His most famous device in this connection is the water screw: a screw located within a pipe that raises water as the screw turns on its axis. But it was realized that such a device could not raise water as high as, say, a thirty-five-foot waterwheel that had buckets mounted on it to bring water from below the ground to a higher level. But building such a huge wooden wheel imposed its own limits. The water-raising technology invented by Archimedes was later passed on to the Arab Middle East by translators.[6] The most famous of the Arab technicians to devise a variety of water-raising devices, water clocks, and such was the late-twelfth-century writer al-Jazari (d. 1206). Little is known about him, except that he was born in Mesopotamia, between the Tigris and Euphrates rivers, toward the middle of the twelfth century. His masterwork, *Book of Knowledge of Ingenious Mechanical*

---

[6] This according to the medieval Islamic thinker Ibn al-Ridwan, in Ahmad Y. al-Hassan and Donald R. Hill, *Islamic Technology: An Illustrated History* (Cambridge: Cambridge University Press, 1986), p. 56.

Devices, contains many fascinating mechanical items and what little is known about him. Though the book contains many ingenious devices, there are no scientific breakthroughs.

Al-Jazari did invent in the thirteenth century a machine for raising water, but it was as much an engineer's toy as a water-lifting device.[7] In his diagram of the machine, water flowed into the mechanism from a lake on the right but then flowed down through the floor to a hidden chamber. The falling water struck scoops mounted on a wheel; its axle then turned a pair of gears, one of the gears mounted on a vertical shaft. Far up this vertical shaft, another pair of gears turned a horizontal shaft, activating a chain of pots that lifted water from the main water chamber, producing a stream of water flowing out to the left of the assembly. However, the machine included a model cow that had no functional importance, thereby adding a level of mystification to the whole assembly.

One of his successors, al-Qazwini (d. 1283), was also a person about whom we know very little. He described a process of raising water from below ground that seems to reflect the fact that mechanical pumps of the Archimedean type could not raise water above thirty-four feet and that building large wooden waterwheels with buckets attached was also very difficult.

Without mentioning the height to which the water was being raised, he describes raising the water to a cistern, whereupon the water is pumped to the next higher level and then again pumped higher. No doubt this was due to the physical limits imposed by the weight of air and the inefficiency of pumps.[8]

But no advances were made in the basic science of hydraulics that would explain why these limits existed. Although the Chinese had long studied various kinds of hydraulics, those studies did not go forward in the seventeenth century to unlock the mystery of why a column of water could not be raised above ten meters and hence to the conclusion that air has weight.[9] From this, we relearn the fact that designing new mechanical devices is one thing, whereas understanding the fundamental principles and structure of nature is another.

---

[7] For a color image of al-Jazari's machine, see http://en.wikipedia.org/wiki/File:Al-Jazari_Automata_1205.jpg.

[8] See al-Hassan and Hill, *Islamic Technology*, pp. 239–40.

[9] Needham alludes to ancient studies of the vacuum created when a hot sphere is plunged into water, but this is a far cry from the sophistication of Europeans studying these phenomena in the seventeenth century. See note 49.

Suction pumps had been around for some time, perhaps back to antiquity. Georg Bauer (1494–1555), known as Agricola, described a series
of suction pumps used in Germany to drain a mine in the sixteenth century, powered by a waterwheel.[10] The height between each of the cisterns
is about twenty-four feet, suggesting the limits to suction pumping (Figure 8.1). Even so, this kind of technology must have been developed some
time before Agricola reported on it. However, Joseph Needham suggests
that the Chinese did not develop or use piston pumps until after the
sixteenth century and the arrival of Europeans and provides no examples
of their use.[11]

At the end of the sixteenth century, patents were still being issued to
European inventors who, in some way, modified the design of Archimedean water pumps. Galileo was issued one such patent for a device
of this type that he invented around 1594.[12] Yet, it was not until the
1630s that Italian and other experimenters began inventing devices that
could bring new light to bear on the problem. In 1630, a very learned
native of Genoa, Italy, a man named Giovanni Battista Baliani (1582–
1666), wrote a letter to Galileo about a siphon problem. It might seem
surprising that Baliani was a mathematician, a lawyer, and a senator in
Genoa as well as an ingenious experimenter. That reflects that humanistic
and scientific education of that period were far more integrated than they
tend to be today, especially in the United States. Baliani had traveled to
Florence to meet Galileo back in 1615. Now he wrote to Galileo about
the failure of a water pump to bring water up a hill about sixty feet high.

---

[10] Agricola's rich description of this "seventh kind of pump" is well worth citing:

> [It was] . . . invented about ten years ago [i.e., ca. early 1540s], which is the most ingenious
> durable, and useful of all, can be made without much expense. It is composed of several
> pumps, which do not, like those last described, go down the shaft together, but of which
> one is below the other, for if there are three, as is generally the case, the lower one lifts
> the water of the sump and pours it into the first tank; the second pump lifts it again from
> that tank into a second tank, and the third pump lifts it into the drain of the tunnel.
> A wheel fifteen feet high raises the piston-rods of all three pumps at the same time and
> causes them to drop together. The wheel is made to revolve by paddles, turned by the
> force of a stream which has been diverted to the mountain.

As cited in Charles J. Singer, ed., *A History of Technology* (Oxford: Clarendon Press,
1954), p. 2:17.

[11] *SCC*, 4/3:666.

[12] Stillman Drake, *Galileo at Work* (New York: Doubleday, 1957), p. 35.

FIGURE 8.1. A triple-lift suction pump powered by a water wheel from the first half of the sixteenth century. From Agricola, *De re Metellica* (1556).

Baliani attributed this failure to the weight of the atmosphere, but Galileo doubted this.[13]

Eight years later, in 1638, Galileo published his last major work, *Discourses Concerning Two New Sciences*, but he continued to assert that the limits imposed on the height of lifting were the result of the limits of the "resistance of the vacuum."[14] Galileo used the analogy of a breaking rope to refer to the column of water breaking whenever the height was greater than "eighteen cubits" (thirty-four feet). The fact of the maximum height was agreed on, but not the explanation. Indeed, one of Galileo's students, an accomplished scientist in his own right who lived with him during his last years, later pointed out that Galileo "does not say that he admits the operations of heavy air, but persists in asserting that nature also concurs in resisting a vacuum."[15] This was Evangelista Torricelli.

The question now became one of devising an experimental apparatus that would measure the weight of a column of water, or perhaps a column of air, if that could be imagined. At this point, another long-term resident of Rome with many scientific talents took the limelight. Gasparo Berti (ca. 1600–43), who had been born in Mantua, was a mathematician and natural philosopher who had come to the attention of the international network of scientists, including the globally famous Jesuit scientist from Germany, Athanasius Kircher (1601/2–80). He was also known to the most prolific correspondent of the seventeenth century, Nicholas-Claude Fabri de Peirsc. These two scholars were a sort of clearinghouse of European scientific life during the first half of the seventeenth century who corresponded with hundreds of scientists and thousands of others. Berti's mathematical skills were such that he was appointed to succeed Bendetto Castelli, Galileo's friend and former student, as professor of mathematics at the College of Sapienza in Rome in 1643. Castelli was also a hydraulics expert.

---

[13] "Baliani, Giovanni Batitsta," in "The Archimedes Project," http://archimedes2.mpiwg-berlin.mpg.de/archimedes_templates/biography.html?-table=archimedes_authors&author=Baliani,%20Giovanni%20Battista&-find; and Mario Gliozzi, "Torricelli, Evangelista," *DSB* 13 (1970–80): 433–40.

[14] Galileo, *Discourses Concerning Two New Sciences*, trans. Henry Crew and Alfonso de Salvio (Buffalo, NY: Prometheus Books, 1991), p. 17.

[15] Evangelista Torricelli, "Letter to Michangelo Ricci Concerning the Barometer" (June 11, 1644, in *A Source Book in Physics*, ed. William Francis Magie (New York: McGraw-Hill, 1935), http://web.lemoynne/~giunta/torr.html.

In the meantime, Berti, perhaps in cooperation with a scholar named Rafael Magiotti, built a water barometer about thirty-six feet high attached to his house. The top was capped by a glass dome or large flask, connected to a lead pipe that emptied into a water basin below, controlled by a valve. He filled this tall pipe and its dome with water, then opened the valve at the bottom to see how tall the column of water would be. After an initial misstep, the height was found to be just ten meters. Because the water level in the pipe dropped to about thirty-four feet (ten meters), a vacuum was created at the top. Berti later opened the top of the flask and witnessed the rush of air moving in to fill the void. Nevertheless, various investigators, including Father Athanasius Kircher and a French scholar from Toulouse who later reported on the event, Emmanuel Maignan (1601–76), doubted that a vacuum had formed. Kircher seems to have been the most opposed to the idea of a vacuum, whereas Maignan, at least later, accepted the idea that a vacuum could form in such a device.[16]

This led to more experiments. True to the medieval roots of curiosity about the nature of a vacuum, and apparently at the suggestion of Kircher, Berti equipped his device with a hammer controlled by a magnet outside the vacuum that would attract the hammer inside striking the bell, providing a test of whether or not *sound* can travel through a vacuum.[17] However, it was concluded that the bell's ringing caused the glass dome to vibrate, nullifying the outcome.[18]

Although measurements connected to this experiment confirmed that water in a pipe will not fall below thirty-four feet (and hence would only rise to such a level), the question of a vacuum's creation in the sealed

---

[16] Maignan's account is translated and printed in W. E. Knowles Middleton, *The History of the Barometer* (Baltimore: Johns Hopkins University Press, 1964), pp. 10–15. Also see Stillman Drake, "Gasparo Berti," *DSB* 2 (1970): 83; Theodore S. Feldman, "Barometer," in *The Oxford Companion to the History of Science*, ed. John Heilbron (New York: Oxford University Press, 2003), pp. 80–81; and Richard Westfall, *The Construction of Modern Science* (New York: Cambridge University Press, 1977), pp. 43–45.

[17] For an illustration of Berti's experimental apparatus, see http://en.wikipedia.org/wiki/Gasparo_Berti. It seems that the experiment was repeated with the bell both inside and outside the potential vacuum chamber. Hearing the sound, some observers concluded that the space could not be a vacuum; Middleton, *History of the Barometer*, p. 14.

[18] Middleton, *History of Barometer*, p. 14; and Luciano Boschiero, *Experiment and Natural Philosophy in Seventeenth-Century Tuscany: The History of the Accademia del Cimento* (Dordrecht, Netherlands: Springer, 2007), p. 137n77.

space at the top of the flask remained undecided. Emmanuel Maignan years later accepted the idea, but Kircher and other Jesuits denied it.[19]

Word of Berti's experiment, which had been witnessed by several scholars, prompted others to contrive additional experiments using different liquids, including wine, honey, and seawater. Because each of these had different specific weights, the height of the column would vary proportionately to the weight of water. But a device based on a column of water more than thirty feet tall was not convenient for scientific experimentation.

In 1644, Evangelista Torricelli (1608–47) hit on the device that would unlock the mystery of the weight of the air. Torricelli had been given a Jesuit education starting in 1624, probably outside Rome, but soon thereafter at the Roman College, the Collegio Romano, where he came under the influence of Galileo's good friend Benedetto Castelli. Torricelli became a gifted mathematician and a believer in the Copernican hypothesis. But after the official condemnation of Galileo in 1633, he shifted his interests to mathematics and other subjects, especially hydraulics.[20] As Galileo's health continued to fail during his last years in Arceti (outside Florence) under house arrest, Castelli recommended Torricelli as a companion for Galileo. After a delay, Torricelli arrived in Arceti in October 1641, and there he remained until after Galileo's death in 1642. He was then appointed court mathematician to the Grand Duke Ferdinando II of Tuscany.[21] He remained in that post until his untimely death in 1647. At the same time, Galileo's last student and first biographer, Vincenzio Viviani, was living with Galileo in Arceti. As a result, Torricelli and Viviani became close friends and worked on the barometer project together.[22]

Soon after Galileo's death, Torricelli began a series of experiments using different liquids, all focused on the problem of the vacuum and the weight of the air that many experimenters thought was responsible for the failing pump problem. Very little of Torricelli's writings were published before he died, but a letter he wrote from Florence to a religious official in Rome, Michelangelo Ricci, in 1644 had become famous for announcing that Torricelli had invented a new scientific instrument: what we call the weather barometer.

[19] Middleton, *History of the Barometer*. The Jesuit position is described in detail by Grant, *Much Ado about Nothing*, pp. 157–63.

[20] J. J. O'Connor and E. F. Robertson, "Evangelista Torricelli," http://www-groups.dcs .st-and.ac.uk/~history/Biographies/Torricelli.html.

[21] Ibid., p. 2.

[22] Boschiero, *Experiment and Natural Philosophy*, pp. 44ff, 120–23.

In the letter, he confirms that a number of "philosophical experiments" had been undertaken in connection with the vacuum and the question of whether such a thing could be produced. He thought it could. But he wanted to design a device "not simply to make a vacuum but to make an instrument which will show the changes in the atmosphere, as it is now heavier and more gross and now lighter and more subtle."[23] This statement is the fundamental assumption that we take for granted whenever we hear the daily weather forecast and reports on barometric pressure. A fall of barometric pressure generally precedes a change in the weather. Torricelli then explains more of what he did and his thinking:

We have made many glass vessels . . . with tubes two cubits [approximately thirty-five inches] long. These were filled with quicksilver [mercury], the open end was closed with the finger, and the tubes were then inverted in a vessel where there was mercury. . . . We saw that an empty space was formed and that nothing happened in the vessel where this space was formed. . . . I claim that the force which keeps the mercury from falling is external and that the force comes from outside the tube.[24]

Earlier in the letter, Torricelli used the metaphor of a "sea of air": we live, he wrote, "immersed at the bottom of a sea of elemental air." Other researchers of that era had estimated that the air surrounding the earth was about fifty miles high. So this means that the sea of air bears down its weight on all the surfaces below:

On the surface of the mercury which is in the bowl rests the weight of a column of fifty miles of air. Is it a surprise that into the vessel, in which the mercury has no inclination and no repugnance, not even the slightest, to being there, it should enter and should rise in a column high enough to make equilibrium with the weight of the external air which forces it up?[25]

With this, Torricelli invented the barometer and explained how it works: the pressure of the outside air, which fluctuates over time, holds the mercury in the tube at twenty-nine inches and falls when weather conditions change, when the air becomes lighter and less saturated with moisture, and it rises to about twenty-nine inches when the air pressure increases. No doubt many others in and around Rome and across Europe repeated this experiment on their own. As we shall see, experimenters in France were intrigued by Torricelli's results.

[23] Torricelli, "Letter to Michangelo Ricci."
[24] Ibid.
[25] Ibid.

The really significant element for the march of science was that Torricelli had formulated a clear idea explaining the nature of air pressure and had invented a new scientific instrument that could precisely measure fluctuations of it. It was relatively compact and portable: theoretically, the device could be taken anywhere. It was another portable laboratory easily used by anyone once it was calibrated. From a modern entrepreneurial point of view, it had all the makings of a household item for affluent citizens who would want one in their houses, just as millions of people across the world have wanted thermometers in their homes.

New Experiments in France

There was one last test that had to be carried out to incontrovertibly establish that air has weight and that the barometer measures that weight: someone had to take a barometer up a mountaintop and then observe the height of the column of mercury. For doubting Thomases, the question was, if the barometer were taken up a mountain top – say, 4,000 feet above sea level – would the height of the column of mercury in the barometer fall? Within four years, that task would be accomplished by an extraordinary young French mathematician, Blaise Pascal.

In 1644, Blaise Pascal was thirty-nine and a maturing young scholar, gifted in mathematics but also in many other areas as well. Although his mother died when he was only three, his father, Etienne, a lawyer, tax collector, and man of wealth, educated Blaise according his own unique system. He personally taught his son grammar, Latin, Spanish, and mathematics, though in the early stages of this tuition, the father had attempted to shield his son from mathematics until age fifteen. But the son managed to undertake his own studies in geometry at the age of twelve, apparently inventing his own mathematical notation, leading his father to relent by giving him a copy of Euclid's *Elements*.[26]

During his early years, Pascal lived with his father in Paris, where the latter was an intimate of the famous Mersenne circle of natural philosophers, discussed previously. By the age of fourteen, Pascal began accompanying his father to these meetings and even presented a one-page set of thoughts containing theorems in projective geometry. Soon thereafter,

[26] J. J. O'Connor and E. F. Robertson, "Blaise Pascal," http://www-groups.dcs.st-and.ac.uk/~history/Mathematicians/Pascal.html; and F. Perier, Preface to *The Physical Treaties of Pascal: The Equilibrium of Liquids and the Weight of the Mass of Air*, ed. and trans. I. H. B. Spiers and A. G. H. Spiers (New York: Columbia University Press, 1937), pp. xv–xx.

he invented a digital calculating machine to aid his father making calculations.[27]

Later in life, Pascal was to become famous for his religious writings, especially his championing an Augustinian branch of Catholicism called Jansenism. Inspired by the Dutch priest Cornelius Jansen, the group turned against the casuistry and convoluted moral arguments of the Jesuits. In his *Provencial Letters* of 1656–57, Pascal made fun of the extreme moral positions (sometimes called moral laxity) that Jesuits got into when justifying the moral position of the individual before the sight of God. Ultimately, this would lead Pascal to make his wager that it was safer to bet that God existed and therefore meet God with a good conscience than possibly to be found wrong and condemned by the eyes of God.

In the meantime, Etienne Pascal moved his large family to Rouen, a city northwest of Paris situated on the river Seine. The city of Rouen was a center of glass-blowing that made possible the construction of very long lengths of glass tubes.

In the 1640s and early 1650s, France became a hotbed of interest in hydraulics, the patterns and principles of fluid motion. According to some accounts, the theologian and philosopher Father Mersenne was the first in France to hear of Torricelli's experiments in 1644. He began spreading the news by word of mouth and his large list of correspondents. Scholars associated with his circle began replicating Torricelli's experiment and, by 1646, Pascal and the chief of fortifications in Rouen, M. Petit, repeated the experiment "just as it had been performed in Italy, duplicating in every particular what had been reported from that country."[28]

During the next two years, Pascal devised a series of ingenious experiments that conclusively demonstrated Torricelli's claims that air has weight and is responsible for holding the mercury suspended at twenty-nine inches, that the "vapor chamber" above the mercury has no effect on the outcome, and that air pressure does indeed vary by altitude.

He began by showing that liquids in long tubes always settle at a level proportional to their weight of air. For water, this was between thirty-two and thirty-four feet; for wine, a bit higher; and so on, with olive oil and other liquids, all depending on location and altitude. Pascal even filled a forty-six-foot glass tube attached to the mast of a ship in the harbor with wine to show that although it was thought to be more vaporous and

[27] O'Connor and Robertson, "Blaise Pascal."
[28] F. Perier, Preface to *The Physical Treaties of Pascal*, p. xvi.

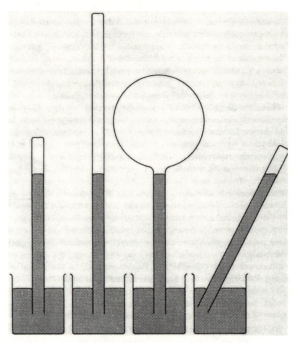

FIGURE 8.2. Proportion of mercury to a vacuum. Pascal filled tubes standing at different angles to show that the height of the column of mercury was always the same, no matter what the incline of the tube or what the shape of the space above the mercury. From Richard Westfall, *The Construction of Modern Science* (New York: Cambridge University Press, 1977), p. 46.

hence more prone to push the column lower, it did no such thing. The level always settled at the same height from the surface of the liquid in the reservoir below. Even tilting the tube of wine so that the wine filled the whole tube, the distance from the top of the reservoir at the base of the tube to the top of the liquid, Perier reports, always remained at thirty-two feet[29] (Figure 8.2).

For Pascal, the empty space at the top of the tube was a vacuum, and it had nothing to do with sustaining the level of the liquid. The shape of the space above the liquid – large, small, domed, or rectangular – had no effect on the height of the liquid's point of equilibrium with the air. That was determined by the weight of the air pushing up on the liquid. The results of these early experiments using many different liquids and

[29] Ibid., pp. xv–xx.

different shaped tubes were printed in a pamphlet and sent off to scholars far and wide.

Puy de Dôme: The Mountain Test

Late in 1647, Pascal came up with the crucial experiment that would show that the height of a barometer varied directly with the weight of air. The task was actually given to his brother-in-law Florin Perier, who lived in Clermont near the mountain called Puy de Dôme in south-central France. In November 1647, Pascal sent him a letter, instructing him to carry a barometer up the 4,800-foot-high mountain and to record the level of the mercury there.

Perier was unable to carry out the experiment until September 1648, but he then executed the test with meticulous precision. Perier recorded the barometric reading at three levels on the way up, noticing also the estimated height at those intervals. At the same time, he placed another barometer at the base and instructed a reliable cleric from the monastery nearby to watch and record the level of that instrument throughout the day. His observations confirmed that the height of the column in the barometer at the base of the mountain remained at twenty-six inches all day.

In Chapter 7, we saw that early- and mid-seventeenth-century natural philosophers were keen to have their scientific results witnessed publicly by reliable witnesses. So, too, in this case, Perier notified "several people of standing" in the town of Clermont to accompany him on this experimental mission. He thought it "proper to carry out [the undertaking] in the presence of a few men who are as learned as they are irreproachably honest."[30] These included several clerics and laymen such as a lawyer attached to a municipal court and a doctor practicing nearby. They were not only professionals but also individuals able in "every field of intellectual interest."[31]

Having stationed one barometer at the base of the mountain with an observer, the team carried their barometric device up the mountain, carefully observing the height of the mercury in the tube as they went, repeating the observations many times over the course of the day. The results tabulated in Table 8.1 show that the height of the column of mercury did indeed vary with the altitude, thereby confirming that our

[30] Ibid., p. 103.
[31] Ibid., p. 97.

TABLE 8.1. *Results of Perier's Puy de Dôme Experiment, September 1648*

	Elevation (feet)	Mercury Level	Difference
Lowest point at Mimins*	–	26″, 3.5 lines	–
7 fathoms higher	42	26″, 3 lines	Half a line
27 fathoms higher	162	26″, 1 line	2.5 lines
150 fathoms higher	900	25″	15.5 lines or 1″ and 3.5 lines
500 fathoms higher	3,000	23″	37.5 lines or 3″ and 1.5 lines

* The monastery in Clermont made famous by the Jansenists; constructed from Perier's letter in Blaise Pascal, *The Physical Treaties of Pascal: The Equilibrium of Liquids and the Weight of the Mass of Air*, trans. I. H. B. Spiers and A. G. H. Spiers (New York: Columbia University Press, 1937).

"sea of air" has weight: the column of air decreases with altitude, and the level of the barometer falls. With this demonstration, Pascal and Perier had shown that the air has weight, the barometer measures it, and that as one rises above the surface of the earth, the weight of the air decreases.

Without any doubt, Pascal's comprehensive experiments established that the weight of air varies under all sorts of conditions and that worries about a vacuum are irrelevant. Although Pascal clearly brought pneumatics fully into the experimental age, it is impossible to pass over the continuity of Pascal's actual experiments with the thought experimenters of the late fourteenth century. A fragment of Pascal's equilibrium experiments from his collected works, but not included in the *Physical Treaties*, reveals Pascal's knowledge of the bellows experiment of Buridan and Albert the Great.

In the fragment, Pascal tells the reader to fashion a bellows only three inches in diameter, hermetically seal it, and then fasten it to the ceiling. From the bottom handle of the bellows, suspend a chain that extends to the floor with extra links, altogether equaling about 120 pounds, not more. He then tells us that we will find that a weight of only 113 pounds is sufficient to open the bellows. Moreover, if you leave the apparatus in place and observe it over time, you will find that as the air becomes heavier, the bellows handle will rise, and it will fall when the air becomes lighter, with more chain links falling onto the floor. For the climatic conditions of Rouen where Pascal lived, the weight of the chain suspended varied between 107 and 113 pounds.[32] This was an elegant little

[32] Ibid., pp. 79–82.

demonstration that any instructor or lab technician would love to show contemporary students.

Once again, Pascal had these results (exclusive of this fragment) printed up and sent to scholars in every direction across France and to others in Sweden, Holland, Poland, Germany, and Italy. For those who look for "the emergence of modern scientific culture,"[33] it had clearly been established by the early decades of the seventeenth century, if not long before. Just as Galileo, Kepler, and other astronomers corresponded with each other across Europe in the first decade of the seventeenth century, so too these students of hydraulics had corresponded regarding these experimental questions since the 1630s and before. Likewise, in Chapter 7, we saw that physicians and anatomists across Europe had been crossing borders to study leading specialists for centuries, and postmortem autopsies had been carried out on human subjects since the thirteenth century.

The history of the barometer and air pressure had begun in Italy, migrated to France in the 1640s, and then shifted to Germany in the 1650s. Later, in the 1660s, it would migrate again, this time to England, in the work of Robert Boyle and the Royal Society, which was not chartered until 1661. Clearly, the "new experimental philosophy" for which the Royal Society later became famous had already been launched in the first half of the seventeenth century.

For the history of humankind and the advancement of economic development, the next step was to find a practical application of this fundamental principle that the air has weight. Within less than a decade, a German scientist and engineer would give Europeans the biggest clues about how to harness this great source of energy that remained tied to the question of what happens when a vacuum is formed.

Stronger Than Horses

Although Pascal's experiments definitively established the existence of air pressure and its variation by altitude, translating that insight into practical applications, especially a new source of energy, had many steps to go. Pioneers such as Gasparo Berti, Torricelli, and Pascal and his associates were convinced that there was no such thing as nature's abhorrence of a vacuum. But old ideas sometimes persist for decades, long after conclusive evidence for the new has been brought forth. Even the great philosopher

---

[33] Stephen Gaukroger, *The Emergence of Modern Scientific Culture* (New York: Oxford University Press, 2006).

René Descartes could not abandon his belief that nature's constitution absolutely ruled out the existence of a vacuum. He even ridiculed Pascal in private correspondence, saying that he had too much "vacuum in his head."[34] But equally, we find that those who hold on to the more incorrigible metaphysical aspects of their ideas can be impelled, perhaps against their will, to undertake experiments that serve to reject their idées fixes.

The metaphysical miasma connected to the idea of a vacuum – that it was impossible – continued for decades after the second half of the seventeenth century. The instrument that did the most to end that belief was the air pump. Yet, it was invented by an engineer and civil servant who held on to many aspects of the medieval view. This was the mayor of Magdeburg, Germany, Otto von Guericke. He had been influenced by the Jesuits and the scholastic view that a plain nothing cannot exist in the void of infinity. He seemed to believe "that imaginary space, vacuum or the Nothing beyond the world, is God Himself."[35] That seemed to rule out a vacuum.

Nevertheless, Guericke was a highly educated university graduate, lawyer, and engineer who was equally comfortable as a public official, city counselor, diplomat, or inventor of the air pump. He came from a wealthy family and from the age of twenty-four served in the town council, assuming increasingly important official positions as he matured.

Some scholars have suggested that the air pump is one of the four most important scientific instruments of the seventeenth century. The others include the telescope, the microscope, and the pendulum.[36] Between 1647 and 1652, von Guericke invented the air pump and began using it for his investigations of the vacuum, air pressure, and one might say the "work capacity" of the vacuum. It seems possible that he carried out many of these investigations before he learned of Torricelli's experiments.[37]

He began by constructing a suction pump for withdrawing the liquid out of a beer cask. But finding that the cask could not be sealed tight enough, even with a surrounding water jacket to prevent air from seeping into the emptying cask, he invented an air pump.

---

[34] Letter to Huygens, as cited in O'Connor and Robertson, "Blaise Pascal."
[35] Grant, *Much Ado about Nothing*, p. 219, and Grant, *A Source Book*, pp. 563–68.
[36] E. J. Dijksterhuis, *The Mechanization of the World Picture* (New York: Oxford University Press, 1961), p. 455.
[37] Otto von Guericke, *New Magdeburg Investigations [so-called] on Void Space (Neue Magdeburg Versuche über den Leeren Raum)*, facsimile edition, ed. Hans Schimak (Dusseldorf, Germany: VDI, 1968), p. 131.

FIGURE 8.3. Otto von Guericke's pneumatic experiment of 1654, demonstrating the enormous power of air pressure on a vacuum-induced metal sphere. Horses pulling in opposite directions were unable to separate the spheres. From von Guericke's *Neue Magdeburger Versuche über Den Leeren Raum* (1672).[38]

This was at the time when von Guericke learned of Descartes's belief that a vacuum was an impossibility, that if one should occur in a confined space, the space would implode. To overcome the leakage von Guericke had experienced earlier with wooden casks, he then ordered the construction of a hollow copper sphere with a petcock for evacuating the air. When the air was pumped out, the sphere imploded. This might have been taken as evidence in favor of Descartes's ideas, but von Guericke apparently thought otherwise and proceeded with more investigations. He ordered a stronger set of copper hemispheres, improved his air pump, and then staged one of the most dramatic scientific tests ever performed. Before many observers in public in Magdeburg, von Guericke carried out the test proposed back in the fourteenth century by Jean Buridan. Instead of ten horses, he used eight on either side of the hemispheres, pulling in opposite directions (Figure 8.3).

Under those conditions, the teams of horses could not pull the two halves apart because of the extraordinary air pressure exerted on the

[38] The conventional date of this demonstration is 1654, but Fritz Krafft, in "Guericke, Otto von," *DSB* 5 (1970): 575, puts the date at 1657 in Magdeburg, not Regensburg in 1654.

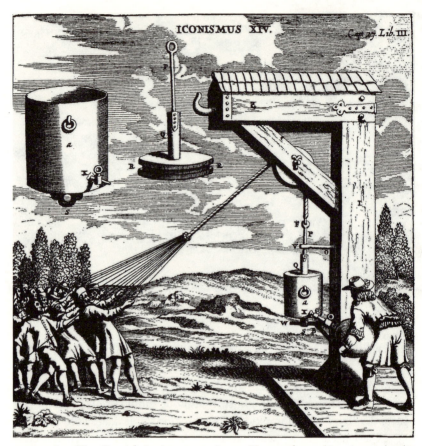

FIGURE 8.4. Von Guericke's piston experiment. First performed in 1654 under the influence of Otto von Guericke, it shows twenty or so men trying to hold a small piston up while air is being pumped out of a cylinder, creating a vacuum. From von Guericke's *Neue Madgeburger Versuche über Den Leeren Raum* (1672).

shells. Some might think that these must have been rather feeble horses,[39] but the demonstration captured the imagination.

However, it was another experiment devised by von Guericke that provided more clues for practical applications (Figure 8.4). In this experiment, a piston is located in a copper cylinder that has a port for evacuating the air. Twenty to fifty men who were supplied with ropes holding the piston up were unable to do so when the cylinder was emptied

---

[39] Donald Cardwell, *The Fontana History of Technology* (London: Fontana Press, 1994), p. 117.

with the air pump. This gave an impressive estimation of the force of air pressure compared to human strength.

In other investigations, von Guericke loaded weights onto a platform attached to the evacuated cylinder to determine just how much weight the vacuum in the cylinder could sustain.[40] What these experiments did was to show that there is an enormous force built into nature that might be put to useful purposes. This second example using a piston in a cylinder may have triggered in some observers a new image of how atmospheric pressure might be harnessed; that is, with a piston attached to a lever, one might be able to capture the force of the atmosphere.

Power from Steam and the Atmosphere

The development of the steam engine toward the end of the seventeenth century is well documented. What has been understated, however, is the degree to which the steam engine, the "atmospheric steam" engine, is the product of seventeenth-century scientific understanding made possible by the many experimenters who carried out decisive experiments throughout the first half of the century. These extend from the Italians such as Giovanni Battista Baliani, Gasparo Berti, and Evangelista Torricelli (and their coworkers) to Pascal and his associates in France to Otto von Guericke in Germany. By the time the results of von Guericke's experiments were published in 1657, the power of the vacuum created by air pressure had been amply demonstrated and measured. By the 1660s, air pumps were being made and used all over Europe. As Steven Shapin and Simon Shaffer have shown, there was extensive communication across Europe between 1659 and 1661 regarding air pumps. News of experiments traveled between London and Paris, Paris and The Hague, London and Florence, London and Würzbug, and Madgeburg and Würzburg. At the same time, and in succeeding years, air pump experimenters from The Hague witnessed demonstrations in London, and vice versa, while others from The Hague witnessed pneumatic experiments in Paris.[41] The inquiry was trans-European.

Still the technical problem was how to create a vacuum quickly and efficiently, faster than could be achieved with an air pump. Christiaan

---

[40] See Abbott Usher, *A History of Mechanical Inventions*, rev. ed. (Cambridge, MA: Harvard University Press, 1954), pp. 337–53; and von Guericke, *Neue Madgeburger Versuche über Den Leeren Raum*, pp. 124–25.

[41] Steven Shapin and Simon Schaffer, *Leviathan and the Air-Pump* (Princeton, NJ: Princeton University Press, 1985), p. 228.

Huygens and his assistant Denis Papin, working in Paris, quickly tried various experiments using gunpowder in a chamber under a piston, leaving a vacuum after the explosion. Clever though this idea was, it was impractical. But the idea of using a piston in a chamber was a crucial step forward on the way to creating a doubling-acting device: a piston pushed in one direction by an explosive force and in the other by the force of atmospheric pressure filling a void created by a vacuum. The other conceptual leap needed here was the realization that the implosive power of steam could be applied to push the piston up before atmospheric pressure forced it back down. This was the brilliant insight achieved by Thomas Newcomen in the first years of the eighteenth century. As one historian of Western technology put it, "the invention of the atmospheric engine in this form was the greatest single act of synthesis in the history of the steam engine."[42] The historical record is still unclear just when Newcomen achieved this insight. There are indications that our ever-present Robert Hooke, shortly before his death in 1703, suggested to Newcomen the possibility that the piston might be operated "by creating a vacuum under it by the 'speedy' condensation of steam."[43] In any event, it seems that the first atmospheric engine designed by Newcomen was built near Dudley Castle, Staffordshire, in the Midlands of England, in 1712.[44]

Prior to this, another Englishman, Thomas Savery (ca. 1650–1715), developed an "engine" for "Raising Water by the Impellent Force of Fire" with no moving parts (except the valves), for which he received a patent in 1698. It did indeed use atmospheric pressure to lift water out of a mine, but it did so by heating a chamber with steam, then creating a vacuum (by cooling) that would draw water from down below up into the chamber. Then the water had to be expelled from the chamber by the application of more steam heat and high pressure. The design proved to be impractical because efforts to improve its operation created more rather than fewer problems. It was soon abandoned.[45]

Newcomen's steam engine, however, with its use of steam and atmospheric pressure to propel a piston up and down, was the authentic item. It harnessed steam and atmospheric pressure to work a suction pump

[42] Usher, *A History of Mechanical Inventions*, p. 347.

[43] Ibid., p. 348.

[44] Richard L. Hills, *Power from Steam: A History of the Stationary Steam Engine* (New York: Cambridge University Press, 1989), pp. 20–23.

[45] Among others, see Usher, *A History of Mechanical Inventions*, pp. 345–47; Hills, *Power from Steam*, pp. 16–20; and Cardwell, *Fontana History*, pp. 118–20.

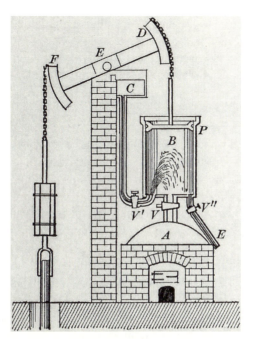

FIGURE 8.5. Newcomen's steam engine. Steam flows up from the furnace and boiler (A) at the bottom, forcing the piston (P) to the top of the cylinder. The injection of cold water into the cylinder (via value V′) causes a rapid condensation of air, creating a vacuum while atmospheric pressure forcefully pushes the cylinder back down. The rocker arm (F–D) attached to the cylinder controls another piston, activating a vacuum pump pulling water up from depths of a mine (from Wikipedia Commons, http://commons.wikimedia.org/wiki/File:Newcomen6325 .png).

that brought water up out of a mine shaft. There is a continuity here between the pump apparatus of Agricola of the early sixteenth century and Newcomen's engine. But now the waterwheel had been replaced by steam and atmospheric power (Figure 8.5).

To this newly discovered agent, "so powerful that it could overcome the strongest horses and evidently rivaled the largest water-wheels and windmills, science and common sense could set no obvious limit."[46] The discovery of the weight of air had now been transformed into a new machine. With it, an almost unbounded amount of work could be accomplished: water could be lifted, wheels turned, and machines making cloth

[46] D. S. L. Cardwell, *Steam Power in the Eighteenth Century* (London: Sheed and Ward, 1963), p. 9.

or other products self-powered. Soon after its initial successful appearance in England in 1712, other models of this engine were built across
Europe in Austria, Belgium, France, Hungary, Germany, and Sweden. In
England alone, 100 models had been built by 1729, when Newcomen's
patent ran out.[47]

Whereas multiple Europeans were developing new models of the steam
engine and exploiting this new source of power across the region, no
similar development took place in the Middle East, Mughal Indian, or
China in the seventeenth or early eighteenth centuries. The Indians of
Mughal times did not have piston-based pump technology, neither had
they learned it from the Europeans.[48]

Joseph Needham speculated about the possibility of using double-
action devices invented by the Chinese in this connection. Although he
presented no clear evidence that such devices originated in China and
came to the West, he imagined that they might have been used to help
control aspects of a steam engine once invented. But such devices would
be of no significance without the scientific discovery of the weight of air,
not to mention Newcomen's leap of the imagination putting all the components together. The Chinese had invented some engineering devices and
had constructed all sorts of waterworks, but there is no evidence that the
Chinese understood atmospheric pressure.[49] There are no reports that
Chinese scholars or inventors ever conducted any experiments comparable to those of Berti (1639–40), Torricelli (1644), Pascal (1648), Otto
von Guericke (1654), or other European experimenters of the seventeenth
century.

Furthermore, Needham observed that the Chinese did not use piston
pumps until after the sixteenth century (coincidently, when the Jesuits
arrived) so that evidence is lacking that the Chinese invented a device like
Agricola's "seventh" piston-based suction pump (see Figure 8.1) with
such family resemblance to the principles of the Newcomen machine.
In "The Pre-natal History of the Steam-Engine," Needham refers to a
Chinese "experiment" in the second century B.C. in which a vessel containing boiling hot water is plunged into a well with a resultant "sound

---

[47] Hills, *Power from Steam*, p. 30.
[48] Ahsan Jan Qaisar, *The Indian Response to European Technology and Culture* (A.D.
1498–1707) (Delhi: Oxford University Press, 1982), p. 32.
[49] For Needham's history of hydraulics in China, see *SCC* (1971), 4/3:sect. 28, pp. 211–378.
There is only a brief reference to pumps, mainly for ships.

like thunder."[50] This is hardly a demonstration of the existence of air pressure, even if one might concede that given our knowledge today, we know that the conditions could have produced a vacuum. This suggests that the claim that the Chinese knew "the basic scientific principle" of steam power is highly exaggerated.[51] A history of the spread of steam technology for pumping water, locomotion, and manufacturing to other parts of the world would tell us a lot more about these different origins of science and economic progress.

[50] Joseph Needham, "The Pre-natal History of the Steam-Engine," in *Clerks and Craftsmen in China and the West* (Cambridge: Cambridge University Press, 1970), pp. 145–46.

[51] Kenneth Pomeranz, *The Great Divergence* (Princeton, NJ: Princeton University Press, 2000), p. 61.

9

Infectious Curiosity III

Magnetism and Electricity

The ideas about magnetism and electricity that began to be widely discussed by natural philosophers at the outset of the seventeenth century take us deep into the mysteries of the fundamental forces of nature. Even at the end of the twentieth century, this part of modern physics had many unanswered questions, including just how to think about the four basic forces of nature: *strong, weak, gravitational,* and *electromagnetic.* Today, perhaps electric and magnetic forces seem the simplest to comprehend, but in 1600, no one had even imagined the existence of "electricity." William Gilbert stumbled onto it while divining the nature of magnetism. Only that innovation paved the way for the continuous study of electric forces throughout the seventeenth century. In the meantime, astronomy was about to be transformed from mere mathematical model-building to philosophical speculation about just what holds our universe together. But before we can approach that great intellectual struggle, we need to consider the discovery of the more subtle forces that bind our world, and that began to be glimpsed in the early seventeenth century.

Holding the World Together

The question of what holds the planets in their orbits was abruptly brought into focus in the late sixteenth century. In 1577, a comet appeared in Europe, seen by many observers, but especially Tycho Brahe. He was then the most accomplished European astronomer. He noticed that the path of the comet was such that it would have crashed through the crystalline spheres that were supposed to hold the planets and fixed stars in their orbits. If this comet on a path through a crystalline sphere did not cause a crash, then those spheres vanished. If the crystalline heavenly spheres were gone from the universe and therefore could not explain why the planets and fixed stars continued in their daily and yearly paths, then cosmological thinkers had to ask themselves if there is not some intrinsic force in nature that attracts objects to each other. This was the deeper background to Kepler's thinking in 1605.

Even to modern ears, this intellectual puzzle of divining the forces holding the universe together and creating a unified celestial and terrestrial science of motion sounds daunting. The challenge was indeed on the minds of many great European thinkers in the seventeenth century, especially Kepler, Descartes, and Newton, but also Robert Hooke, Sir Christopher Wren, and Edmund Halley. Even before them, an English physician by the name of William Gilbert had approached the problem from another direction at the very beginning of the seventeenth century. That was the exploration of magnetism and electricity.

Magnetism and Electricity

William Gilbert's landmark book on magnetism (*De Magnete*) appeared in 1600, before Galileo's telescopic discoveries and Kepler's new astronomy launched the full flowering of the scientific revolution. In that sense, it was truly in the vanguard of the scientific revolution of the seventeenth century. Partly because of Gilbert's own language, and partly because of Francis Bacon's stilted conception of the kind of "great instauration" (new beginning) that was required to launch the new science,[1] Gilbert's achievement has often been undervalued or overlooked as an example of the new experimentalism that was to become the hallmark of the seventeenth century. The sad fact is that Bacon could not find much to admire in the science of his era and thus could not appreciate the accomplishments of one of his contemporaries.[2] More broadly, he missed that Europe was

---

[1] Francis Bacon, *The Great Instauration*, in *The New Organon*, ed. Fulton H. Anderson (Indianapolis, IN: Bobbs-Merrill, 1960). For the classic study of that idea and its implementation in England in the seventeenth century, see Charles Webster, *The Great Instauration: Science, Medicine, and Reform 1626–1660* (New York: Holmes and Meier, 1976); and the essays in *Puritanism and the Rise of Modern Science: The Merton Thesis*, ed. I. Bernard Cohen, with K. E. Duffin and Stuart Strickland (New Brunswick, NJ: Rutgers University Press, 1990).

[2] John Heilbron, *Electricity in the 17th and 18th Centuries* (Berkeley: University of California Press, 1979), p. 175.

already poised for its great run of experimental inquiries, many of which we have already considered.

The rhetorical language of casting anything before the seventeenth century into the mold of "scholastic" hair-splitting, Aristotelian paralysis, or the dreaming of "light-headed metaphysicians" (Gilbert's term) serves to obscure the continuity between the era of Gilbert's work and his predecessors. Although Gilbert used rhetorical language suggesting that he was turning over a great new page in natural philosophy, moving on from the "sophists and spouters," it should be remembered that Gilbert himself was a product of the much-maligned universities from which "reason" had allegedly been excluded.[3] He had received three degrees, including his MD in 1569 from St. John's College at the University of Cambridge. His scientific imagination had not been damaged by the experience.

No doubt the intellectual world of the sixteenth century was changing faster than universities could be reformed. Nevertheless, Gilbert's penchant for exaggerated comment obscures the extraordinary fact that Gilbert's opening pages of experimentalism were inspired by the writings of the thirteenth-century pioneer of magnetism, Peter Peregrinus (fl. 1269). The newly stimulated interest in navigating the world's oceans in the sixteenth century also brought greater interest in the use of the compass and magnetic directionality. Furthermore, Gilbert's achievement was accomplished by standing on the shoulders of such sixteenth-century students of magnetism as Girolamo Fracastoro (1478–1553), Girolamo Cardano (fl. 1561–76), Georg Hartmann (1489–1564), Robert Norman (fl. 1576, the second discoverer of magnetic "dip"), and his immediate contemporary William Barlowe (fl. ca. 1597).[4]

It had been known for centuries that lodestones have powerful magnetic effects. A number of pioneer investigators of magnetism from the twelfth century through the sixteenth had written about magnetism so

---

[3] At the time of Descartes's reburial in France in 1667, a writer by the name of Nicolas Boolean sponsored a prank by talking about the "unknown person called Reason" being unable to enter the university in the seventeenth century to overthrow Aristotle; as cited in Heilbron, *Electricity*, p. 36. But Aristotle's reason had long ago served its purpose in educating the great figures who carried forth the scientific revolution, including Descartes himself.

[4] See A. Crichton Mitchell, "Chapters in the History of Terrestrial Magnetism: [1] The Discovery of Directionality," *Journal of Terrestrial Magnetism and Atmospheric Electricity* 37, no. 2 (1932): 105–46; "The Discovery of Declination," *Journal of Terrestrial Magnetism and Atmospheric Electricity* 42 (1937): 241; and "The Discovery of Dip," *Journal of Terrestrial Magnetism and Atmospheric Electricity* 44 (1939): 77.

that Gilbert's great work was really a classic synthesis of all that preceded him. In a sense, his book was the first important scientific textbook
of the seventeenth century, based on experimental methods designed to
illuminate magnetic and electric attraction, while separating the two in
important ways.[5]

It was also known since the Greeks that amber when rubbed produces
an attractive force that draws chaff to it. In the present context, Gilbert's
clear separation of electric from magnetic forces, apart from establishing an exemplary model of experimental inquiry, was his most significant accomplishment. For that conceptual separation based on experimental evidence gave birth to the whole new field of electricity. That
discovery of invisible microforces different from magnetism (though ultimately related) would eventually give the world a limitless revolutionary
new source of power. No glimmer of that field was to be found outside
Europe for centuries more. In China, Joseph Needham pointed out that
"there was no real advance" in the area of electromagnetism and, indeed,
there is no indication that the Chinese had discovered "electrostatic phenomena" at all apart from amber. Conversely, Gilbert created a whole
new field. Although Joseph Needham wrote a hundred pages on various
aspects of magnetism, geomancy, and related magnetic phenomena, he
wrote barely a page on "electrostatic phenomena," including a review of
ancient sources.[6]

Peter Peregrinus

The first systematic treatise written on magnetism in the Western world
was that of the French scholar, Peter Peregrinus (fl. 1269), also known
as Pierre de Maricourt. Most probably, it was the first systematic treatise

---

[5] See Heilbron, *Electricity*, pp. 169–79; Suzanne Keller, "William Gilbert," in *Dictionary of Scientific Biography* 5 (1972): 396–401; Duane Roller and Duane H. D. Roller, *The Development of the Concept of Electric Charge*, Harvard Case Studies in Experimental Science, 8 (Cambridge, MA: Harvard University Press, 1954); and for a useful overview of Gilbert's book and purpose, H. H. Ricker III, "William Gilbert: Founder of Terrestrial Magnetism," http://www.wbabin.net/science/ricker6.pdf. Another view is in Edgar Zilsel, "The Origin of Gilbert's Scientific Method," *Journal of History of Ideas* 2, no. 1 (1941): 1–32. Zilsel's fastidious analysis of Gilbert comes up with an ambiguous portrait, both praising and criticizing Gilbert, implausibly faulty for not using quantitative analysis. At the same time, he fails to see Gilbert's work as part of an experimental tradition that had long been practiced in Europe.

[6] Joseph Needham, *SCC*, 4/1:236–37.

written anywhere.[7] He had studied at the University in Paris and was highly praised by other scholars such as Roger Bacon. According to him, Peregrinus was "a master of experiment" and "he knows by experiment natural history and physic and alchemy" as well as optics and "all things in the heavens and beneath them."[8] Magnetic effects and the use of the compass were known in other parts of the world, especially in China, but Peregrinus's "letter" is a remarkably clear and coherent description of the magnet, magnetic phenomena, and how to test for magnetic effects. It appears that he invented the *terrella*, the spherically shaped lodestone that stood as an analogue to the earth. This was the source of Gilbert's use of that device reported in *De Magnete*.

The Chinese scholar Shen Kua in 1088 speaks of a magician rubbing a needle so that it becomes magnetized by a lodestone, which then "is able to point to the south. But it always inclines slightly to the east, and does not point directly south."[9] It may be that discovery of the directionality of the lodestone and its use as a compass occurred about a century earlier in China than in Europe. In the absence of written records of that use in the eleventh century, as well as the absence of any documents showing the transmission of knowledge of the compass from China to the West,[10] it is doubtful that Peregrinus, or before him Alexander Neckham, was influenced by Chinese sources. Neckham described magnetic forces (ca. 1190) and refers to European mariners using the compass at that time but does not indicate that it was something new.[11]

In any event, Peregrinus's letter is a systematic treatise on the magnet that attempts to explain magnetic attraction. It also includes the invention of the compass-card and the dry-pivot compass not known in China until the arrival of Europeans in the sixteenth century.[12] In China, and before the time of Peregrinus, compasses were based on either a lodestone

---

[7] Although Needham wrote quite a few pages about magnetism in China, it is surprising how much that account is based on isolated comments, inscriptions, and widely scattered sources, especially what he calls the "bewildering" arrangement of geomantic documents. The most authoritative and first reliable written record describing magnetic directionality and its use for finding the polar directions comes from Shen Kua (d. ca. 1095). He was a great Chinese scholar, but his writings have often been characterized as rather mixed and muddled. "Notices of the highest originality stand cheek by jowl with trivial didacticism, court anecdotes, and ephemeral curiosities," providing little insight. Nathan Sivin, "Shen Kua," *DSB*, 12 (1975): 374.

[8] Roger Bacon on Petrus Peregrinus in Edward Grant, ed., *A Source Book in Medieval Science* (Cambridge, MA: Harvard University Press, 1974), p. 824.

[9] In Needham, *SCC*, 4/1:249.

[10] Ibid., 4/1: 330.

[11] Mitchell, "Chapters in the History," pp. 105–46.

[12] Needham, *SCC*, 4/1: 289.

or a magnetized needle floating in water to find the polar directions. Peregrinus, however, after telling his reader how to make a wet compass, gives directions for performing experiments and how to construct a dry compass with the needle resting on a pivot set within a shallow box. In addition, he invented the compass card that consists of the cardinal directions (north, south, east, west) all placed on the circumference of a compass, with further subdivisions into ninety parts. Thus, the compass card was eliminated.[13]

The other pioneering aspect of Peregrinus's work was the invention of the spherically shaped lodestone, the terrella. This gave him an experimental device for observing magnetic effects. With it, Peregrinus discovered that if you lay a needle on the surface of the stone, the needle will orient itself to the north and south poles of the stone. Peregrinus did just that: he placed a needle on the surface of the lodestone, then drew lines on the surface to mark the direction of the needle. He did this until he could see all the lines of direction mapped out on the terrella's surface intersecting at the poles of the stone. Although he demonstrated in this way the fact that magnets have complementary poles, he was not able to make the leap to the idea that the earth, too, is just like a lodestone, with poles at the north and south that cause the directionality of the compass needle.

Apart from the fact that these discoveries became permanent additions to our stock of knowledge about lodestones, the importance of the treatise by Peregrinus is reflected in the fact that many copies of it were produced in the Middle Ages, with the first printed edition appearing in 1558. Following that, a French priest by the name of Jean Taisner published another edition in 1572 that he passed off as his own.[14] It was from this work that Johannes Kepler first learned about magnetism as a potential force regulating objects in the universe. Remarkably, many more editions of Peregrinus's treatise on magnetism were printed across Europe into the nineteenth century.[15]

De Magnete and Demonstration

From the outset, Gilbert made his readers aware that his study was not just spouting and hair-splitting but was based on experiments "carefully,

---

[13] "The First Systematic Description in Europe of the Properties of the Lodestone," written by Peregrinus (fl. 1269), in Grant, *A Source Book*, esp. pp. 374–75.

[14] "Pierre de Maricourt," *Catholic Encyclopedia*, http://www.newadvent.org/cathen/12079e.htm.

[15] See the annotations to Grant, "Peter Peregrinus," *DSB* 10 (1974): 532–40.

skillfully and deftly" designed and performed. He challenges the reader to "make the same experiments" and reassures one that there is nothing in the book that "has not been investigated again and again and repeated under our eyes."[16]

In the third chapter of Book I, he introduces Peregrinus's device of the terrella and tells us how to determine the poles of any lodestone by laying a needle on the surface and then tracing out the directional lines that always intersect at the poles. Then he proceeds to demonstrate the fact that the opposite poles of any two lodestones attract each other, while the same northern or southern poles of two magnets repel each other.

The plan of the book is to investigate experimentally each of the major aspects of the magnet, the obvious aspects of magnetic *attraction* (that he calls "coition"), the *directionality* of the compass needle, the *variations* from true north, the "dip" (or *declination*) of the compass whereby the needle is deflected downward from the horizontal, and what Gilbert calls the "verticity" of lodestones that results in a wire standing vertically when brought near the poles. Using his terrella, when possible, as an analogue to the earth, Gilbert performed numerous experiments, showing which materials are attracted by a magnet and which are not.

In all this, Gilbert is developing what he calls the *magnetic philosophy* that assumes that all the matter of the earth "is brought together and held together by itself electrically."[17] He declares that the earth is a giant lodestone and the reason for the compass needle pointing to the north pole is not because of the attraction to the northern pole star but because of the magnetic force of the earth. Gilbert also believed that the magnetic characteristics of the earth explained why it rotates daily on its axis.

Gilbert's "Electrics"

The most revolutionary discovery of Gilbert's work concerns his "electrics." In 1600, when he published *On the Magnet*, the idea of gravitational pull had not been grasped, neither had electricity been discovered. What Gilbert did was to insist that apart from magnetism, there is another force in nature that is generated differently from magnetism and that lots of materials, "electrics" he called them, have the capacity

---

[16] William Gilbert, *William Gilbert of Colchester, on the Lodestone and Magnetic Bodies, and The Great Magnet Earth. New Physiology, Demonstrated with Many Arguments and Experiments*, trans. P. Fleury Mottelay (New York: John Wiley, 1893) (hereinafter *De Magnete*), p. xlix.

[17] Ibid., p. 97.

FIGURE 9.1. Gilbert's versorium, or electroscope, designed to detect an electrical charge from his *De Magnete* (1600).

to generate a charge. Amber (*elektron* in Greek) was the classic substance that was known to produce an attractive force when rubbed. But now Gilbert, after conducting all sorts of experiments himself, concludes that many other kinds of materials have this same capacity. It is from this word, *electrics*, that we get our word *electricity*. With that insight, Gilbert created the completely new field of electric studies.

This was accomplished in Book II, chapter 2, where he investigated amber and a great variety of objects and substances that were found to have electric properties similar to amber. He was convinced that there is a "vast difference between this [amber] and the magnetic actions."[18] For example, interposing a piece of paper between an electric and a detection device or another electric would cancel the effect, whereas magnetic forces still work. Likewise, moisture enveloping an electric would disperse its change but not magnetic attraction.

As with his chapters on magnetism, Gilbert proceeds in an experimental fashion, telling the reader what he found out and encouraging him or her to undertake similar investigations. Through long and meticulous work, he had discovered that this mysterious class of electrics "not only draw to themselves straws and chaff, but all metals, wood, leaves, stones, earths, even water and oil."[19] Here, too, he introduced a new instrument, the electroscope, probably the very first of such instruments for detecting an electric charge. He called it a "versorium" (a turning indicator; Figure 9.1).

Although use of the needle appeared in an earlier work by Girolamo Fracastorio, Gilbert certainly deserves credit for incorporating it into electric studies. To understand all these electric effects "from experience," Gilbert wrote, "make yourself a rotating needle (electroscope-versorium) of any sort of metal, three fingers long...poised on a sharp needle."[20] Then bring near to one end of the device a piece of amber, a gem, or

---

[18] Ibid., p. 75.
[19] Ibid., p. 78.
[20] Ibid., p. 79.

other polished object that has been rubbed, and the needle will move in response to the charge.

In short, Gilbert had taken a major leap into the experimental age; he had devised a new scientific instrument that could detect subtle electric forces. It would be the forerunner of all such devices invented by electricians. And yet, electric charges were so subtle and mysterious that they would remain poorly understood until roughly the time of Franklin at the end of the eighteenth century.[21] Nevertheless, Gilbert had opened the door to this new investigation, and by the end of the seventeenth century, other experimenters would be able to generate static currents that could be transmitted hundreds of feet. As John Heilbron put it, notwithstanding all the limitations of Gilbert's pioneering classic, it "defined the subject matter of electricity, identified its problems, and provided the framework for their resolution" for centuries to come.[22]

The Magnetic Philosophy and Kepler

Before turning to later seventeenth-century developments in electricity, it should be noticed that Gilbert's magnetic philosophy had both followers and detractors in the first half of the seventeenth century. The most extraordinary and surprising of such followers was Kepler. From the moment that he discovered Taisner's plagiarized rendition of Peregrinus's letter on magnetism, Kepler came to believe that the magnetic forces of nature might provide an explanation for the coherence of the celestial domain. In his *New Astronomy* of 1609, Kepler struggled with the problem of the forces necessary to explain the attraction between physical bodies and their circular motions around the sun. He saw the need for a "true doctrine concerning gravity" for, according to him, "gravitation consists in the mutual bodily striving among related bodies toward union or connection; (of this order is also the magnetic force)."[23] Gerald Holton refers to Kepler's thinking here as a "premonition of universal gravitation"[24] that seems to be suggested in this statement from Kepler:

If two stones were placed in a certain place of the world, near each other and outside the orb of the virtue of a third cognate body, these stones, in a manner

---

[21] Hirshfeld, *The Electric Life of Faraday* (New York: Walker & Company, 2006).
[22] Heilbron, *Electricity*, p. 179.
[23] *New Astronomy*, as cited in Gerald Holton, "Johannes Kepler's Universe: Its Physics and Metaphysics," in *Thematic Origins of Scientific Thought: Kepler to Einstein* (Cambridge, MA: Harvard University Press, 1973), p. 73.
[24] Ibid.

similar to magnetic bodies, would meet in an intermediate place, each nearing the other by an interval proportional to the bulk [moles] of each other.[25]

Kepler's ideas about gravity were clearly ambivalent, leading him to acknowledge an attraction between some bodies (such as the earth and the moon), but in the general case denying gravitational attraction between the planets.[26] Although Kepler is not able to reach the inverse square rule of attraction between physical bodies that Newton developed, he did suggest that one could observe the gravitational pull of the moon on the ocean tides of the earth – a view that Galileo disparaged, as did Jean Cassini years later. In the end, Kepler had to remain content with his three laws of celestial motion without the principle of universal gravitation. He opted instead for the efficacy of magnetic attraction, as argued by Gilbert. In his *Epitome of Copernican Astronomy* of 1621, Kepler tried to dispense with the idea of a governing intelligence or soul within planetary bodies. He argued that the sun has an internal virtue that attracts other bodies and causes the planets to revolve around it. When he tries to explain how this is so, he calls up the magnetic powers of the lodestone and the compass needle. The planets and the sun are like the "lodestone and the iron pointer, which has been magnetized by the lodestone and which gets magnetic force by rubbing. Turn the lodestone in the neighborhood of the pointer; the pointer will turn at the same time."[27] (See Figure 9.2.)

Kepler guessed that the sun rotates on its own axis (which was later confirmed by sunspot activity) and that the sun's rotation had a paddlelike effect: the rays of force emanating like spokes out from the sun served to push the planet forward, while the magnetic forces pulled it toward the sun.

But there is a problem in that the sun seems to attract and repel moving bodies, for Kepler has to explain the principle of the polar forces of magnets. In his model, however, the points of attraction and repulsion appear to be in the middle of the celestial bodies, unlike Gilbert's clear delineation of the polar location of maximum magnetic force. Kepler's model assumes that positive and negative forces are located at the middle of the sun and seem to reside on opposite sides of his equator.

This is an imaginative attempt to employ Gilbert's magnetic philosophy, but it failed to win the assent of other astronomers. The attempt to

---

[25] As cited in Alexandre Koyré, "Copernicus and Kepler on Gravity," in *Newtonian Studies* (Cambridge, MA: Harvard University Press, 1965), p. 174.
[26] Koyré's discussion is helpful on this; ibid., pp. 173–75.
[27] Kepler, *Epitome of Copernican Astronomy and Harmonies of the* World (Amherst, NY: Prometheus Books, 1995), p. 57.

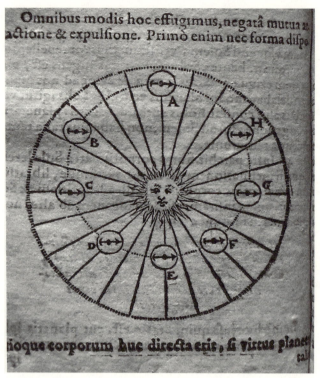

FIGURE 9.2. Kepler's diagram of the magnetic attraction between the sun and a planet from his *Epitome of Copernican Astronomy* (1618–21). Courtesy of Owen Gingerich.

build on magnetic attraction in the celestial sphere proved to be another garden path, though some astronomers continued to invoke magnetic forces until the end of the century. Nevertheless, Kepler's law of elliptical orbits and his two additional laws remained intact and were later incorporated in modified form into Newton's synthesis (on which more later).

Phosphorus and Electrostatic Machines

It is not my purpose here to spell out all the details of the advances and setbacks in the field of electrical studies in the seventeenth century. It is enough for us to catch the main drift of the new electrical inquiries in which European scientists and natural philosophers were engaged during the seventeenth century, keeping in mind the absence of such progressive inquiry outside Europe.[28]

Although Gilbert's work served as the basic framework for electrical studies for the next century, not everyone agreed with his conclusions, and rightly so. Between 1600 and 1799, there were at least 210 significant electricians who advanced the field and whose work has been studied in detail by John Heilbron, Park Benjamin, Edmund Whittaker, and other historians of science.[29] No doubt there were other students of electricity who explored basic issues in electrical studies but whose work either did not get published or was forestalled because their conclusions were similar to published results. But the 200-odd pioneers whose works have been studied serve as a sure indicator of the spirit of curiosity that animated this field across Europe. Later, they would have major implications for scientific and technological development. The discovery of electricity obviously had momentous implications for all humanity. It is the fundamental natural force whose understanding unlocks the possibilities of electric lighting, electric-powered machines, the telegraph, the telephone, and myriad electrical devices of the twenty-first century. But without the centuries-long preparation for its harnessing that began in 1600 with Gilbert's classic work, no peoples on this globe could suddenly jump ahead of the curve. The intellectual capital, harvested from the study of electrics as well as the emerging new science of motion, accumulated like gold in the Western world from 1600 onward. Eventually, it would have an overwhelming impact on the world of the future, including the balance of power between the societies and civilizations of our globe.

For the advancement of electricity, a number of conceptual and mechanical hurdles had to be surmounted. Once electric charge had been discovered, it was necessary to determine that such a charge could be

---

[28] Efforts to suggest that the Muslim world of the Middle East was not totally isolated and might have contributed to the "Republic Letters" are in Sonja Brentjjes, "The Interests of the Republic of Letters in the Middle East, 1550–1700," *Science in Context* 12, no. 3 (1999): 435–68; and "Western European Travelers in the Ottoman Empire and Their Scholarly Endeavors (Sixteenth–Eighteenth Century)," in *The Turks*, vol. 3, ed. Kemal Cicek et al. (Ankara: Yeni Turkiye, 2002), pp. 795–803. These are not convincing, and one finds no reports of experimental activity anything like that of Europeans with the telescope, microscope, air pump, or electrical studies.

[29] See Heilbron, *Electricity*, pp. 135–36, for his tabulations. For more details of the period, see Park Benjamin, *A History of Electricity (The Intellectual Rise of Electricity from Antiquity to the Days of Benjamin Franklin)* (New York: Arno Press, 1975) as well as Duane Roller and Duane H. D. Roller, *The Development of the Concept of Electric Charge: Electricity from the Greeks to Coulomb* (Cambridge, MA: Harvard University, 1954), among others.

transmitted from a generating device over a thread or wire to another place or object. To do that, a way had to be found to produce electric charges of sufficient force and quantity to be transmitted more than a few inches, such as Otto von Guericke discovered in the 1660s with his mineral-laced sulfur globe. Eventually, electricians would have to discover the difference between conductors and insulators as well as the fact that electricity involves both positive and negative charges that flow from one part of an object to another. That would explain in part why electric charges can be both mutually attracting and repulsive. All these aspects of electricity were explored and sometimes glimpsed in the seventeenth century, but clearly another century would have to elapse before scientists and engineers would gain the capacity to produce electricity in sufficient quantity for useful purposes.

As discussed in Chapter 8, the discovery that a vacuum is possible in the 1640s and the invention of the air pump in the 1650s made it possible to study systematically a whole new range of physical phenomena that would otherwise fall beyond the limits of human capacity. Nevertheless, medieval Europeans had asked hypothetical questions about whether sound or light can travel through a vacuum if it were possible for one to exist. So, too, in the study of electricity, questions were asked about the role of air in the production of electrical effects, whether heating a piece of amber and watching the attraction of a piece of paper or chaff was due to the supposed *effluvia* emanating from the amber, or whether this was really the result of warm air currents flowing from the amber outward and then circling back, bringing the chaff with it. This was the theory put forth by the Italian natural philosopher, Niccolò Cabeo. His treatise of 1629 on the magnet (*Philosophia Magnetica*) was the first by an Italian but remained influential throughout the century. Testing for the effects of air with regard to electric charges, however, depended on the discovery that a vacuum is possible. That first came about with Torricelli's invention of the barometer (in 1643), with its empty chamber above the mercury. Using that chamber as an experimental space was both a challenge and enticement for early-seventeenth-century European experimenters. Soon after the founding of the Accademia del Cimento – the Italian Academy of Science – in 1657, its members interested in electricity made a first attempt to study an electric charge in a vacuum in the early 1660s. They hoped to chafe a piece of amber inside the Torricelli vacuum chamber that would generate an electric charge and attract a piece of paper. The amber was to be reached either by a leather sleeve, fastened tightly around a wrest, or by a wooden rod inside a similar chamber that could massage the amber and generate an electric charge. Both attempts failed.

In the meantime, the mayor of Madgeburg, Germany, Otto von Guericke, carried out experiments with a sulfur ball laced with a concoction of minerals that von Guericke thought better approximated the earth's composition than Gilbert's spherical lodestone, the terrella. He poured his sulfurous porridge into a glass sphere the size of a large grapefruit and waited for it to cool and congeal. Then he broke away the glass, leaving his mineral-enhanced sulfur ball. With a finesse that apparently only von Guericke's toughened dry hands could produce by rubbing, he generated a charge on his sulfur ball that caused a feather to float while he walked around a room. He also noticed that a string touching his globe transmitted a charge to an object next to it. Most surprising of all, but little noticed elsewhere at the time, in the dark, von Guericke's sulfur ball glowed. Here was a clue of immense importance, yet it would have to be reinvented decades later.[30]

Shortly after this, a French priest and very able astronomer by the name of Jean Picard happened to notice in 1675 that when he moved his mercury barometer about at night, the space in its Torricelli vacuum at the top seemed to emit sparks and sometimes glowed. This discovery stimulated considerable curiosity, but no one could reproduce the effect, much less explain it. With hindsight, this event indicated that the movement of mercury on the surface of a glass tube caused its electrification and subsequent glow.

Still another curious discovery was made at this time in Germany. In about 1669, a physician by the name of Hennig Brandt (or Brand) accidentally extracted phosphorus from urine.[31] When the news that this new substance glowed in the dark was noised about in a pamphlet of 1678, it became one of the elixir rages of the day. The existence of this glowing substance, secreted in cow and human urine, led many to think that perhaps impurities in mercury were of a phosphorous nature, leading to the study of "phosphorus luminosity."

By this time, the Royal Society of London was in full flower. Yet, it was not until 1703 that its members suggested that it should take up investigation of this phosphorus and mercurial luminosity. The society was then under the direction of Isaac Newton. Robert Boyle had carried out a number of electrical investigations in the mid-1670s, but they did little to advance the field. Boyle agreed that electrics, when heated, sometimes emitted some kind of "effluvia" or, more reliably, did so when

---

[30] Heilbron, *Electricity*, pp. 216–17.
[31] John Emsley, *The Shocking History of Phosphorus: A Biography of the Devil's Element* (London: Macmillan, 2000); Benjamin, *History of Electricity*, p. 454.

rubbed. Beyond that, he had little to offer. He disagreed with Descartes's alternative theory that suggested some kind of particles were being forced through "pores or crevices" in the glass. He found Descartes's theory so daunting that he referred the reader to the original source.[32]

The next decade of the 1680s was to be a momentous one as Newton fashioned his grand synthesis of astronomy and the science of motion. Consequently, the publication of his *Principia Mathematica* in 1687 created a great stir, while drawing attention away from electrical studies that were poised to make great discoveries.

Hauksbee Sheds New Light

The renewal of those studies within the Royal Society coincided with Isaac Newton's assumption of the presidency of the Royal Society in 1703 and his appointment of Francis Hauksbee as the demonstrator of experiments. At the time, Hauksbee was an artisan of about fifty. Although he was little schooled in the formal studies of his day, he was a deft experimenter whose skills had gained him admittance to the Royal Society while winning Newton's trust.

Hauksbee's first appearance as demonstrator for the Royal Society in December 1703 was a great success marked by a shower of sparks. He had invented a new air pump and put it to good use in his new position. He had chosen for his first experiments the luminosity of mercury when it runs down the sides of a glass vessel. This was a clever laboratory simulation of the conditions within the Torricelli chamber of a barometer. It required two glass vessels, one inside the other, the outer one about eighteen inches tall. The external vessel was exhausted of air, creating a vacuum around the inner chamber, which had a domed top, down which Hauksbee caused mercury to run. When this happened, sparks began to fly. Not only could sparks be seen on the crown and sides of the enclosed vessel, "from the crown of the [enclosed] glass were darted frequently flashes resembling lightning, of a very pale color."[33] For sure, these displays were a winning success.

Hauksbee was a most ingenious experimenter whose cleverness would reveal many more secrets of electricity. His next move was to show that

[32] Boyle, "Experiments and Notes about the Mechanical Origins or Production of Electricity," vol. 8, p. 512, of *The Works of Robert Boyle* (online edition of Past Masters).
[33] Francis Hauksbee, "Several Experiments on the Mercurial Phosphorus, Made before the Royal Society, at Gresham College," *Philosophical Transactions* 24 (1704–5): 2134.

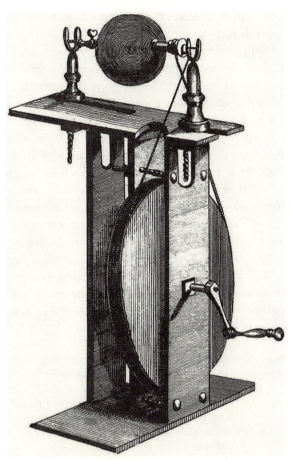

FIGURE 9.3. Francis Hauksbee's electrostatic machine. With this device, Hauksbee demonstrated that a static electric charge could be generated simply by friction on a glass container. From Hauksbee, *Physico-Mechanical Experiments* (1719).

a glow could be produced by amber being rubbed mechanically against wool inside an air-exhausted glass chamber, all mounted in a new machine that Hauksbee had invented. This demonstration also succeeded as the glow appeared.

Hauksbee's next bold step was to invent a machine that would produce a glow simply by friction applied to the outside of it. This landmark invention has become known as the prototype of the electrostatic generator (Figure 9.3). With this new device, Hauksbee proceeded in his 1705–06 experiments to demonstrate that a luminous glow could be produced directly by a charged vacuum-sealed glass sphere. Based on a nine-inch globe that could be evacuated of air, mounted on a rotating shaft, the device was capable of producing a charge of electricity that could be felt by bringing a finger near the apparatus. Using this new machine, Hauksbee showed that when the evacuated globe is rotated with a hand gently resting on it in a dark room, an electric glow is generated. It created so much light that it could be seen on a wall ten feet away, and the large letters in an open book nearby could be seen.[34]

As Hauksbee put it, "the light produced has been so great that a large print without much difficulty might be read by it: And at the same time, although in a pretty large room, the whole became sensibly illuminated."[35] Put in such an understated way, this extraordinary discovery of electric illumination might escape one's attention. But there it was in 1706, and Hauksbee demonstrated it several times and in a variety of configurations. Using his electric machine, but also employing a long, thirty-inch tube of about an inch in diameter, vacuum sealed, Hauksbee demonstrated still another effect: electric influence at a distance. With the electric machine spinning, Hauksbee brought the gently rubbed tube in his other hand toward the spinning and electrified globe; as he did so, the tube itself began to glow. In Hauksbee's words:

After this I took a long glass tube, which had lain by me exhausted of its air for more than six months. This glass having been rubbed little by my hand to expel the humidity on its outside, I held over the unexhausted glass in motion, which at the same time was rubbed by my hand. It would now and then (for it was not constant) be very surprising to see what large flashes of light would be produced in the long glass tube without its touching the glass in motion or itself being either moved or provoked by any immediate attrition [stroking].[36]

In related experiments, Hauksbee experienced himself a kind of "electric wind" that grazed his face and raised the hairs on the back of his hand when he brought an electrified tube near.

Although Hauksbee did not speak of it directly, clearly he had hit on the earliest form of electrification that would grace cities all over the world

---

[34] Heilbron, *Electricity*, p. 230; and Hauksbee, "An Account of an Experiment Made before the Royal Society at Gresham College, Together, with a Repetition of the Same, Touching the Production of a Considerable Light upon a Slight Attrition of the Hands on a Glass Globe Exhausted of Its Air; with Other Remarkable Occurrences," *Philosophical Transactions* 25 (1706–7b): 2277–82.

[35] Hauskbee, "An Account of an Experiment," p. 2278.

[36] Hauksbee, "Several Experiments Shewing the Strange Effects of the Effluvia of Glass, Produceable on the Motion and Attrition of It," *Philosophical Transactions* 25 (1706–7c): 2377.

in the future. Of course, there were a great many additional questions to be studied and answered. Historians of science have wanted to determine exactly when Hauksbee recognized that what he had accomplished was electrification and not described in some other, more obscure terms. For us, it is enough to recognize that Hauksbee had taken a giant step forward in the understanding and use of electricity. He had varied the amount of air in the globe of his electrostatic machine, finding that the glow increased in intensity as the amount of air decreased; and, conversely, the level of attraction between the surface of the globe and experimental brass leaf inside dropped as the air decreased. Electrification took place at the expense of attraction.[37] In the experiment with the second tube, Hauksbee had kept the maximum amount of air in the globe of his generating machine because that increased the charge in that device and hence had greater power to influence the long, thin tube. Earlier in these same experiments, Hauksbee had discovered that as air was let into the globe, the charge became so great that anyone bringing a finger within inches of the apparatus would feel an electric charge. Within a decade, this ability to "shock" unsuspecting others became a parlor game all over Europe and America.

This sketch should not be taken to suggest that Hauksbee was the only pioneer making these advances at the time. Indeed, Stephen Gray, in a famously unpublished letter to the secretary of the Royal Society in 1708, had both replicated many of Hauksbee's results and gone beyond them.[38] In the next few years, he would demonstrate that electric charges could be transmitted over hundreds of feet. Electric transmission had been discovered.

The evidence of the distinct differences between magnetic and electric charges identified by Gilbert in 1600 had now been turned into a rich field

---

[37] Hauksbee, "An Account of an Experiment before the Royal Society at Gresham College, Touching the Extraordinary Elistricity [*sic*] of Glass, Produceable on a Smart Attrition of It; with a Continuation of Experiments on the Same Subject," *Philosophical Transactions* 25 (1706–7a): 2327–35, esp. p. 2330. In this experiment, Hauksbee had only to wave his hand to transfer the electric glow to another evacuated tube:

Now the attrition [rubbing] of the tube being made in the dark, it was very observable that when the glass became warm, a light would continually follow the motion of the hand, backward and forward; and at the same time, if another hand was held near the tube, a light would be seen to break from it with noise, much like that of a green leaf in the fire, for smartness, but nothing so loud: although when the experiment has been very silently made, I have heard several cracks at 7 or 8 feet distance or more.

[38] The full letter with commentary is in R. A. Chipman, "Unpublished Letter of Stephen Gray on Electrical Experiments, 1707–08," *Isis* 45, no. 1 (1954): 33–40.

of experimental studies. For theoretically inclined minds, especially Newton, the new challenge was to imagine how the forces of mechanics, the phenomena of optics, and electric and magnetic forces could be brought into a unified theory. As we shall see, Newton did have something to say on that subject.

10

Prelude to the Grand Synthesis

Before we look at the new synthesis of astronomy, the science of mechanics, and other forces, we should recall the scientific context outside Europe, especially in the Muslim world, regarding astronomy and the science of motion.

Earlier, in Chapter 5, I outlined developments in optics, astronomy, and the science of motion in the Muslim world up to the end of the seventeenth century. We saw that when the telescope arrived in Mughal India (1615), in the Ottoman Empire (ca. 1630), and the broader Middle East, there was no response triggering an upsurge in astronomical activity. No new telescopes were designed, no new observatories were built, and no new astronomical observations were compiled using the telescope.

Most important of all, there was no revolution in astronomy, the great metaphysical shift to a heliocentric system. This is the more puzzling when, as some writers have suggested, Copernicus was the last of the great astronomers in the Arab-Muslim tradition of astronomy, attempting to work out the defects of the Ptolemaic system.[1] Even more poignant is the fact that the Copernican system is at least partially "a formal transformation of the Ptolemaic theory."[2] For although Copernicus brought forth the heliocentric system by such a transformation, the Arab-Muslim astronomers not only did not arrive at the new system but also either

[1] Noel Swerdlow and Otto Neugebauer, *Mathematical Astronomy in Copernicus's 'De Revolutionibus'"* (New York: Springer, 1984), p. 295.

[2] O. Neugebauer, "On the Planetary Theory of Copernicus," *Vistas in Astronomy* 10 (1968): 103ff.

rejected or ignored it for centuries thereafter, all the way through the writings of Copernicus, Galileo, Kepler, and Newton.[3]

Breaking with the Islamic Past

There were a number of philosophical problems in the Ptolemaic system. The first was the fact that the planetary models composed of circles and epicycles did not preserve the Aristotelian principle of uniform circular motion. Second, although it was not a defect, the planetary motions were irregular, and sometimes the planets appeared to go backward – that is, to retrogress. Those movements could not be explained by the philosophical assumptions of the Ptolemaic system that expected uniform and circular motion to prevail throughout. Yet, it is true that seen from the earth, the planets do go backward because the earth overtakes other planets as it revolves around the sun.

Another problem was that the centers of Ptolemy's eccentric circles were not located at the geometric center but at a point off center called the *equant*. This was the center around which uniform motion was maintained, but it was not the geometric center. The leading Muslim astronomers, especially in the twelfth century and later, labored assiduously to remove Ptolemy's equant and to preserve uniform circular motion. The closest that the Maragha school came to achieving those objectives was the work of Ibn al-Shatir, the time keeper in the mosque in Damascus, who died in 1375. Some aspects of Ibn al-Shatir's models were incorporated in some of the Copernican models, though no one has shown a path of transmission between Ibn al-Shatir and Copernicus.

But removing the equant (the off-center point of Ptolemy's models) did not solve the great problem of modern astronomy and it had, after all, been invented to establish a point around which uniform motion was preserved. The central problem was to restore uniform circular motion and to arrange the planets in a coherent relational pattern. From the time of Ptolemy and before, there were separate models for the so-called superior planets (Mars, Jupiter, and Saturn) and another set of models for the inferior planets (Venus and Mercury), the whole lacking a systematic arrangement.[4] Moreover, the variety of mechanisms, deferents, epicycles,

[3] Ekmeleddin Ihsanoglu, "The Introduction of Western Science to the Ottoman World: A Case Study of Modern Astronomy (1600–1800)," and other essays in his *Science, Technology, and Learning in the Ottoman Empire* (Aldershot, UK: Ashgate, 2003).

[4] Among others, see Noel Swerdlow, "Astronomy in the Renaissance," in *Astronomy before the Telescope*, ed. Christopher Walker (London: British Museum), pp. 187–230; Olaf Pedersen, "Astronomy," in *Science in the Middle Ages*, ed. David C. Lindberg (Chicago:

and homocentric models used by others led Copernicus to claim that the astronomical ensemble was a horror,

just like someone including in a picture hands, feet, head, and other limbs from different places, well painted indeed, but not modeled from the same body, and not in the least matching each another, so that a monster would be produced from them rather than a man.[5]

With regard to the equant problem, one might be able to devise astronomical models similar to those of Ibn al-Shatir that remove the equant and yet the heliocentric orientation would be missing. Likewise, the coordinated system that Copernicus discovered establishing a standard relationship between the distances of the planets from the sun and the periods of their revolution around the sun was missing in Ibn al-Shatir's equant-free models. The ultimate form of any new system had to be heliocentric, and that was nowhere to be found in Islamic astronomy.

Equally important, the physical basis of the whole system, the actual principles of physics or natural philosophy, had to be discovered, and they were missing in all Middle Eastern astronomical thought.[6] In actual fact, it was the genius of Kepler that led him to first propose the idea of a "celestial physics based on causes," but this leap could only have been made – as it was by Kepler – on the assumption that the sun is the approximate center and motive force of our universe.

So we should recall that throughout this period of late medieval and early Renaissance history, there was no coherent picture of the whole system, and astronomers were simply mathematical model builders: they were not supposed to opine about the real structure of the universe; that was the job of natural philosophers. Copernicus broke that assumption by proposing a new astronomical system, yet he said nothing about the physical laws that might govern this new conception. He found an amazing correspondence between the size of a planet's orbit and the period it took to complete a circuit around the sun but remained silent about the forces holding the system together.

University of Chicago Press, 1978), pp. 303–37; and William Donahue, "Astronomy," in *The Cambridge History of Science*, vol. 3, *Early Modern Science*, eds. Katherine Park and Lorraine Daston (New York: Cambridge University Press, 2006), pp. 564ff.

5 *Copernicus on the Revolutions of the Heavenly Spheres: A New Translation from the Latin by A. M. Duncan* (North Vancouver, BC: David and Charles, 1976), p. 25.

6 For a discussion of the nature of the study of the heavens (*hay'a*) in the Arab-Muslim tradition, see A. I. Sabra, "Configuring the Universe: Aporetic, Problem Solving, and Kinematic Modeling as Themes in Arabic Astronomy," *Perspective on Science* 6, no. 3 (1999): 288–330; George Saliba, "Arabic versus Greek Astronomy: A Debate over the Foundations of Science," *Perspectives on Science* 8, no. 4 (2001): 328–41; and A. I. Sabra, "Reply to Saliba," *Perspectives on Science* 8, no. 4 (2001): 342–45.

Although we still do not know the exact route by which Copernicus arrived at his heliocentric view, the central achievement of this revolution involved arranging the planets around the sun in eccentric circles. Mathematically considered, that meant transforming planetary epicyclic models centered on the earth to eccentric circles centered near the sun. In large measure, this meant transforming the Ptolemaic system into a heliocentric system, with many features of Ptolemy's system still intact. The mechanism for that transformation was found in the work of Regiomontanus, the great astronomer who also systematized for Europeans the Middle Eastern heritage of trigonometry.

Between 1510 and 1514, Copernicus drafted his "little commentary," that is, the *Commentariolas*, which first set out the heliocentric worldview. Using the parallelogram device that geometrically changes epicyclical models into eccentric models, he was able to reorient the system so that the planets revolve around the sun, as did the earth (Figure 10.1). In a critical as well as technical sense, the insight that the earth is a planet that revolves around the sun had to be the starting point for Copernicus's "proofs" that if the other planets were arranged in a heliocentric pattern based on the earth-sun distance as a basic unit, then indeed the whole system would conform to the mathematical parameters derived from the planetary tables that all astronomers in Europe and the Middle East used but were transposed into heliocentric orbits. These were the so-called Alphonsine tables constructed in Spain between 1252 and 1270.[7] Furthermore, the apparent retrograde motion of the planets as seen from the earth could easily be explained by the motion of the earth as it overtakes slower-moving planets in their circuits around the sun. The tables were, however, entirely based on the parameters used by Ptolemy centuries earlier.[8]

Working with this earth–sun standard, Copernicus discovered that the distance of each of the planets from the sun corresponded to the length of time that each planet takes to make a complete circuit around the sun.[9] The farther from the sun each planet is, the longer it takes for it to make a complete cycle. For Saturn, the farthest out in the hierarchy, it took thirty years for a complete revolution around the sun; for Mercury, the closest to the sun, it took but eighty-eight days. This arrangement also resulted in a vastly larger universe than had been imagined before Copernicus.

---

[7] See Alfonsine tables: http://en.wikipedia.org/wiki/Alfonsine_tables.
[8] See note 33.
[9] This is what Copernicus said and has been pointed out by a number of historians of science, including Edward Rosen, *Three Copernican Treaties* (New York: Octagon Books, 1959).

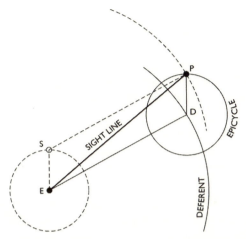

FIGURE 10.1. The geometrical transformation device suggested by Regiomontanus and used by Copernicus. In this diagram, angles EDP and ESP are equal so that the same observational results are obtained regarding the orbit of planet P at point E or S – that is, a geocentric or solar-centric point. From Owen Gingerich and James MacLachlan, *Nicolaus Copernicus. Making the Earth a Planet* (New York: Oxford University Press, 2005), p. 58; with permission. A more detailed account of this transformation appears on pp. 58–59 of that work.

This outcome was indeed pleasing to Copernicus and brought the whole system together as never before. As Copernicus wrote:

We find, then, in this arrangement the marvelous commensurability of the universe, and a sure linking together in harmony of the motion and the size of the spheres, such as could be perceived in no other way.[10]

At last our solar system could, within reasonable limits of error, be represented as a set of nested eccentric circles centered slightly off center of the sun. There is no hint of such a reorientation of the planets in such a coordinated fashion around the sun in the work of Ibn al-Shatir, 'Ali al-Qushji, or any other Muslim astronomer whose writings have been studied by historians of Islamic science.[11]

There is another feature of Copernicus's deep understanding of our universe that should be noticed. That appears in his very first postulate

---

[10] *Copernicus on the Revolutions of the Heavenly Spheres*, p. 50.
[11] The similarities between Ibn al-Shatir's geocentric and Copernicus's heliocentric model (that included the Tusi couple) have been known for more than a half-century, since Victor Roberts pointed it out. Also see E. S. Kennedy and Roberts, "The Planetary Theory of Ibn al-Shatir," *Isis* 50 (1959): 227–35.

of the *Commentariolus*: "there is no one center of the celestial spheres" and, consequently, the "center of the universe is *near* the sun" (emphasis added), not directly at the sun.[12] This vague fact represents something that Copernicus could not have known and only Kepler discovered nearly a century after Copernicus's innovation: that the orbits of the planets are ellipses with one focus at the sun. But that is a development for the next chapter.

In the meantime, it now seems certain that the transformation device that Copernicus adopted from Regiomontanus to establish eccentric circles with a heliocentric orientation had been found earlier by the Muslim astronomer ʿAli Qushji, of whom we have spoken.[13] In a very short commentary, ʿAli al-Qushji worked out the same proof of the equivalence of the epicyclic and eccentric models for the lower planets. But, as noted, neither he nor any other Muslim astronomer has been found to champion the heliocentric view, though ʿAli al-Qushji maintained that there is a *possibility* that the earth moves.[14] Very strong condemnations of the view that the earth moves (which is necessitated by the heliocentric view) were regularly heard in the Muslim world long after Copernicus, Galileo, and Kepler.[15] Furthermore, ʿAli al-Qushji wanted to substitute the will of God for the regularities of nature and the principles of natural philosophy.[16] He clearly had a different orientation than Copernicus, who worked out his heliocentric system within about thirty-five years of ʿAli Qushji's death in 1474.

What all this suggests is that astronomers in the Arab-Muslim world had all the technical and mathematical tools available to them to move astronomy forward to the heliocentric system, the one that in fact became foundational for subsequent astronomy and physics. It was the foundation for the construction of the modern, unified science of celestial and

---

[12] Swerdlow, "The Derivation and First Draft of Copernicus's Planetary Theory," *Proceedings of the American Philosophical Society* 117 (1973): 436.

[13] See F. Jamil Ragep, "ʿAli Qushji and Regiomontanus: Eccentric Transformations and Copernican Revolutions," *Journal for the History of Astronomy* 36, pt. 4 (2005): 359–71. Neugebauer, however, refers to the "familiar parallelogram" as something "known to all astronomers at least since Apollonius"; Neugebauer, "On the Planetary Theory," p. 93. This suggests that it was not first invented either by al-ʿUrdi, as claimed by Saliba, *Islamic Astronomy*, pp. 152–54, or ʿAli al-Qushji.

[14] Ihsan Fazoglu, "Qushji: Abu al-Qasim ʿAli ibn Muhammad Qushci-zade," *Bibliographical Encyclopedia of Astronomers* 2 (2007): 946–48.

[15] Anton M. Heinen, *Islamic Cosmology: A Study of As-Suyuti's "al-Hayʾa as-saniyafiʾ l-hayʾa as-suniya"* (Beirut: Franz Steiner, 1982).

[16] See the discussion in Chapter 5 and Fazoglu, "Qushji."

terrestrial motion worked out by Newton. The Muslim world, however, stopped far short of that, and as we have seen with regard to the arrival of the telescope all across the Muslim world in the seventeenth century, no burst of intellectual curiosity followed its arrival.

Given this sketch of the Renaissance renewal in Europe, it is a stretch to suggest that Arabic-Islamic astronomy was anything more than a broad background condition leading to the development of modern Western astronomy and the scientific revolution.[17] Copernicus's great innovation was entirely his, and when we move on to Galileo, Kepler, Newton, and all the others in between, there is little evidence of an Islamic astronomical influence. Indeed, we have seen earlier that outside the science of astronomy, there was a great poverty of scientific advancement there from the thirteenth century onward. In earlier times, the natural philosophy of the Greeks was greatly enriched by the work of Arab and Muslim scholars, and that tradition was embedded in the European universities founded in the twelfth century and later. In contrast, that Aristotelian tradition of natural philosophy so vigorously embraced by medieval European scholars was intentionally excluded from the madrasas. Consequently, when we get to the modern scientific developments of the sixteenth and seventeenth centuries, the influence of non-Western sources is virtually absent.

Mathematics from the Middle East

Nevertheless, the most plausible case for an Arab-Muslim *facilitating* influence can be found in mathematics. The Hindu-Arabic numeral system was invented in India sometime before or during the seventh century and transmitted to Syria in 662 by Severus Sebokt. Soon thereafter, the system of ten numerals, including a zero, was adopted by the Persian scholar al-Khwarizmi around 825.[18] It came to Europe in the late tenth century. Slowly, over several hundred years, it was adopted for

[17] The most recent attempted defense of an Islamic influence on Renaissance Europe is in George Saliba, *Islamic Science and the Making of the European Renaissance* (Cambridge, MA: MIT Press, 2007). My review of it is in *The Middle East Quarterly* 15, no. 4 (2008): 77–79, http://www.meforum.org/2005/islamic-science-and-the-making-of-the-european.

[18] Paul Kunitzsch, "The Transmission of Hindu-Arabic Numerals Reconsidered," in *The Enterprise of Science in Islam*, ed. Jan P. Hogendijk and A. I. Sabra (Cambridge, MA: MIT Press, 2003), pp. 3–21. He writes, "All the oriental testimonies speak in favor of this line of transmission, beginning from Severus Sebokt in 662 through the Arabic-Islamic arithmeticians themselves and to Muslim historians and other writers" (p. 4). For earlier discussions of these origins, see Louis C. Karpinski, "The Hindu-Arabic Numerals," *Science* 35, no. 900 (1912): 969–70; Florian Cajori, "The Controversy on the Origins

use by Europeans, although a mix of Roman and Arabic numerals continued to be used through the time of Copernicus and, indeed, in his *De Revolutionibus*.[19] However, the prime operators of plus and minus (+, −) were added to the system in the fifteenth century by Europeans, and the equal sign (=) was introduced by Robert Recorde around 1557.[20] Because Copernicus did not know the modern algebraic notation, his "notation and expressions are entirely geometrical, following the Greek sources."[21] Likewise, though Newton had invented some elements of the calculus in the 1660s, his full system did not emerge until after his *Principia* had been published. Consequently, his masterwork is a work in geometry, not algebraic calculation, and contains no equal signs.[22]

of Our Numerals," *Scientific Monthly* 9, no. 5 (1919): 458–64; and A. I. Sabra, "Ilm al-Hisab," *Encyclopaedia of Islam*, 2nd ed., 3 (1979): 1138–41.

[19] I was alerted to this practice by Owen Gingerich. A. M. Duncan, the translator of a recent edition of *De Revolutionibus*, points out that Copernicus used Latin (i.e., Roman) numerals for angles and dates, numbers of years and days, but Arabic numerals (which he called "Indian") for lengths of measurement; Duncan in *Copernicus on the Revolutions of the Heavenly Spheres*, p. 320n33. Similarly, Florian Cajori remarked on the continuing mix of Arabic and Roman numerals in written works in the sixteenth and seventeenth centuries, and even more examples in Spanish writings up to the nineteenth century. See Florian Cajori, in a review of *The Hindu-Arabic Numerals* by David Eugene Smith and Louis Charles Karpinski (Boston: Ginn, 1911) in *Science* 35, no. 900 (1912): 503. Also see Richard Lemay, "The Hispanic Origin of Our Present Numeral Forms," *Viator* 8 (1977): 435–59.

[20] For some of this background, see Florian Cajori, *A History of Mathematical Notation* (La Salle, IL: Open Court, 1928), pp. 1:128, 1:230–31, 1:235. For a table listing the dates of the first European use of the four basic operators (+, −, ×, ÷, or addition, subtraction, multiplication, and division, respectively), see Frank J. Swetz, *Capitalism and Arithmetic: The New Math of the 15th Century* (La Salle, IL: Open Court Press, 1987). For a useful review of the European reception of this, see Alfred W. Crosby, *The Measure of Reality: Quantification and Western Society, 1150–1600* (New York: Cambridge University Press, 1994), chap. 6.

[21] *Copernicus on the Revolutions of the Heavenly Spheres*, p. 321n34.

[22] Many recent scholars have deflated the calculus myths as regards the *Principia*; see A. R. Hall, *Isaac Newton: Adventurer in Thought* (New York: Cambridge University Press, 1992), pp. 212–13; I. B. Cohen, "A Guide to Newton's *Principia*," in I. B. Cohen and Anne Whitman, *The Mathematical Principles of Natural Philosophy: A New Translation* (Berkeley: University of California Press, 1999), pp. 49, 50, and sect. 5.4, pp. 114–15, where the question of whether the *Principia* was written in "the manner of Greek Geometry" is discussed. For an analysis of Newton's revolutionary insights in the use of geometry to measure *force*, see Francois de Gandt, *Force and Geometry in Newton's Principia*, trans. Curtis Wilson (Princeton, NJ: Princeton University Press, 1995); J. Bruce Brackenridge and Michael Nauenberg, "Curvature in Newton's Dynamics," in *The Cambridge Companion to Newton*, eds. I. B. Cohen and George E. Smith (New York: Cambridge University Press, 2002), 85–137. For ways in which Newton's mathematics went beyond the classic geometric principles, see D. T. Whiteside, "The Mathematical Principles Underlying Newton's *Principia Mathematica*," *Journal of the History of Astronomy* 1 (1970): 116–38.

At the same time, Arab and Muslim mathematicians made significant contributions to the development of trigonometry. The consensus among historians of Arabic-Islamic science is that the Arabs "were indisputably the founders of plane and spherical trigonometry, which properly speaking, did not exist among the Greeks."[23] Likewise, E. S. Kennedy agrees that trigonometry, the study of the plane and spherical triangle, "was essentially a creation of Arabic-writing scientists."[24] What this meant was that calculations of the sizes and dimensions of planes and spheres, triangles, and related figures were greatly simplified. By the tenth century, the mathematical functions known as the sine, cosine, and tangent, along with their tables of application, had been invented in the Middle East. With the great translation movement of the eleventh and twelfth centuries, these mathematical advances were brought to Europe, along with translations of Ptolemy's astronomy, known as the "greatest book," the *Almagest*, by the Arabs. Even so, just when and from whom Europeans learned of the rich details of trigonometry remains somewhat obscure.[25]

The most important European consolidator and transmitter of the new trigonometry was the German mathematician and printer Regiomontanus (also known as Johann Müller, 1436–76), discussed earlier. He was educated and later taught mathematics at the University of Vienna, though he also traveled around Europe a great deal. But it was his teacher, Georg Peurbach (1423–61), who launched the detailed study and assimilation of Ptolemy's *Almagest*. Unfortunately, Peurbach died at the young age of thirty-eight, passing on to Regiomontanus the first six chapters of the work that would become the most important European work in astronomy of the Renaissance, the *Epitome of the Almagest*.[26] However, the

---

[23] Carra de Vaux, "Astronomy and Mathematics," in *The Legacy of Islam*, 1st ed. (Oxford: Oxford University Press, 1955), p. 276.

[24] E. S. Kennedy, "The Arabic Heritage in the Exact Sciences," *Al-Abhath* 23 (1970): 337. Also see Kennedy, "The History of Trigonometry: An Overview," in *Studies in the Islamic Exact Sciences*, eds. E. S. Kennedy et al. (Beirut: American University of Beirut Press, 1983), pp. 3–29.

[25] Among the familiar Arabic or Persian names whose trigonometric works were studied by Europeans, we find al-Battani, al-Farghani, al-Biruni, Abu l'Wafa, Az-Zaqali, and Nasir al-Din al-Tusi. See John David Bond, "The Development of Trigonometric Methods down to the Close of the XVth Century," *Isis* 4, no. 2 (1921): 295–323; Barnabus Hughes, Introduction to *Regiomontanus on Triangles*, trans. Barnabus Hughes (Madison: University of Wisconsin Press, 1967), pp. 3–13; and J. L. Berggren, "Trigonometry in the Islamic World," in *Episodes in the Mathematics of Medieval Islam* (New York: Springer, 2003), pp. 127–53.

[26] C. Doris Hellman and Noel Swerdlow, "Peurbach," *DSB* 15 (2008): 473–79; Michael Shank, "Regiomontanus," *DSB* 11 (2008): 216–19; and "Regiomontanus," http://www-history.mcs.st-andrews.ac.uk/Biographies/Regiomontanus.html.

Epitome was not a simple translation of Ptolemy's masterwork but rather a very high-level commentary and exposition of the fundamentals of the Ptolemaic system that Peurbach and Regiomontanus deeply explored. It remained the most widely used work on Ptolemaic astronomy throughout this era. Accordingly, Regiomontanus's *Epitome* (finished in 1461 or 1463 but not printed until 1496[27]) was a work that Copernicus studied carefully.

What is also significant in the present context is the fact that Regiomontanus wrote a work on triangles (completed in 1464 but not published until 1533 in Nuremberg) that consolidated a large portion of what was known of the trigonometry that had been brought to Europe through various routes since the twelfth-century Renaissance.[28] That work also influenced Copernicus, who introduced trigonometric material from *De Triangulis* into his *De Revolutionibus*.[29] Although this facilitating role is to be seen in the work of Copernicus, it does not give us the key to the latter's great leap to the heliocentric worldview discussed earlier.

A great deal of ink has been spilt with regard to the "Tusi couple," the "crank mechanism" of the thirteenth-century Persian astronomer Nasir al-Din al-Tusi. It was invented to generate a straight-line motion from a pair of embedded circles. The similarities between al-Tusi's couple and a diagram in Copernicus's book focusing mainly on the lettering was first pointed out by Willy Hartner.[30] Although this mechanism appears in Copernicus's book, no one has shown that the mechanism had anything to do with Copernicus's path to heliocentrism. Furthermore, there were other non-Arab sources from which Copernicus could have derived his model. There were two earlier sources of the reciprocating device, one from Proclus (a fifth-century A.D. Greek philosopher) and the other from

---

[27] Michael H. Shank, "Regiomontanus, Johannes," *DSB* 11 (2008): 216.

[28] Hughes, *Regiomontanus on Triangles*.

[29] Book I, chap. 12, of *De Revolutionibus* contains an explanation of the method of calculating the sines of angles.

[30] Willy Hartner, "Copernicus, the Man, the Work, and His Achievement," *Proceedings of the American Philosophical Society* 117, no. 6 (1973): 413–22. For the long debate, see the following: Noel Swerdlow, "The Derivation and First Draft of Copernicus's Planetary Theory"; I. N. Veselovsky, "Copernicus and Nasir al-Din al-Tusi," *Journal for the History of Astronomy* 4 (1973): 128–30; Mario Di Bono, "Copernicus, Amico, Fracastoro and Tusi's Device: Observations on the Use and Transmission of a Model," *Journal for the History Astronomy* 26 (1995): 133–54; Jerzy Dobrzycki and Richard L. Kremer, "Peurbach and Maragha Astronomy? The Ephemerides of Johannes Angelus and Their Implications," *Journal for the History of Astronomy* 27 (1996): 187–237; F. Jamil Ragep, "'Ali Qushji and Regiomontanus: Eccentric Transformations and Copernican Revolutions," *Journal for the History of Astronomy* 36, pt. 4 (2005): 359–71; Victor Roberts, "The Planetary Theory of Ibn al-Shatir: A Pre-Copernican Copernican Model," *Isis* 48 (1957): 428–32; Swerdlow and Neugebauer, *Mathematical Astronomy*.

Eudoxus (a slightly older Greek astronomer-mathematician), that were known in Europe in Copernicus's time.[31]

Perhaps more important, historians of Islamic science have pointed out that there are a number of similarities between Ibn al-Shatir's models and those of Copernicus. Although that is true, Ibn al-Shatir's models are located in a geocentric framework, whereas Copernicus had transformed the whole system into a heliocentric arrangement. Furthermore, during the more than four decades since the similarities between Ibn al-Shatir's models and those of Copernicus have been known, no documentary evidence of borrowing has been found.

What has apparently been forgotten in these discussions is the fact that since the time of Ptolemy, astronomers have worked with the same mathematical tools and the same Ptolemaic parameters for planetary orbits. Here again, the historian of mathematics, Otto Neugebauer, stressed the point that comparing the tables of parameters for the five planets (Saturn, Jupiter, Mars, Venus, and Mercury) in Ptolemy and Copernicus, "the Copernican tables will produce practically the same results as the Ptolemaic ones."[32] Essentially, no new observations were added, even in the time of the reconstruction of the Alfonsine tables (ca. 1252–70). Although Copernicus did make some observations of his own, in addition to eclipses, this was done mainly to confirm the accepted values from sources such as the Alfonsine tables. Otherwise, the identity of the Alfonsine and Ptolemaic parameters has been shown.[33] Consequently, the parameters for all the planets, not just Mercury and the moon, were based on Ptolemaic values. From this, it should not be surprising that models of planetary orbits of Copernicus and Ibn al-Shatir converged and were sometimes identical.

[31] Indeed, I. N. Veselovsky believes that the device Copernicus used was derived from Proclus because a passage in Copernicus's *De Revolutionibus* is found in Proclus. See Veselovsky, "Copernicus and Nasir al-Din al-Tusi." For illustrations from al-Tusi's work, see http://en.wikipedia.org/wiki/Tusi-couple and also "Tusi's couple" under "images" in Google.

[32] Neugebauer, "On the Planetary Theory," pp. 92, 97.

[33] The identity of the tabular data in Ptolemy and the Alfonsine tables has been tested and reported by Owen Gingerich, "'Crisis' versus Aesthetic in the Copernican Revolution," in *Copernicus: Yesterday and Today*, Vistas in Astronomy, 17, eds. Arthur Beer and K. A. Strand (Oxford: Pergamon Press, 1975), p. 88; and in *The Book That Nobody Read* (New York: Walker, 2004), p. 57. For discussions of the exact origins of the tables, see Emmanuel Poulle, "The Alfonsine Tables and Alfonso X of Castile," *Journal for the History of Astronomy* 29 (1988): 97–113; and José Chabás and Bernard Goldstein, *The Alfonsine Tables of Toledo* (Boston: Kluwer Academic, 2003). Also see Owen Gingerich, "Commentary: Remarks on Copernicus's Observations," in *The Copernican Achievement*, ed. Robert S. Westman (Berkeley: University of California Press, 1975), pp. 99–107.

If we assume that Copernicus was tackling the same problems in astronomy as the Arab and Persian astronomers in thirteenth-century Maragha, was pursuing the same astronomical objectives, and was using the same methods and the same data (from the Ptolemaic-Alfonsine tables), "it is by no means remarkable," the historian of astronomy Mario Di Bono has suggested, that Copernicus "obtains results very similar to those of his predecessors."[34] Furthermore, the "reciprocation device," often referred to as the Tusi couple, "could equally well have been derived from an independent reflection on these same problems."[35] Otherwise, the orientations of the two systems – one heliocentric and one geocentric – are radically different. In the meantime, no one has shown that Copernicus actually had access to Arabic documents (or others) that contained the Tusi mechanism or Ibn al-Shatir's models. Likewise, no one has shown that having access (hypothetically) to this material had anything to do with Copernicus's thinking that led him to his heliocentric models.

Astronomy and the Science of Motion

This brings us finally to the centrality of mechanics and the science of motion, without which the great leap to Newton's synthesis would not have been possible. Physics and the science of motion had been central to Aristotelian physics since before Aristotle's time. Conversely, astronomy in Aristotle's conception of the sciences was located among the mathematical, not the natural, sciences. It was for that reason that astronomy was seen as a mathematical model-building enterprise, not as a natural science legislating the real shape of the world. What that meant is that in the great long run of scientific inquiry leading to the modern scientific revolution, those two sciences had to be brought together, as they ultimately were in Newton's *Mathematical Principles of Natural Philosophy*.[36]

In the extended interval before that accomplishment, progress had to be made in the science of mechanics as well as astronomy. In the Muslim world, very limited progress was made throughout the so-called golden age. It ended altogether in the early twelfth century. The last significant contributor to the science of motion was Ibn Bajja, the Andalusian Muslim philosopher of the early twelfth century. There are indications that

---

[34] Di Bono, "Copernicus, Amico, Fracastoro and Tusi's Device," p. 147. Indeed, George Saliba produced a diagram for the upper planets showing the same point for planet P in Ptolemy, al-Urdi, Ibn al-Shatir, Copernicus, and al-Khafri; see Saliba, *Islamic Science*, p. 206, figure 6.6.

[35] Di Bono, ibid.

[36] This broad line of development is very nicely traced out by Edward Grant, *A History of Natural Philosophy* (New York: Cambridge University Press, 2007).

some of his thinking did aid medieval Europeans.[37] But after him, historians of science have failed to find any successors of his caliber. Neither the Ottomans nor the Mughals wrote significant treatises on the science of motion in the years between Ibn Bajja and Galileo or Kepler.[38]

Consequently, neither the Mughals, Ottomans, nor Persians were on the path to the Newtonian synthesis. That meant that for peoples without high levels of achievement in the science of motion as well as astronomy, the way to the modern scientific conception of a unified celestial and terrestrial science was blocked. Jumping over all the missing ingredients that flowed uniquely from the work of Copernicus, Galileo, Kepler, Descartes, and Newton to modern science does not seem a realistic possibility. Discontinuities within delimited fields of science are possible, but great leaps of the sort considered here are no more possible than leaping to microparticle physics without first discovering electric charge and electricity.

China's Deficits

Those who look for clues regarding possible Chinese influences are likewise short of evidence. Joseph Needham pointed out with regard to physics and the science of motion that there were no Chinese scholars equivalent to such leading figures in the West as Philoponus, Jean Buridan, Thomas Bradwardine, or Nichole d'Oresme[39] and, hence, no

---

[37] For an analysis of Ibn Bajja's work and its influence on medieval European natural philosophers, see Marshall Clagett, *The Science of Mechanics in the Middle Ages* (Madison: University of Wisconsin Press, 1959); and the essays of Ernest Moody, "Laws of Motion in Medieval Physics," *Scientific Monthly* 72 (1951): 18–23; Moody, "Galileo and Avempace: The Dynamics of the Leaning Tower Experiment," in *Roots of Scientific Thought*, eds. Philip P. Wiener and Aaron Noland (New York: Basic Books, 1957), pp. 176–206; as well as John E. Murdoch and Edith D. Sylla, "The Science of Motion," in *Science in the Middle Ages*, ed. David C. Lindberg (Chicago: University of Chicago Press, 1978), pp. 206–64.

[38] A major reason for this is that none of Aristotle's natural books on physics and the science of motion were taught in the madrasas. For an overview of Ottoman science, see Ekmeleddin Ihsanoglu, "The Introduction of Western Science to the Ottoman World: A Case of Study of Modern Astronomy (1600–1800)," in *Science, Technology and Learning in the Ottoman Empire* (Aldershot, UK: Variorum, 2004). Apart from some histories of astronomy in the Mughal Empire in the sixteenth and seventeenth centuries, no studies of fragments of the science of motion or magnetism in India in this period seem to exist. For the history of seventeenth-century astronomy in India, see the references in Chapter 5 and the following: S. M. Razaullah Ansari, "Introduction of Modern Western Astronomy in India during the Eighteenth and Nineteenth Centuries," in *History of Astronomy in India*, eds. S. N. Sen and K. S. Shula (New Delhi: Indian National Science Academy, 1985), and Rajesh Kochhar, "Pre-telescopic Astronomy in India," in *History of Indian Science, Technology, and Culture* AD *1000–1800* (Oxford: Oxford University Press, 2000). For Jai Singh's observatories, see Chapter 5.

[39] Joseph Needham, *SCC*, 4/1:1.

one similar to Ibn Bajja in the Muslim world. Furthermore, the arrival of the telescope in China in 1618, along with Jesuit scientists, failed to elicit major advances in telescopy, astronomy, or the science of mechanics.

The path to the capstone achievement of a new science of mechanics centered in a unified celestial and terrestrial physics evolved out of an Aristotle-based science of mechanics and the tools of Euclidean geometry firmly located in a spherical universe. All these elements were missing in China. As the discussion in Chapter 4 revealed, when Matteo Ricci arrived in China in the 1580s, he discovered that the Chinese still believed in a flat earth and that the whole apparatus of spherical geometry was missing. That was why the first of the scientific books that he chose to translate into Chinese was Sacrobosco's *On the Sphere*. Without that conceptual apparatus, little understanding of the spherical nature of the universe was possible. The second major work was Euclid's *Elements of Geometry*, which was translated by Xu Guangqi and published for the first time in Beijing in 1607. Although various myths have grown up about Newton's use of calculus that he began to construct in the 1660s, his system was not completed until after the appearance of the *Principia Mathematica*; consequently, that work is rooted in a deep tradition of geometry that could hardly be fully understood without knowledge of geometry and, in some cases, would be better understood with the tools of infinitesimal calculus, not to be found outside Europe. Whether it was a mere "convenience," Newton laid out his introductory pages in the *Principia* using the classic Euclidean pattern of definitions, theorems, propositions, axioms, and lemmas.[40] The Chinese had no such tradition and were quite taken by the methods of proof contained in Euclid's *Elements*, as Xu Guangqi exclaimed in his preface to the 1607 edition.[41] Although some scholars associated with the Chinese Bureau of Astronomy knew of this new mathematics, it was not made part of the Chinese educational experience. Verbiest in the 1680s hoped to reform Chinese educational practice with his recommendations for study submitted to the emperor, but the latter refused even the permission to publish the recommendations.[42] In a word, the tools needed to fashion the modern scientific revolution culminating in the work of Newton were entirely absent in China.

---

[40] See note 22.
[41] See Peter M. Engelfriet, *Euclid in China* (Boston: Brill, 1998).
[42] Noël Golvers, "Verbiest's Introduction of Aristoteles Latinus (Coimbra) in China: New Western Evidence," in *The Christian Mission in China in the Verbiest Era: Some Aspects of the Missionary Approach*, ed. N. Golvers (Leuven, Netherlands: Leuven University Press, 1999), pp. 33–51.

I I

The Path to the Grand Synthesis

As astronomy went through its revolutionary transformation from the time of Copernicus to Newton, the ground shifted from mathematical modeling to deep probings of the structures of the universe. We have seen already that seventeenth-century European natural philosophers had stumbled onto the mysterious forces of magnetism and electricity. Solving the problem of the orbits of the planets was not just a mathematical problem based on observational parameters for the seven planets. Sooner or later, astronomers would be released from the confines of geometry to the soaring world of *philosophers* of the universe such as Galileo wished to be. That meant grasping the forces of nature, both large and small.

Philosophers of the Universe

This was to be the new age of cosmology. Inevitably, it required working toward a unified science of terrestrial and celestial physics. Kepler was the first of these new philosophers of the universe to propose a new astronomy based on physical causes, something missing from Copernicus's great work. Yet, even he did not envision a unified terrestrial and celestial physics, as Newton did. He had a grand vision for the shape of astronomy based on physical causes, but just what that meant in Kepler's time, nobody could say. He laid out that vision in an insight from 1605 that was not published until the appearance of his *New Astronomy* of 1609:

I am much occupied with the investigation of the physical causes. My aim in this is to show that the celestial machine is to be likened not to a divine organism

267

but rather to a clockwork...insofar as nearly all the manifold movements are carried out by means of a single, quite simple magnetic force, as in the case of a clockwork all motions [are caused] by a simple weight. Moreover I show how this physical conception is to be presented through calculation and geometry.[1]

To propose a machinelike universe animated by a single force was audacious. Galileo was a committed Copernican, and his extraordinary visual exploration of the heavens using the telescope yielded the discovery of the cratered surface of the moon, the satellites of Jupiter, and the phases of Venus, all of which supported the Copernican hypothesis as he saw it. Yet, he did not have a grander vision of celestial physics beyond the success of the Copernican system.

Nevertheless, the hypothesis of a single force regulating the heavens quickly turned into a real question: Is there a singular force in the universe that holds all the planets in their orbits and propels them around their circuits? This question about the paths of planetary motion was, in fact, a question at the heart of the science of motion; that is, it was a question about how and why all objects on our earth or in the whole universe move the way they do. The Aristotelian view was that objects "naturally" travel in a circular motion but, on the terrestrial level, they do so only as long as they are propelled by an impressed force. Another assumption was that all objects sought their "natural place" in the world, which meant that most objects fell straight toward the center of the earth, whereas light things floated up.

Galileo began with such conceptions in his early years, but his study of freely falling objects, whether dropped from the tower of Pisa or the mast of a ship, took him to the idea that provided a key to the puzzle of celestial motion: inertia. Although Galileo never used that term, he has, rightly, been credited with working it out. For, clearly, he gave an example of it in his *Dialogue Concerning the Two Chief World Systems* of 1632.

Using the character of Salviati as his mouthpiece and Simplicio as the follower of Aristotle and traditional authority, Galileo maneuvers the conversation into what would happen if a heavy object, a stone, were dropped from the mast of a moving ship. This analogy was intended to evoke the idea of a turning earth as proposed by Copernicus. The argument against the Copernican view starts with the assumption that the earth travels around the sun and turns fully around on its own axis

---

[1] Johannes Kepler, as cited in Gerald Holton, "Johannes Kepler's Universe: Its Physics and Metaphysics," in *Thematic Origins of Scientific Thought: Kepler to Einstein* (Cambridge, MA: Harvard University Press, 1973), p. 73.

in twenty-four hours. If it were true, that the earth actually moves that fast, then between the time when a stone was dropped from a tower and landed at the base, the earth would have traveled many feet, perhaps yards, leaving the stone behind. Of course, balls dropped from towers always do land at the foot of towers. But Galileo refused to take this as proof of the immobility of the earth.

He proposed another thought experiment: What if a stone were dropped from the mast of a moving ship – would the heavy object fall, relative to the mast, *behind* as the ship moved forward? In the *Dialogo*, Simplicio agrees with ancient authorities that this must be so. Galileo, still using the language of "impressed forces" that would be imparted to the stone atop the mast, suggests that in fact the ball would keep moving and hence land at the base of the mast, just like the ball in front of the tower on land. Having asked Simplicio if he had ever performed such an experiment and gotten a negative answer, Galileo declares that anyone who performs the experiment will find just the opposite of what traditional authorities averred. Whether or not Galileo had performed the experiment, he declares that such an experiment "will show that the stone always falls in the same place on the ship, whether the ship is standing still or moving with any speed you please."[2] In other words, the stone, just like the ship on which it is carried, has received a force (what we would call momentum) from the ship's movement equal to that of the ship so that when it is released from the mast, it continues to move forward by what we now call *inertia*. So the stone will not fall behind, even though the ship is moving.

The French philosopher René Descartes, in the 1640s, was the first to reformulate the idea of inertia so that it implies that objects continue in their paths unless otherwise diverted. In other words, objects continue in a state of motion (or rest) unless moved by another force or object. For modern physics, this was a very big jump forward, preparing the way for Newton's grand synthesis. That would take another half-century and the work of many scientifically inclined investigators to accomplish. The list of those contributors to the new science of mechanics is much longer than indicated here. It is sufficient for us to recognize that the whole development of the new astronomy and unified mechanics of the seventeenth century was a Europewide movement that included scholars from England, France, Germany, Holland, Italy, Poland, and Scandinavia.

---

[2] Galileo, *Dialogue Concerning the Two Chief World Systems*, trans. Stillman Drake (Berkeley: University of California Press, 1967), p. 144.

Kepler's Platform

Within that broad background of scholars paving the way for the new science, Johannes Kepler's achievements remained a bedrock starting point for Newton's new endeavor that merits special attention. Kepler's contribution to the revolutionary reconstruction of astronomy and the science of motion is remarkable in a number of ways. In the history of astronomy, it was the very first work based on a sun-centered (Copernican) astronomy that employed new astronomical observations while at the same time proposing physical causes that were absent in the Copernican system. The new observations were those meticulously compiled by Tycho Brahe at his Uraniburg observatory off the coast of contemporary Sweden (then controlled by Denmark) between 1584 and 1601, when he died. The most difficult to understand of those observations concerned Mars and were given to Kepler by Brahe to see what order he could bring to them. It took Kepler from 1601 to 1605, during which time he was the imperial mathematician for King Rudolph II in Prague, to bring order to the data. He discovered that the orbit of Mars is elliptical, which was a major departure in the history of astronomy. That discovery of the elliptical orbit was a major blow to the Aristotelian and Ptolemaic systems, which depended on the idea of perfectly circular motion. Kepler assumed but did not prove that all planetary orbits were elliptical. Although Kepler had some followers, not until Newton did elliptical orbits become a cornerstone of modern astronomy. Consequently, the ellipticity of planetary orbits is what we now call Kepler's first law.

Kepler's second law of motion states that a line extending from the sun to a planet sweeps out equal areas in equal times. In Figure 11.1, the departure of Kepler's ellipse from a circular orbit is exaggerated. Nevertheless, one can see the principle of equal areas swept out in equal times in the geometric figures of the diagram. When Newton was prodded to work out the path that a body would follow under an inverse-square rule of attraction, the first order of business was to demonstrate that equal areas *are* traversed in equal times in an ellipse *under any force* impelling a body toward a center (more on which to follow).[3]

But that was not all. Kepler's so-called harmonic law would figure strategically in Newton's new synthesis. According to that rule, the ratio of the distance cubed (from the sun) over the time squared (the interval

---

[3] Newton refers to Kepler's rule throughout the *Principia* as the rule of 3/2 but sometimes stated as the *sesquialterate* rule.

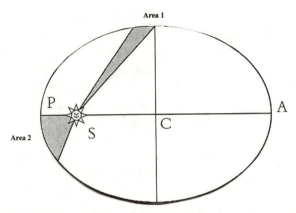

FIGURE 11.1. Kepler's second law of planetary motion, in which equal areas are traversed by a planet in equal times.

required to travel around the sun) was the same for all planets. This was a magical discovery that Kepler apparently stumbled on.[4] But it added immensely to his mystical notions about the harmony of the universe. Surely, one does feel a sense of awe and mystery imagining a universe with the sun at its center and the six planets (Mercury, Venus, Earth, Mars, Jupiter, and Saturn) all revolving around that center and all regulated by this simple rule: the cube of the (mean) distance divided by the square of the period is the same, a constant for all the planets. This discovery was a more mathematically precise formulation than Copernicus's simple but profound discovery that the periods of the planets are proportional to their distances from the sun. Even so, Kepler's rule would prove to be too imprecise in Newton's system, yet it was a bedrock starting point for Newton.[5]

Surprisingly, Kepler did not attempt to demonstrate the rule and failed to find the force of gravity that works (universally) according to the inverse square of the distance between planetary centers. However, a table of values extracted from Kepler for the distances and periods of the planets was published by the astronomer and historian of science, Owen Gingerich, from which one can see that the relationship between distances and times holds rather well.[6]

[4] Owen Gingerich, "Johannes Kepler," in *General History of Astronomy*, v. 2a, edited by R. Taton & C. Wilson (New York: Cambridge University Press, 1989), p. 72.
[5] For a discussion of the path from Kepler's law to Newton's, see Bernard Cohen, *The Newtonian Revolution* (New York: Cambridge University Press, 1980), chap. 5.
[6] Gingerich, Ibid.

Likewise, one cannot resist thinking that there must be a force that causes this perfect harmony of the spheres. To get beyond the limits of his system, Kepler needed the concept of inertia and inertial motion, but the concepts had not yet been published before he died.

By the 1680s, the problem of understanding gravitational attraction and the curves that planets follow when circling the sun under such rules became a widely shared concern across Europe. Galileo discovered that a pendulum takes the same amount of time to swing back and forth no matter how far out its lateral distance might be. Huygens, however, later discovered that beyond certain limits of the swing, the time elapsed was not uniform. In the meantime, Newton found experimentally that no matter what the material concerned (e.g., sand, wood, metal), a swinging pendulum always took the same amount of time to complete its cycle of swings, holding the pendulum length constant.

Likewise, Galileo had demonstrated the law of free fall, and Newton set out to apply it. Already in 1665–66, when Newton was only about twenty-three or twenty-four years of age, he applied the idea to the hypothetical "fall of the moon" as it circles the earth[7]; that is, Newton reasoned that instead of continuing in a straight line, the moon fell toward the earth, thereby remaining in its orbit, and the hypothetical fall from a straight line conformed to his rough calculations of how a body would fall in such a situation. In other words, even at that early date in Newton's career, he apparently assumed that gravitational forces operated between the earth and moon; later, he would definitively define that force as the inverse-square rule of attraction. The attempt to work out all these elements of the great celestial puzzle was a pressing concern among the leading figures of the Royal Society of London. Insofar as Newton's particular interests in this problem were concerned, three encounters drew him out of his self-induced silence into the analysis that produced the grand synthesis. The first of these was his correspondence with Robert Hooke.

The Hooke–Newton Correspondence

Robert Hooke, born in 1635, was undoubtedly one of the most brilliant scientific minds of all times. He was by age and experience nearly two

[7] Among others, see I. B. Cohen, "Newton's Discovery of Gravity," *Scientific American* 244, no. 3 (1984): 166–79, and Cohen, "A Guide to Newton's *Principia*," in Isaac Newton, *The Principia: Mathematical Principles of Natural Philosophy* (Berkeley: University of California Press, 1999), pp. 64–65.

decades Newton's senior.[8] Although the eventual conflict between the two men resulted in Hooke's reputation being tarnished for centuries, he has recently been showered with encomiums. Many now refer to him as "England's Leonardo" or the "man who measured London."[9] He is credited with many original scientific insights in both the life and mechanical sciences. He was a constant inventor, technician, and engineer who made improvements on air pumps, microscopes, and telescopes; fashioned one of the first pocket watches driven by a coiled spring and an escapement mechanism; and experimented with odd dietary prescriptions.[10]

We met Hooke earlier in connection with the publication of his landmark *Micrographia*, the single work most responsible for stimulating a broad interest in microscopy from 1665 onward. It was Hooke who was called on to determine whether Antoni Leeuwenhoek's "little animals" really could be detected in pond water and in the pepper-infected concoction in 1677. But the ubiquitous Mr. Hooke was to be found everywhere, tinkering with air pumps, inventing new instruments, staging experiments for the Royal Society, serving as chief surveyor of London after the Great Fire of 1666, and working out mathematical problems in astronomy. Although he was self-taught in many areas, he received two degrees (MA and MD) from Oxford in about 1662. His acquaintance with Robert Boyle at Oxford led to his work assisting Boyle, constructing air pumps, and conducting experiments. His participation in the circle of natural philosophers who founded the Royal Society made him part of the society from the outset, when he was appointed curator of experiments in 1661.

When Hooke took over the position of secretary to the Royal Society in 1679, he was asked to initiate discussions with Isaac Newton, who was living in Cambridge at the university. Owing to professional criticism of

---

[8] On the dispute between Hooke and Newton, see, among others, Alexandre Koyré, "An Unpublished Letter of Robert Hooke to Isaac Newton," *Isis* 43 (1952): 312–37, reprinted in *Newtonian Studies* (Cambridge, MA: Harvard University Press, 1965), pp. 221–60; A. Rupert Hall, *Isaac Newton: Adventurer in Thought* (New York: Cambridge University Press, 1992), pp. 204ff; Cohen, "Guide to Newton's *Principia*," pp. 15ff; Michael Nauenberg, "Hooke, Orbital Motion, and Newton's *Principia*," *American Journal of Physics* 62, no. 4 (1994): 331–50; Patri J. Pugliese, "Robert Hooke and the Dynamic of Motion in a Curved Path," in *Robert Hooke: New Studies*, eds. Michael Hunter and Simon Schaffer (Woodbridge, UK: Boydell Press, 1989), pp. 181–205.

[9] Lisa Jardine, *The Curious Life of Robert Hooke: The Man Who Measured London* (New York: HarperCollins, 2004); Allan Chapman, *England's Leonardo: Robert Hooke and the Seventeenth-Century Scientific Revolution* (Bristol: Institute of Physics, 2005).

[10] For more on this, see Jardine, *Robert Hooke.*

Newton's revolutionary work in optics, his experiments revealing the multicolored composition of light, Newton became excessively sensitive to scholarly criticism of his writings. This was in 1672 and 1675, when Hooke also joined in the scholarly critique of Newton's optical work. Indeed, Hooke had pressed another of his excessive claims of plagiarism that left a residue of bitterness on Newton's part. It was back then that Newton had written his famous "I Desire to Withdraw" letter announcing his retreat into silence in Cambridge.[11]

Hooke's friendly overture to Newton came in a letter written on November 24, 1679, inviting Newton to rejoin the philosophical discussions in progress at the Royal Society. In the process of extending his invitation, Hooke broached a technical subject concerning the path that an object would describe if it were deflected from its straight path toward a center. Newton responded by discussing the fall of an object under uniform gravity from a great height toward the center of the earth while the latter turned at a uniform rate. This was a technical challenge that Newton fully understood and could resolve, though he and Hooke fell into a dispute over how to interpret a rough diagram that Newton included with his reply. Despite the disputes that came out of the affair, it is fair to say that Hooke's query put Newton onto a new train of thought. Indeed, it seems to have led Newton to radically rethink his notions about physical forces and how they operate.

Unfortunately, the exchange of ideas led eventually to a monumental falling out between Hooke and Newton. It had to do with how much Hooke actually knew about the inverse-square law that became a cornerstone of the new mechanics. Previously, in 1674, Hooke had published a paper that contained three important ideas: the notion that all the planets possess a mutual gravitational attraction; that celestial bodies continue to travel in a straight line unless otherwise constrained; and that gravitational forces are stronger the closer their centers come together.[12] He did not insist on Kepler's law of elliptical orbits, neither did he articulate the inverse-square law, though he came very close.

---

[11] The *Correspondence of Isaac Newton*, eds. H. W. Turnbull, J. F. Scott, A. Rupert Hall, and Laura Tilling (Cambridge: Cambridge University Press, 1959–77), March 8, 1673. This was well explored in Dale Christianson, *In the Presence of the Creator* (New York: Free Press, 1984), chap. 8.

[12] Hooke's lecture, "An Attempt to Prove the Motion of the Earth," is published in G. T. Gunter, *Early Modern Science at Oxford*, vol. 8 (Oxford: Oxford University Press, 1931), esp. pp. 27–28.

Much earlier in his career, Newton had already worked out the case of the moon's attraction to the earth, the principle that the moon's fall toward the earth is a function of the inverse-square rule, though this result was only approximate.[13] But that train of thought began with the assumption that there is a centrifugal force impelling the moon off into space, while the attraction of gravity prevented that outward motion. The mathematics of this centrifugal motion had been worked out by Huygens in 1673, something that remained in Newton's mind. Now, in 1684, Newton was approaching the problem from the point of view of centripetal force – that is, the force causing a moving object to fall toward a center, not flee from it.[14]

Whether Hooke deserves credit for working out a very close approximation to Newton's law that planetary attraction varies inversely with the square of the distance between objects, the task was to prove it, and only Newton was able to accomplish that. Nevertheless, the correspondence with Hooke led to a radical shift in Newton's thinking about the forces of gravity and how they work. Whatever progress Newton made on this subject in the months of 1680 he kept to himself. It required two more events to galvanize Newton's pen into action. One of these was a natural event.

Beware of Comets

A startling celestial event occurred on November 14, 1680.[15] A German astronomer by the name of Gottfried Kirch spotted a new comet, passing surprisingly close to the earth, with the use of a telescope.[16] Soon it would be seen across the world in the northern hemisphere, causing worry and alarm for ordinary folks (Figure 11.2). For laymen, this apparition was an awesome and frightening sight, perhaps suggesting divine displeasure. The first Dutch settlers of Esopus, New York, a small town on the Hudson about 100 miles north of Manhattan, left this record:

On the 9th of December 1680, there appeared an extraordinary comet, which caused very great consternation throughout the province, with forebodings of

---

[13] Cohen, "Guide to Newton's *Principia*," pp. 64–69.

[14] Ibid., pp. 76ff, and Cohen, "Newton's Discovery of Gravity," pp. 170–72.

[15] Eric G. Forbes, "The Comet of 1680–81," in *Standing on the Shoulders of the Giants: A Longer View of Newton and Halley*, ed. Norman J. W. Thrower (Berkeley: University of California Press, 1990), p. 314.

[16] Eric G. Forbes, "The Comet of 1680–81," in Thrower, *Standing on the Shoulders of the Giants*, pp. 312–23.

AN
ALLARM
TO
EUROPE:
By a Late Prodigious
COMET
feen November and December, 1680.

With a Predictive Difcourfe. Together with fome preceding and fome fucceeding Caufes of its fad Effects to the *Eaf* and *North Eaftern* parts of the World.

Namely, *ENGLAND*, *SCOTLAND*, *IRELAND*, *FRANCE*; *SPAIN*, *HOLLAND*, *GERMANY*, *ITALY*, and many other places.

By *John Hill* Phyfitian and Aftrologer.

The Form of the *COMET* with its Blaze or Stream as it was feen *December* the 24th: Anno 1680. In the Evening.

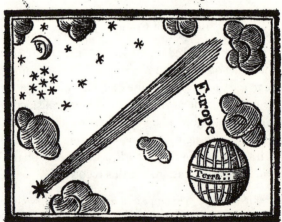

London Printed by *M. Brugis* for *William Thackery* at the Angel in Duck-Lane;

FIGURE 11.2. An engraving in a pamphlet about the comet seen across Europe in November and December 1680. From John Hill, *An allarm to Europe, by a late prodigious comet seen November and December, 1680 with a predictive discourse* (1681).

dreadful happenings and divine punishments. It is described, in a letter dated January 1st, 1681, as having *"appeared in the Southwest on the ninth of December last, about two o'clock in the afternoon, fair sunshine weather, a little above the sun, which takes its course more northerly, and was seen the Sunday night, right after about twilight, with a very fiery tail or streamer in the west, to the great astonishment of all spectators, and is now seen every night with clear weather. Undoubtedly, God threatens us with dreadful punishments if we do not repent."* (emphasis in original)[17]

No doubt many thought prayer and repentance were in order. The sheer size of the comet and its brightness, visible even during daylight, added an extraordinary dimension. As it turned out, either the same or another comet of the same magnitude appeared later in December, causing some to think that there were two comets. Whereas the reaction of the untutored was alarm, natural philosophers in England, especially Edmond Halley, Newton, and others connected to the Royal Society, saw the event as a most puzzling scientific problem. A half-dozen or more new observatories had been built all around Europe since the initial one in Paris in 1667. Most were now equipped with telescopes, along with the newly invented micrometer with cross hairs in the field of vision that permitted more precise astronomical observations than ever before. These technical improvements remained absent in the Ottoman Empire and Mughal India, where no new observatories equipped with telescopes were built until the early nineteenth century. Neither did they make significant modifications to the telescope, which arrived in the late 1620s.

In China, the great observatory in Beijing had been recently re-equipped with models of Tycho Brahe's pretelescopic instruments under the direction of Ferdinand Verbiest, but the Chinese made no advances regarding the design and manufacture of telescopes themselves. Nevertheless, as long ago as A.D. 837, the Chinese had observed the odd phenomenon of the tails of comets pointed away from the sun whether they were approaching or receding from it.[18]

For European astronomers, the question was what kind of path this or other comets follow as they flare through the heavens. The first step was to gather enough accurate observational data on which to base calculations of the comet's path. As Newton later reported in the *Principia*,

[17] From Marius Schoonmaker, *The History of Kingston* (1888), p. 70, as cited in J. Werner, "The Great Comet of 1680 – Kirch's Comet," http://jwwerner.com/history/Comet.html.
[18] Joseph Needham, *SCC*, 3:432. This reversal is caused by solar winds composed of charged particles produced by the sun blowing the gaseous material of the comet ahead of it as it recedes from the sun. This was discovered in the early twentieth century.

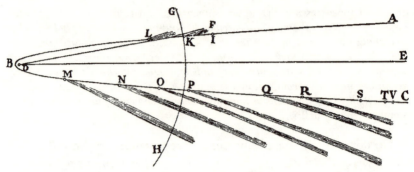

FIGURE 11.3. Newton's diagram of the path of the comet of 1680–81, with its tail preceding the head as it recedes from the sun in the lower half of the diagram. From the *Principia* (1687).

he used a seven-foot telescope equipped with a micrometer to make his own observations of the comet in February 1681 in Cambridge.[19] In the meantime, in late fall 1680, Newton's later collaborator and future eponymous legatee of "Halley's comet" was on his way to Paris when the second sighting of the comet occurred. Consequently, he gathered as much information as he could from other European astronomers on the comet while in Europe. Jean Cassini, the director of the Paris observatory, generously gave Halley a compilation of observations gathered from a number of continental observers that Halley brought back to England, but not until January 1682.[20] By the time Newton put all his thoughts on comets together in the *Principia*, he had a dozen or more observations, some of better accuracy than others. These came from across Europe and even America. He plotted them in a diagram showing their position relative to the sun, with the tail preceding the comet's head as it receded from the sun (Figure 11.3).

To naive and experienced observers alike, the path of a comet could look like a straight line, as it did to Kepler when he published *Three Tracts on Comets* in 1619. Indeed, there were all sorts of theories about the likely trajectory of comets. In 1678, Robert Hooke had published lecture notes that contained questions about comets that seemed to sum

[19] Isaac Newton, *The Mathematical Principles of Natural Philosophy*, trans. I. B. Cohen and Anne Whitman, assisted by Julia Budenz (Berkeley: University of California Press, 1999), p. 905.
[20] David W. Hughes, "Edmond Halley: His Interest in Comets," in Thrower, *Standing on the Shoulders of the Giants*, p. 341.

up all the possible theories and explanations that had been set forth by others over the last half-century. Among such questions was:

what kind of motion [is] it...carried with? Whether in a straight or bended line? And if bended, whether in a circular or other curve, as elliptical or other compounded line, whether the convex side of the curve were turned towards the Earth? Whether in any of those lines it moved equal or unequal spaces in equal or unequal times? Whether it ever appears again, being moved in a circle or be carried clear away and never appear again, being moved in a straight or paraboloidical line?[21]

The flashy comet of November 1680 stirred up all these questions once again. Flamsteed, at the Royal Observatory, thought that the second cometary event was just the return of the earlier comet, but he imagined that it had turned around right in front of the sun. Newton thought this sudden reversal of direction was improbable.

By early 1681, Newton was convinced that the comet appearing in December 1680 was actually the same comet seen in November (as Flamsteed suggested), but the question remained: Had this comet reversed course around right in front of the sun, or had it gone around it and then made a sharp detour back toward earth? Either way (and the fact was that the comet of 1680 had gone around the sun before circling back), the question was, what exact path (straight line, parabola, hyperbola, or ellipse) did such celestial objects follow? More important, what physical force could have caused this heavenly body to deviate from its straight path off into infinity and instead to turn in a sharp arc around the sun and then return in the direction from whence it came? Flamsteed had actually proposed that the magnetic fields of the sun and comet opposed each other, thereby redirecting the comet's path.[22]

For centuries, philosophers and astronomers had talked about gravity, especially the force that causes objects to fall from heights and to roll down inclined planes. But how could something like gravity affect distant objects such as the sun, planets, and maybe comets? Was not suggesting attraction at such a distance invoking occult properties – properties that were now banned by the corpuscular theory, according to which all motion is just the movement of small particles banging into each other and mechanical actions? The demonstrable effects of magnets showed at least that such attraction was possible. Certainly, navigators could reliably follow their compass needles (with small variations that sailors and

---

[21] Hooke, *Lectures and Collections*, as cited in Hughes, "Edmond Halley," pp. 334–35.
[22] Hughes, "Edmond Halley," pp. 338–39, with a diagram of the sun's proposed path.

geographers like Halley had charted), which were somehow attuned to the earth's magnetic pull. Still, it was known that magnetic forces sharply decrease as the distance from a magnet increases. Kepler's attempt to explain the Copernican system by proposing magnetic forces failed, yet Flamsteed had proposed such an explanation for the comet.

In a word, the appearance of the comet of 1680, along with all the related discussions going on among members of the Royal Society, created considerable interest in resolving all these cosmic problems. Hooke, Sir Christopher Wren, Edmond Halley, and a host of other seventeenth-century astronomers wondered what precise curve a planet would follow under an inverse-square rule. For Newton's mathematically obsessed mind, the question was, what *exact curved* path did the comet of 1680–81 follow? For privately, he already believed (after 1679) that the planets followed elliptical orbits. To be sure, it seemed likely that the path of a comet was some form of a conic section, but whether it was an open (e.g., parabolic) or closed curve (an ellipse) was a momentous question that, Newton thought, had to be answered, despite the lack of adequate observational data. For if the curve were closed as an ellipse, then the comet of 1680 (and another from 1664) would someday return, something never before shown to be the case. Little wonder that the *Principia* would contain page after page on comets.

Halley's Visit

Just what transpired in Newton's mind between 1681 and 1684 is unknown. Newton probably had worked out in the early months of 1680 the solution to the problem that Hooke had proposed in a letter of January 17, 1680. In modernized terminology, the question was "If a central attractive force causes an object to fall away from its inertial path and move in a curve, what kind of curve results if the attractive force varies inversely as the square of the distance?"[23] Whatever Newton's response to this question was, he did not send a reply to Hooke. It took

---

[23] This is I. B. Cohen's restatement in "Newton's Discovery of Gravity," p. 169. Hooke's letter reads, "It now remains to know the proprietys of a curve Line (not circular nor concentricall) made by a centrall attractive power which makes the velocitys of Descent from the tangent Line or equall straight motion at all Distances in a Duplicate proportion reciprocally taken"; Hooke to Newton, January 17, 1680, in D. T. Whiteside, ed., *The Mathematical Papers of Isaac Newton, 1684–1692* (Cambridge: Cambridge University Press, 1974), 6:13.

the surprise visit of the young Edmond Halley, then twenty-eight, with Newton in Cambridge to bring the latter's silent cogitation to public light.

Halley has been portrayed as a tall, handsome young man with a pleasing demeanor. His father owned a soap factory in London with a preferment that meant that Halley would one day inherit the family wealth. He was a gifted mathematician and apparently had memorized every star in the sky and could quickly spot on a celestial map or sphere any missing or misplaced star.

We know that Halley, from the time of the appearance of the comet of 1680 and his visit to Paris, was fascinated by cometary patterns and that he continued to labor over calculations of scarce observational data on them. But in January 1684, Halley met with Robert Hooke and Sir Christopher Wren, probably in a local London inn, where they discussed celestial orbits and the forces impelling them. All three seemed to agree that the planetary force needed to explain the heavenly patterns was the inverse-square law. In his typically forceful manner, Hooke declared "that upon that principle all the laws of the celestial motions were to be demonstrated and that he himself had done it."[24] Halley, however, confessed that he had not been able to do it, whereupon Wren, who also admitted that he could not do it, "said that he would give Mr. Hooke and me two months time to bring him a convincing demonstration thereof." Sir Christopher sweetened the challenge by offering to award the successful demonstrator a book worth forty shillings to go along with the honor of the accomplishment. Hooke, in his hubris, claimed that he already had the answer but that he would withhold it for a while "so that, others, trying and failing, might know how to value it when he should make it public."[25]

When nothing was forthcoming from Hooke seven months later, Halley made a decision to visit Newton in Cambridge, where he met with him in August. The report of that meeting was recorded by the French mathematician, Abraham de Moivre, who was friend to both Newton and Halley. According to de Moivre's notes, Halley and Newton met that August in Cambridge and conducted a long conversation, toward the end of which Halley posed his burning question; that is, he wanted to know "what he [Newton] thought the curve would be that would be described by the planets supposing the force of attraction towards the

---

[24] *Correspondence of Isaac Newton*, 2:442.
[25] This correspondence is also reported in Christianson, *In the Presence of the Creator*, pp. 283–84.

Sun be reciprocal to the square of their distance from it. Sir Isaac replied immediately that it would be an ellipsis," which de Moivre claimed struck Halley "with joy & amazement."[26] When Halley asked Newton how he knew this, Newton replied, "I have calculated it." This was startling news. But when Newton rummaged through his papers looking for this singularly important result, he was unable to find it – or as some historians think, Newton was not yet willing to part with his conclusion without a further review of his calculations.[27]

During this eventful meeting, Newton promised that he would provide Halley with a copy of his results once he had retrieved them. The proof that Newton promised Halley was not forthcoming until November 1684. When it did arrive, it appeared in a short revolutionary treatise called *On the Motion of Bodies in Orbit* (*De Motu corporum in gyrum*). When Halley read this little treatise, a bare eight pages in Latin (twenty-two pages in English translation), he realized at once that this was a revolutionary work that created a foundation for a new general science of mechanics. The short treatise of 1684 was far from perfect, with only one of Newton's eventual laws listed as an "hypothesis,"[28] but it was unquestionably pathbreaking.

Although Edmond Halley had obtained Newton's permission to publish the little *De Motu* in *Philosophical Transactions*, that was not to be for Newton was engaged in the huge struggle to work out every detail in all the applications and deductions that came to his attention as he reworked *De Motu* into his magnum opus, the *Mathematical Principles of Natural Philosophy*. It was finally delivered to Halley in April 1686.

In the meantime, the portrait of Newton as "never at rest" that emerges from the notes of Humphrey Newton, a distant relative employed as an assistant, seems to apply to Newton during this period and probably before:

I never saw him take any Recreation or Pastime, either in Riding out to take the air, Walking, Bowling, or any other Exercise whatever, Thinking all Hours lost, that was not spent in his Studies, to which he kept so close, that he seldom left his Chamber.... So intent, so serious ... that he eat very sparingly, nay, oftimes he has forget to eat at all, so that going into his Chamber, I have found his mess untouched, of which when I have reminded him, would reply Have I; & then

[26] Ibid.
[27] The latter view is that of Richard S. Westfall, *Never at Rest: A Biography of Newton* (Cambridge: Cambridge University Press, 1993).
[28] *Mathematical Papers*; and see De Gandt's discussion of them in *Force and Geometry*, trans. Curtis Wilson (Princeton, NJ: Princeton University Press, 1995), esp. pp. 18ff.

making to the Table, would eat a bit or two standing, for I cannot say, I ever saw Him sit at Table by Himself. . . . He very rarely went to Bed, till 2 or 3 of the clock, sometimes not till 5 or 6, lying about 4 or five hours.[29]

From Revolutionary Treatise to Magnum Opus

The little work *On Motion* revealed for the first time just what Newton had been thinking over the last half-dozen years and what the fruit of his 1679–80 correspondence with Hooke had been.[30] The most radical shift in Newton's thinking came from his abandonment of the framework of centrifugal forces elaborated earlier by Christiaan Huygens. In place of that view, Newton coined the concept called *centripetal* force – that is, a force causing a moving object to fall toward a center, not flee from it. This turned out to be another key for a general science of motion.[31] Rather than seeing the problem (as he did earlier in the case of the moon) as one of a moving body flying off in a tangent to a circular orbit propelled by a centrifugal force, Newton now thought of a body moving in a straight line continuously but being pulled by a centripetal force back toward a center. Once he did that, all the pieces of the new theory of planetary motion, and hence the new mechanics, began to fall into place.[32]

The treatise began with a proof of Kepler's area rule and some of the implications that follow from it (see Figure 11.4). The intent was to show the impact of a force on a moving body and to show that the resultant path was an ellipse. But it also demonstrated that a centripetal force (later to be called *gravity*) was operative. In Newton's thinking, there were two components of the force: an inertial component propelling the body forward and a centripetal force attracting the body toward a center, pulling it back from a straight path. This was the new idea from Hooke, that of "compounding the celestiall motions of the planets [out] of a direct motion by the tangent & an attractive motion towards the central body."[33] The correspondence with Hooke thus altered Newton's view with the result that now, in 1684, he spoke of an "inertial force"

---

[29] As cited in Christianson, *In the Presence of the Creator*, p. 305.
[30] This intense expansion of the *De Motu* in the *Principia* is very nicely explained in Westfall, *Never at Rest*, chap. 10.
[31] Cohen, "Guide to Newton's *Principia*," pp. 76ff, and Cohen, "Newton's Discovery of Gravity," pp. 170–72.
[32] Cohen, "Newton's Discovery of Gravity," pp. 170–72.
[33] Hooke correspondence to Newton, November 24, 1979, as cited in Cohen, "Newton's Discovery of Gravity," p. 167; and *Correspondence of Isaac Newton*, 2:297.

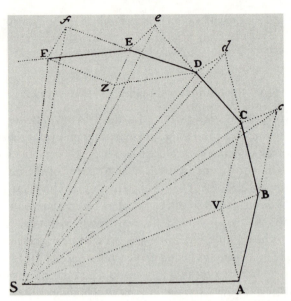

FIGURE 11.4. Newton's proof that the areas ASB, BAC, CAD, and so on, are equal under any force deflecting a body along the arc AF toward the center. The proof first appears in *De Motu*. From the *Principia*, Proposition 1, Theorem 1.

(*vis insita*)[34] propelling an object in a straight line and a centripetal force of attraction between all bodies, which together produce the elliptical orbits. In Figure 11.4, we can imagine a moving body being knocked successively from its straight path to a point on the curve. The force diverts the object passing from B to C so that it arrives at C instead of c. Moreover, the difference between those two points (C − c) might be taken as a measure of the acting force (that of gravity). The curve that results is not a circle but rather an ellipse. What Newton had done was to prove mathematically that Kepler's area law is indeed true – that it is more than an hypothesis.[35]

The second part of Newton's revolutionary advances involved linking the law of areas to the inverse-square law of attraction. Students of Newton's early development point out that Newton had already, in 1665–66, in his own words, "deduced that the forces which keep the Planets in their Orbs must [be] reciprocally as the squares of their distances from the

34 Cohen, *Newtonian Revolution*, p. 190.
35 I. B. Cohen, "Newton's Laws: Stages of Transformation Leading to Universal Gravitation," chap. 5 in *Newtonian Revolution*, esp. pp. 229–30.

centers about which they revolve."[36] He did this by combining a measure of uniform circular velocity and Kepler's 3/2 law.[37] But it was only when Newton abandoned the centrifugal (fleeing motion) frame of reference after 1679 and adopted Hooke's centripetal perspective that the inverse-square law could be set in place. It became the law of attraction between all objects at a distance based on a force proportional to the product of their masses and inversely to the square of their distance apart. Hence, it was in *De Motu*, and three years later in the *Principia*, that Newton arrived at the idea of universal gravitational attraction.[38]

In *De Motu*, Newton laid out the connection between Kepler's third law and the inverse-square law of attraction, which is contained in his Corollary 3: "If the squares of the periodic times are as the cubes of the radii [Kepler's law], the centripetal forces are inversely as the squares of the radii, and conversely."[39] Then, in a scholium (a sort of mathematical aside demonstrating a point), Newton claims that astronomers are "agreed" that Kepler's third law applies to all the planets[40] and, therefore, the inverse-square law of attraction applies universally.

With this, Newton had achieved something monumental. As we saw earlier, Kepler's third law postulates a universe with the sun at its center around which the six planets (Mercury, Venus, Earth, Mars, Jupiter, and Saturn) all revolve in such a manner that the cube of the distance divided by the square of the period (time) is a constant. Here was cosmic harmony as Kepler envisioned it, a celestial system regulated by an amazingly simple mathematical proportionality.

But now Newton goes further to assert that Kepler's third law is agreed to by astronomers and, given that regularity, the force holding this system together is a centripetal force (gravity) that operates according to the inverse-square rule: the attractive force decreases inversely with the square of the distance between bodies. This formulation was the centuries-old realization of the deeply rooted Western belief that God is a geometer,

---

[36] As cited in Cohen, "Guide to Newton's *Principia*," p. 65. This comes from an early Newton autobiographical sketch. Cohen goes on to say that the two documents referred to by Newton with these calculations have been found, one in the so-called *Waste Book*. The other was found in a Cambridge Library collection referred to as ULC MS Add. 3968; ibid., p. 66.

[37] Ibid.

[38] Cohen, "Newton's Discovery of Gravity," p. 172.

[39] *De Motu*, as cited in De Gandt, *Force and Geometry*, p. 29.

[40] As cited in ibid., p. 29, and Cohen, "Newton's Discovery of Gravity," who suggests that astronomers had not been agreed that this was an empirical fact before Newton.

that the world is made according to "number, weight and measure,"[41] for which there is no Islamic or Asian equivalent. Or, in Galileo's exuberant phrase, the book of nature is written in the language of mathematics; "its characters are triangles, circles, and other geometric figures, without which it is impossible to understand a single word of it."[42] In the *Principia*, Newton had used all those tools, symbols, and propositions to arrive at the mathematical principles that regulate the cosmos. The result was Kepler's wish come true.

As Newton worked through a range of problems in *De Motu* and introduced the principle of (universal) attraction based on the inverse-square rule, he discovered that large planets such as Jupiter and Saturn, especially when coming into conjunction, would have a mutual attraction that would perturb their orbits, altering their uniformly elliptical paths. Because of this *mutual* attraction and the fact that the celestial system is a many-body system, "there are as many orbits to a planet as it has revolutions, as in the motion of the Moon, and the orbit of any one planet depends on the combined motion of all the planets, not to mention the actions of all these on each other."[43] Newton could not but realize at the same time that "to consider simultaneously all these causes of motion" in such an interactive planetary system, "and to define these motions by exact laws allowing of convenient calculation exceeds, unless I am mistaken, the force of the entire human intellect."[44]

Throughout this great undertaking, Newton constantly reviewed the body of published astronomical observations and requested more precise data from Flamsteed at the Royal Observatory to confirm his calculations about the "real world." Clearly, there were many calculating problems in this new world system that needed to be worked out. For readers who got access to the *De Motu* manuscript, however, it was clear that Newton had solved the big riddle posed by Hooke and Halley regarding celestial motion: the curve followed by bodies gripped by the inverse-square rule and the way in which gravity affects all particles and bodies in the universe. Nevertheless, because the force of gravity was an empirical rule that

---

[41] See Benjamin Nelson, *On the Roads to Modernity*, ed. Toby E. Huff (Totowa, NJ: Rowman and Littlefield, 1981), p. 159, where he discusses a famous painting used as the frontispiece of Riccioli's *Almagestum novum* of 1651, depicting the hand of God revealing "number, measure, and weight" at the ends of his fingers.

[42] Galileo, "The Assayer," in *The Discoveries and Opinions of Galileo*, trans. and with an introduction by Stillman Drake (New York: Doubleday, 1957), p. 238.

[43] Newton's *De Motu*, as cited in Cohen, "Newton's Discovery of Gravity," p. 172.

[44] Ibid.

lacked any explanation, it always retained a problematic status. Consequently, it was many hundreds of pages into the *Principia* before Newton declared the equivalence of "centripetal force" and "gravity" as a universal rule. This follows Newton's demonstration (with new observations from John Flamsteed) that Jupiter's satellites as well as Saturn's satellites conform to the same centripetal forces in their orbits around their central bodies as those bodies follow in their paths around the sun. With this conclusion, he offers the following declaration:

> Hitherto we have called "centripetal" that force by which celestial bodies are kept in their orbits. It is now established that this force is gravity, and therefore we shall call it gravity from now on. For the cause of the centripetal force by which the moon is kept in its orbit ought to be extended to all the planets.[45]

This was another way of stating the *universal law* of gravitational attraction: all particles in the universe attract each other.

In *De Motu*, Newton asked whether comets could be brought within the compass of the inverse-square rule and parabolic or elliptical orbits. In the *Principia*, he shows that they could and worked out a method for calculating the elliptical orbits of comets. Indeed, the *Principia* contains dozens of pages on comets, making it a formidable scientific text on comets in its own right. Following Newton's lead, Edmond Halley set to work studying the historical records of comets until he found the path of a comet that had appeared three times earlier (in 1531, 1607, and 1682) and predicted that it would return in 1758, seventy-five years after its last appearance.[46] This was "Halley's Comet," which arrived on time, although Halley had died sixteen years earlier, in 1742.

Terrestrial Applications

The fact that the *Principia* announces principles for the "System of the World" often obscures the fact that Newton's book also contains numerous applied problems in terrestrial dynamics, including fluid mechanics as well as the refraction and reflection of light. He worked out such problems for cases with and without resistance.

It was one of Newton's more impressive accomplishments to show that the ocean tides are caused by the gravitational force of the moon and also the sun in combination. In that connection, he posed the problem

---

[45] Isaac Newton, *The Principia: Mathematical Principles of Natural Philosophy*, new trans. I. B. Cohen and Anne Whitman (Berkeley: University of California Press, 1999), p. 806.
[46] Hughes, "Edmond Halley," p. 351.

"to find the force of the moon to move the sea." He then draws on the observations of a Samuel Sturmy, who reported on the tides at the mouth of the river Avon below Bristol in the spring and autumn, which is "approximately 45 feet, but in the quadrature is only 25 feet."[47] Newton then proceeds to show how one can calculate the gravitational forces of the sun and moon, jointly and separately, and their tidal effects. These are surely earthly applications.

Indeed, there are dozens of problems laid out by Newton that could have a variety of industrial and commercial applications. In Book 2, Proposition 3, Problem 1, the task is "to determine the motion of a body which, while moving straight up or down in a homogeneous medium, is resisted in proposition to the velocity, and which is acted on by uniform gravity." Proposition 4, Problem 2, of the same book states, "Supposing that the force of gravity in some homogeneous medium is uniform and tends perpendicularly toward the place of the horizons, it is required to determine the motion of a projectile in that medium, while it is resisted in proportion to the velocity."

All such problems as those are complements to the many other theorems and problems that concern the movements of the celestial bodies, the arcs of the moon, and the problem of determining the "unequal motions of the satellites of Jupiter and of Saturn from the motions of our moon" (Book 3, Proposition 23, Problem 5)[48] or "to find and compare with one another the weights of bodies in different regions of our earth" (Book 3, Proposition 20, Problem 4). At the same time, many of these problems lay out physical conditions that are clearly incompatible with Descartes's notion of "vortices" carrying all matter around. The idea that the world is full of extended matter and whirling vortices does not seem compatible with celestial bodies in motion according to Kepler's 3/2 law and the fact that different celestial bodies travel at different rates of speed and move in contrary directions. Comets, for example, seem to revolve in many different directions and cannot be imagined to revolve according to a singular direction, as postulated by Descartes's vortices.

Folklore has it that a student once said as Newton passed by, "There goes the man that writt a book that neither he nor anybody else understands."[49] But the historical fact is that the *Principia Mathematica*

---

[47] Newton, *Principia*, new trans., p. 875.
[48] Ibid., p. 834.
[49] As cited in Christianson, *Isaac Newton and the Scientific Revolution* (New York: Oxford University Press, 1996), p. 83.

was a treasure trove of applied mathematics that was soon mined by Europeans for a great range of applications in ballistics, mining, surveying, machine making, and many other applications.[50]

Roots of a Final Theory

The great mind that has been described as "never at rest" was but forty-five years old when the *Principia* appeared. That meant that Newton was bound to make more contributions to science in the seventeenth and early eighteenth centuries. Newton had, it goes without saying, an extraordinary capacity to work out the mathematical details of physical problems coupled with an ability to grasp the larger forces and structures that govern the natural world. After the *Principia* was published, Newton went through an emotional bad patch, until he moved to the London Mint in 1695. There he served as second in command (warden), though the frequent absence of the master of the mint left Newton in charge. When he began to venture outside of Cambridge in the years after the publication of the *Principia* in 1687, Newton began to interact more with fellow natural philosophers and the Royal Society. Finally, in 1703, he was elected to the post of president of the society, where he served until his death in 1727.

The Royal Society of London had had an impressive run of extraordinary scientific undertakings during the more than forty years of its existence. When Newton assumed the presidency, membership had fallen off, and its members often did not show up for meetings. Of course, this distressed Newton and galvanized him to launch another new beginning. He drafted a new scheme for the Society that was to focus on "discovering the frame & operations of Nature, reducing them (as far as may be) to general Rules or Laws, establishing those Rules by observations & experiments, and then deducing the causes & effects of things."[51] Newton had already stated his belief along these lines in the preface to the *Principia*. There he spoke of the basic task of philosophy "to discover the forces of nature from the phenomena of motions and then to demonstrate the other phenomena from these forces."[52] Then he went on to hope that

---

[50] For an historical account of these, see A. E. Musson and Eric Robinson, *Science and Technology in the Industrial Revolution* (Toronto, ON: University of Toronto Press, 1969), among others.

[51] As cited in Christianson, *Isaac Newton*, p. 119.

[52] Newton, *Principia*, new trans., p. 192 (preface of 1687).

"we could derive the other phenomena of nature from mechanical principles by the same kind of reasoning!" Still forging ahead, he avers, "For many things lead me to a suspicion that all phenomena may depend on certain forms by which the particles of bodies, by causes not yet known, either are impelled toward one another and cohere in regular figures, or are repelled from one another and recede."[53] In the final paragraph of *General Scholium* (2nd ed., 1713), he added the following thoughts:

A few things could now be added concerning a certain very subtle spirit pervading gross bodies and lying hidden in them; by its force and actions, the particles of bodies attract one another at very small distances and cohere when they become contiguous; and electrical [i.e., electrified] bodies act at great distances, repelling as well as attracting neighboring corpuscles.[54]

We may at this point recall the earlier work of Francis Hauksbee and the many other electricians who pioneered electrical studies in the time of Newton. Moreover, given the scholarly attention to these comments by Newton in the *Principia* as well as the unpublished addendum that Newton drafted for the second edition of the book, there is little doubt but that Newton had the forces of electricity in mind when he wrote those lines.[55] In 1704, the year after Hooke died, Newton published this landmark study of optics that he had carried out back in the 1670s but had refused to publish because of the intense criticism of Hooke and others. But now, with Hooke gone from the scene, he decided to bring it out. A Latin edition of the book was published two years later, in 1706. There, Newton laid out the framework of a grand scheme, the skeleton of a final theory that included a complex set of natural forces:

Have not the small particles of Bodies certain Powers, Virtues, or Forces by which they act at a distance, not only upon the Rays of Light for reflecting, refracting, and inflecting them, but also upon one another for producing a great Part of the Phaenomenon of Nature? For it's well known, that Bodies act one upon another by the Attractions of Gravity, Magnetism, and Electricity . . . and it is not improbable but that there may be more attractive Powers than these.[56]

53 Ibid., pp. 382–83.
54 Ibid., p. 944.
55 The addendum is published in full in Newton, *Principia*, new trans., pp. 287ff. This is preceded by a discussion of "Newton's 'Electric and Elastic' Spirit," sect. 9.3. An earlier discussion of Newton's thinking about these issues appears in the preface to the Dover edition of Newton's *Optics*, based on the London 3rd ed. (New York: Dover, 1951).
56 As cited in Westfall, *The Reconstruction of Modern Science* (New York: Cambridge University Press, 1977), pp. 141ff.

In a word, Newton had laid the foundations for thinking about a final theory. Whether this might be possible or not, "Dreams of a Final Theory," authored by Steven Weinberg, the contemporary physicist, were rooted in Newton.[57]

There is no question but that the scientific revolution unfolding in Europe in the seventeenth century bequeathed an enormous legacy of intellectual insight into the forces of nature. That legacy was a product of the long study of the natural books of Aristotle that had been ensconced in the universities of Europe since the twelfth century. Conversely, no comparable naturalistic study was undertaken outside the West. In the decades and centuries after the seventeenth-century revolution, that intellectual legacy (both the medieval and the modern), some might say, would be worth all the tea and silk of China, all the gold of Africa, and all the cloth and spices of India and the Middle East. The world had entered an entirely new phase: old ways of thought and social organization would be insufficient for the scientific, technological, and commercial revolutions that were to follow.

---

[57] Steven Weinberg, *Dreams of a Final Theory: The Search for the Fundamental Laws of Nature* (New York: Pantheon Books, 1992).

12

The Scientific Revolution in Comparative Perspective

The Revolutionary Grand Synthesis

The achievement of the modern scientific revolution, most elegantly put forth in the work of Sir Isaac Newton, was the outcome of a joint European adventure. It brought together extraordinary advances in optics, astronomy, and the science of motion, all governed by the law of universal gravitation. Whether we consider Newton's new unified system of terrestrial and celestial physics of 1687, or his even grander vision of that system augmented by particle attractions, magnetic, electric, and other forces acting "at a great distance," the result is undeniably revolutionary.[1]

The seventeenth century also witnessed great strides in pneumatics and electrical studies: advances in the former field would bring the steam engine, whereas those in the latter would bring electrification and an unimaginable new source of energy: electric power. It is difficult to imagine the Industrial Revolution without steam power and our modern digital world without electricity and its harnessing. Neither could any other part of the world get us there without first discovering and harnessing electric forces.[2]

---

[1] This contrasts with Steven Shapin, who claims, "There was no such thing as the Scientific Revolution, and this is a book about it," in *The Scientific Revolution* (Chicago: University of Chicago Press, 1996), p. 1. Likewise, he asserts that "historians now reject even the notion that there was any single coherent cultural entity called 'science' in the seventeenth century to undergo revolutionary change" (p. 3). My view has greater affinity with John Henry, *The Scientific Revolution and the Origins of Modern Science*, 3rd ed. (New York: Palgrave Macmillan, 2008).

[2] Jack Goldstone has claimed that "quantum theory and chaos theory both argue that nature is not continuous, and that sudden and dramatic 'jumps' can develop from slight

Although medical research and advances in microscopy were not theoretically integral to the grand synthesis of the new physics, they represent nonetheless a simultaneous burst of creative inquiry that was revolutionary in its own way and remained nonpareil throughout the world in the seventeenth century.

On this high note of theoretical understanding across a number of fields, it is difficult to deny that there was indeed a major intellectual upheaval in the seventeenth century and that it merits the label of a scientific revolution. Of course, it did not happen all at once, or in one place in the European landscape, or in just one discipline. Yet, it is evident that vast areas of naturalistic thought were radically different, let us say, in 1710 than they were in 1600. Only with the revolution in chemistry and then biology (nineteenth-century possibilities) would even the thought of biochemistry be imaginable in the twentieth century. Those revolutions also were destined to appear only in the West. Multiple Europeans were indeed fashioning a new intellectual edifice and worldview that in future years would shape modern thinking, as Herbert Butterfield suggested, across the globe.

The great departure of intellectual paths between East and West becomes starkly evident if one compares the intellectual apparatus of late-seventeenth-century Europe with intellectual thought in China, Mughal India, or the Ottoman Empire in the same period. If one were to make a roster of outstanding contributors to the leading edge of the scientific transformation in Europe and seek counterpart achievements in other parts of the world, there would be no equivalent to the advances of Galileo, Kepler, Descartes, Huygens, or Newton; no William Gilbert, Otto von Guericke, or Francis Hauksbee; no Torricelli, Blaise Pascal, Robert Hooke, Robert Boyle; or any counterpart to William Harvey, Marcello Malpighi, Regnier de Graaf, Jan Swammerdam, or Antoni Leeuwenhoek. This is the very short list of stellar scientific pioneers, but it makes the point.

The absence of any clearly definable inputs into the Newtonian achievement from outside Europe will sound jarring to our early-twenty-first-century egalitarian sensibilities. Nevertheless, the same was true with

tips or deviations in underlying functions or relationships." He takes this to mean that underdeveloped societies can suddenly "jump" ahead. See Goldstone, "Capitalist Origins, the Advent of Modernity, and Coherent Explanation: A Response to Joseph M. Bryant," *Canadian Journal of Sociology* 33, no. 1 (2008): 119–33. Bryant, however, suggests that such a view is incoherent; see Joseph Bryant, "A New Sociology for a New History? Further Thoughts on the Eurasian Similarity and Great Divergence Theses," *Canadian Journal of Sociology* 331, no. 1 (2008): 149–67.

regard to optics, electrical studies, microscopy, and pneumatics. This outcome ought to force us to rethink efforts to understand the deep grounding of the cultural, scientific, and metaphysical foundations of the various civilizations (Europe included) that occupy our shared planet. All cultural roads may not lead to the same mental space. Be that as it may, the legacy of the scientific revolution coming out of the seventeenth century was indeed a world-changing transformation, for better or worse.

Roots of Experimental Science

Searching for the roots of the scientific advances of the seventeenth century requires looking even further back in the European tradition than many observers have wished. The emergence of scientific culture in Europe, even its experimental form, is far older than the glories of the Royal Society of London, founded in 1661.

Ever since the appearance of the Harvard Case Studies in Experimental Science in the 1950s that recounted specific experimental advances in electrical studies, Robert Boyle's pneumatics, the discovery of oxygen, and the atomic-molecular theory of matter, students have been directed to inquiries that flowered in the late seventeenth century and thereafter. That makes considerable sense in the unfolding case of the discovery of electric charge and the extraordinary laboratory work of Francis Hauksbee and others.[3] Although there was in that case study a nod in the direction of William Gilbert's seminal discoveries regarding magnetism and electric charge, the general impression left is that all the ideas about experimental manipulation of natural conditions originated in the late seventeenth or early eighteenth century. This is wide of the mark.[4]

[3] Duane Roller and Duane H. D. Roller, *The Development of the Concept of Electric Charge*, Harvard Case Studies in Experimental Science, 8 (Cambridge, MA: Harvard University Press, 1954).

[4] The existing histories of experimental endeavors from the late medieval period onward suffer from lack of attention to the pioneers who actually carried out experimental procedures, however inexact they may have been. The classic work that may need renewed attention, especially in comparative perspectives, is A. C. Crombie, *Robert Grosseteste and the Origins of Experimental Science, 1100–1700* (Oxford: Clarendon Press, 1962). The recent discussions of the putative "torture of nature" falsely attributed to Francis Bacon sheds no light on the present context; cf. Carolyn Merchant, "'The Violence of Empediments': Francis Bacon and the Origins of Experimentation," *Isis* 99, no. 4 (2008): 731–60; and Peter Persic, "Proteus Rebound: Reconsidering the 'Torture of Nature,'" *Isis* 99, no. 2 (2008): 304–17. More useful is Peter Dear, "The Meaning of Experiment," in *The Cambridge History of Science*, vol. 3 (New York: Cambridge University Press, 2006), chap. 4.

Magnetism and Electrical Studies

For the operative idea that a variety of experimental conditions had to be tested, even with a variety of alternative substances, using controlled interventions is clearly evident in Gilbert's *De Magnete*, published in the very first year of the seventeenth century. Gilbert, in fact, tested a great range of materials in search of magnetic charge, trying to find when it is strongest, what attenuates it, and what possible role it might play in the constitution of our earth and universe. In all this, he was the first to clearly distinguish between magnetic and electric charges. He even developed the prototype of the first electrical measuring device, the *versorium* or electrical charge indicator, to detect magnetic or electric charges. As I stressed in Chapter 9, Gilbert himself was building on the fertile legacy of his *sixteenth-century* peers and predecessors. That means that this experimental strand of scientific culture was widely diffused in late-sixteenth-century European culture.

But not only that: one of Gilbert's models was the thirteenth-century polymath Peter Peregrinus (fl. 1269), whom Roger Bacon called "a master of experiment" and who "knows by experiment natural history and physic and alchemy" as well as optics and "all things in the heavens and beneath them."[5] It was Peregrinus who devised the spherical lodestone for purposes of experimentation and who wrote the first systematic treatise anywhere in the world on the properties of the lodestone and how to test them.

Seen in this light, Gilbert's seminal study of magnetism that first identified electric phenomena, published in 1600, was the first scientific textbook using experimental methods, crude as they were. At the same time, it was the culmination of an experimental tradition in natural philosophy that, with the publication of Gilbert's book, established the starting point for electrical studies. Furthermore, Gilbert's "magnetic philosophy" had a major impact on Johannes Kepler and many other scholars of the first half of the seventeenth century.

Astronomy

At the same time, no historian of science will overlook the fact that astronomers have been making experimental predictions for centuries

---

[5] Roger Bacon on Peter Peregrinus in Edward Grant, ed., *A Source Book in Medieval Science* (Cambridge, MA: Harvard University Press, 1974), p. 824.

regarding solar and lunar eclipses, the expected zodiacal locations of the five planets, and their conjunctions with each other. That tradition obviously goes back to the Greeks and Babylonians, and though it passed through Islamic civilization, Muslim astronomers were averse to compiling ephemerides – tables of daily, weekly, or monthly celestial appearances – because of their astrological interference with Islamic thought, claiming that "only God knows" the future. In short, an experimental approach that was based on observation and the making of predictions was integral to the astronomical tradition for centuries, though Islamic scholars were averse to doing this on a weekly or monthly basis.

The Europeans, however, got to be quite proficient at this. As we saw in Chapter 4, the Jesuits, on numerous occasions, challenged Chinese astronomers to make astronomical predictions based on their system both of solar eclipses and planetary conjunctions. In more than a dozen such tests, the Chinese system proved less accurate than the Western system.

Optics

In optics, Ibn al-Haytham is rightly credited with articulating an experimental approach in the eleventh century, but that tradition did not get embedded in the madrasas so that it lost its power within Islamic civilization. Nevertheless, both Europeans and Islamic scholars did develop an explanation of the rainbow (see Chapter 5), and the tradition of conducting optical experiments persisted to modern times.[6]

Given the very high level of analysis in optics achieved by Ibn al-Haytham in the eleventh century that was so influential on Roger Bacon and other European opticians, one might have expected scholars in Islamic culture to have invented those superbly useful aides to vision: eyeglasses. But as we have seen, that was a Florentine invention of 1286 that uniquely put Europeans on the path to inventing the telescope and microscope.

However one articulates the sources and origins of experimentalism in Western culture, it is clear that the Aristotelian version of naturalistic inquiry based on the search for the causes of things, and their explanation

---

[6] For some historical background on the optical tradition and its experimental side, see Eileen Reeves, *Galileo's Glass Works: The Telescope and the Mirror* (Cambridge, MA: Harvard University Press, 2008) as well as Vincent Ilardi, *Renaissance Vision from Spectacles to Telescopes* (Philadelphia: American Philosophical Society, 2007).

through logical analysis, was solidly ensconced in European universities from the twelfth century onward. That institutional setting proved to be a fertile ground within which scientific culture incubated all the way to the seventeenth century and beyond.

Pneumatics

Another illustration of this, apart from the influence of Peregrinus, and this a triple one, shows the influence of that nascent experimental culture on seventeenth-century science looking back to the fourteenth century. It concerns pneumatics and the possibility of creating a vacuum. As pointed out in Chapter 8, two outstanding seventeenth-century experimenters, Blaise Pascal and Otto von Guericke, were inspired by early-fourteenth-century natural philosophers who were asking what-if questions about nature: What if a vacuum were actually created (against all medieval expectations): Would it be possible to pull the surfaces surrounding the void apart? What if a vacuum were created – would it be possible for sound to travel through it?

Albert of Saxony (ca. 1316–90) posed just this sort of question, as did Jean Buridan, who was the rector at the University of Paris in 1328. Albert, known also as Albert the Great because of his prodigious knowledge and intellectual output, entertained a variety of thought experiments that were meant to test critical ideas. As a result, he turned to a sort of experimental approach with regard to the vacuum: If all the openings in a bellows were stopped up so that no air whatsoever could enter, what would happen? Albert opines that the forces that would come into play would be virtually invincible: "no power could raise one handle from the other unless a break occurred somewhere through which air could enter.... This [experience] seems to be a sign that nature abhors a vacuum."[7]

As if Albert's challenge is not daunting enough, Jean Buridan posed the question with a nuanced conclusion: "Not even horses could do it if ten were to pull on one side and ten on the other; they would never separate the surfaces of the bellows unless something were forced or pierced through and another body could come between the surfaces."[8]

---

[7] Albert of Saxony, "Questions on the Physics of Aristotle," in Grant, *A Source Book*, p. 325.
[8] Buridan, "Questions on the Eight Books of the Physics of Aristotle," in Grant, *A Source Book*, p. 326.

By the 1640s, the study of air pressure and the invention of the barometer allowed ingenious experimenters such as the thirty-eight-year-old Blaise Pascal to conduct a variety of experiments that could resolve the challenge of Albertus Magnus and Buridan. Because Pascal had commissioned his brother-in-law to carefully measure the weight of air with the barometer in the Puy de Dôme Mountain tests, he knew that the weight of air varies daily depending on humidity and climatic conditions. He also knew that the empty chamber at the top of a barometer, whether using mercury or another liquid, was a vacuum; hence, such a thing is not impossible. So what if you stopped up a bellows, fastened it to the ceiling, and then suspended a chain from the bottom handle of the bellows so that it extended to the floor with extra links? What one would find, as Pascal did, is that air pressure fluctuates so that a weight of only 113 pounds is sufficient to open the bellows as air pressure diminishes. Moreover, if you leave the apparatus in place and observe it over time, you will find that as the air becomes heavier, the bellows handle will rise, and it will fall when the air becomes lighter, with more chain links falling onto the floor.

This was an elegant, naturalistic demonstration, and when combined with his many other experiments weighing air, he could conclude, "I have demonstrated," against the idea that a vacuum is impossible, "on the contrary, by absolutely convincing arguments that the weight of the mass of the air is [the] real and only cause" of those phenomena previously attributed to nature's alleged abhorrence of a vacuum.[9] The same was true, of course, regarding the opening and closing of the suspended bellows.

The second dramatic experimental demonstration in the mid-seventeenth century, apparently inspired by fourteenth-century scholars discussing the vacuum, was that of Otto von Guericke, the mayor of Magdeburg, Germany. He devised an apparatus that would pit the strength of horses against atmospheric forces. This experiment had to wait for the invention of the air pump, which von Guericke did invent. He then instructed clever craftsmen to manufacture a strong copper sphere, divided into two halves, the whole being capable of being evacuated by an air pump to create a vacuum. Then horses could be attached to each hemisphere so as to pull against each other, but actually against the force

---

[9] From *Physical Treatises of Pascal: The Equilibrium of Liquids and the Weight of the Mass of the Aid*, as printed in Grant, *A Source Book*, p. 329.

of the atmosphere holding the two halves together. When the experiment was carried out in 1654, von Guericke's report confirmed the power of nature to hold the spheres together and the limits of horsepower (see Chapter 8).

Third, we saw in Chapter 8 that Italian experimenters in the early 1640s, having discovered the empty vacuum chamber at the top of the barometer, set about experimentally testing whether sound could be transmitted through it using a specially designed apparatus. We note again that all these pneumatic experiments took place in France, Italy, or Germany before the founding of the Royal Society.

Anatomy and Medicine

Finally, there is still another experimental tradition that is often over-looked: that of medicine and anatomical dissection. Since the twelfth century, Europeans had been carrying out dissections of human and animal corpses, whereas that tradition either atrophied in the Islamic world of the Middle East, as well as India, or did not escape official banning in China. As was shown in Chapter 7, from the thirteenth century onward, there was in Europe no official religious opposition to postmortem examinations, the "look for yourself" of autopsies. Indeed, from then onward, the practice was made part of the curriculum of leading universities and medical faculties across Europe, but especially in Italy. The pinnacle of this early modern research agenda was given concrete form in the early sixteenth century with the publication of the unsurpassed anatomical illustrations contained in Vesalius's *On the Fabric of the Human Body* of 1543.

In a word, the experimental scientific culture of Europe took many forms from the twelfth-century renaissance onward so that it was broadly established for centuries prior to the founding of the Royal Society. As the founders of that institution looked back for suitable models and emblems of their enterprise, one of those proposed was a depiction of Galileo's discovery of the satellites of Jupiter in 1610.[10] All of this seems further evidence of the wide and deep embedding of an infectious *ethos of scientific curiosity* across Europe that remained unmatched outside Europe during the seventeenth and later centuries.

---

[10] See the coats of arms produced by John Evelyn for the Royal Society, in Michael Hunter, *Establishing the New Science* (Woodbridge, UK: Boydell Press, 1989), p. xiv.

Divergent Ways

From this sketch, we see that the scientific and experimental point of view was much more deeply and broadly embedded in Western thought than has generally been realized. Conversely, that orientation, especially the experimental dimension, was much less practiced in other civilizations around the world.

I began this inquiry with the invention of the telescope in 1608 because it is linked to a burst of creativity and scientific curiosity that flowered across Europe in the seventeenth century. Second, the telescope was quickly taken around the world in the early seventeenth century. Yet, the arrival of the telescope in China, India, and the Ottoman Middle East did not produce a similar upsurge in scientific curiosity. We can now see that a comparative approach starting with a different field – say, pneumatics or microscopy – would have produced the same result: extraordinary advances in Europe but few or no parallel advances outside Europe.

Nevertheless, the failure of the telescope to trigger an exciting new burst of scientific creativity, especially in astronomy, around the world serves best to highlight a deficit in scientific curiosity that seems to have prevailed outside Europe from before the seventeenth century all the way to the end of the twentieth century. That is an extraordinary record of cultural disparity. It surely calls for more probing assessments of alternative modernities than have been produced hitherto. It remains for us, however, to consider both the impact of the scientific revolution on the rise of industrialism and the interacting cultural features of Europe that reinforced the emergence of the new scientific culture and propelled the Western world into an ascendancy of centuries.

Epilogue

Science, Literacy, and Economic Development

Given the extraordinary achievements seen in the scientific revolution and the huge cultural and technological advantages that those advances conferred on the Western world, it is surprising that so little has been written about it by those concerned with economic development. Major writers who have claimed either the parity or superiority of China to the West economically prior to the eighteenth century have been almost entirely silent about the European scientific revolution, its long history, and its significance.[1]

If we credit Herbert Butterfield's claim set out in the introduction to this study, then it is clear, as the last chapter has shown, that there was a great transformation of thought regarding our understanding of the forces governing the natural world. That mental transformation uniquely

---

[1] Among many authors who fall into this camp who have denied or ignored the importance of the scientific revolution, see Andre Gunder Frank, *ReOrient: Global Growth in the Asia Age* (Berkeley: University of California Press, 1998); Jack Goldstone, "Efflorescence and Economic Growth in World History: Rethinking the 'Rise of the West' and the Industrial Revolution," *Journal of World History* 12, no. 2 (2002): 323–89; Kenneth Pomeranz, *The Great Divergence: China, Europe and the Making of the Modern World Economy* (Princeton, NJ: Princeton University Press, 2000); R. Bin Wong, *China Transformed: Historical Change and the Limits of European Experience* (Ithaca, NY: Cornell University Press, 1997). The origins and significance of the European scientific revolution are also misconstrued in John Hobson, *The Eastern Origins of Western Civilization* (New York: Cambridge University Press, 2004), pp. 178–81. Joel Mokyr, *The Gifts of Athena: Historical Origins of the Knowledge Economy* (Princeton, NJ: Princeton University Press, 2002), does acknowledge the importance of "propositional knowledge," but this is not the same argument as presented in this study.

unfolded in the West during the last phases of the scientific revolution.[2] This means that Max Weber's question about "what combination of circumstances" were responsible for the great ascendance of the West must include those of the revolutionary new scientific point of view that infused the whole gamut of seventeenth-century natural scientific inquiry, not just astronomy. Put differently, the question of why the West can only be answered by bringing together the great conceptual transformation of the scientific revolution and the effects of the Protestant Reformation that had been noted by Weber. That path of cultural synthesis must consider the facilitating effects of religion along with the emergence of the new print media, the crystallization of a public sphere, and the rising rates of literacy. Indeed, as a sociological factor, the unparalleled rising rates of literacy in Europe were a major contributor to the great ascendance and divergence that set Europe off socially and economically from other parts of the world. Furthermore, the rise of literacy in Europe must be traced back at least to the early sixteenth century, when there was no parallel development in China, Asia broadly, or the Muslim world. At the same time, those developments have to be read against the long developmental background from the late Middle Ages.

Rising Rates of European Literacy

The singular legacy of the seventeenth-century scientific revolution quickly found its way into European patterns of education. A major reason for that absorption was the *educational revolution* of the sixteenth and seventeenth centuries. For the great advances in scientific knowledge that took place in the seventeenth century were the fruit of long-standing European educational traditions that encompassed both higher education and grammar schools.

The dramatic rise in the founding of universities across Europe was accompanied by rising levels of literacy in the general population. By the eighteenth century, there were nearly 150 universities located across Europe.[3] Their steady rise from the thirteenth century to the end of the eighteenth century meant that a growing population of university-educated citizens was multiplying in Western Europe.

[2] Herbert Butterfield, *The Origins of Modern Science 1300–1800*, rev. ed. (New York: Free Press, 1954), pp. 7–8.

[3] For a sociological tabulation of the trajectory of university foundings, see Phyliss Riddle, "Political Authority and University Formation in Europe, 1200–1800," *Sociological Perspectives* 36, no. 1 (1993): 45–62.

Speaking of the general impact of higher education in the sixteenth and seventeenth centuries, the late British-American historian Lawrence Stone hit the right note when he observed that "what is so striking about this period is not the appearance of individual men of genius who may bloom in the most unpromising soil, but rather the widespread public participation in significant intellectual debate on every front."[4]

As Stone and others have shown, the trajectory of literacy, aided and abetted by the rising tide of Puritanism in England, reveals an education revolution that took place in that island kingdom between 1580 and 1640. Summarizing the trend gathered from scattered sources, Stone found that by 1640,

> over half the male population of London was literate, that a high proportion of the one third of adult males who could sign their names in the home counties could read, and that $2\frac{1}{2}$% of the annual male seventeen-year-old age-group was going on to higher education.[5]

These signs reflect a major demographic shift in literacy in England, a "quantitative change of magnitude that...can only be described as a revolution." England was not alone in this transformation, as the Reformation's sponsorship of popular education also swept across the Continent, especially in Protestant areas, where reading the Bible became an essential religious duty for every man and woman.

Confining our attention to the case of England, Stone's first inquiry regarding the literacy revolution from 1580 to 1640 was followed by a second covering the period from 1640 to 1900. The results of a number of other scholars who took up the issue reveal the same conclusion that Stone had reached.[6] What they found was an extraordinary rise in literacy across England over the three centuries, culminating in a drop to less than 10 percent illiteracy by the eve of the twentieth century.[7] From the point of view of an input into an economic system, this burgeoning expansion of *human capital* gave Europeans a huge advantage over other parts of

---

[4] Lawrence Stone, "The Educational Revolution in England, 1560–1640," *Past and Present* 68, no. 28 (1964): 80.

[5] Ibid., p. 68.

[6] Donald Cressy, *Literacy and the Social Order* (Cambridge: Cambridge University Press, 1980); Rosemary O'Day, *Education and Society, 1500–1800* (New York: Longman, 1982); R. A. Houston, *Literacy in Early Modern Europe: Culture and Education, 1500–1600* (London: Longman Group, 1988).

[7] See Cressy, ibid., p. 177, for a chart showing the precipitous drop in illiteracy from the sixteenth to early twentieth centuries.

the world, where literacy rates continued to lag all the way to the end of the twentieth century.

The argument from the human capital point of view is that "educated people make good innovators, so that education speeds the process of technological diffusion" and obviously quickens economic development.[8] As such, it is a major contributor to the creation of wealth in any country and hence played a major role in the Industrial Revolution. I shall return to the argument of economic historians who have emphasized the great importance of human capital formation for economic development.

But let us notice that the revolution in literacy was far from being just an English phenomenon: it took place all across Europe, especially Northern Europe. Other researchers have shown that literacy across Western Europe, especially the Low Countries, northern France, and the western part of Germany, spiraled up from 1500 to the 1750s.[9] Remarkable progress was made in the sixteenth century so that around 1800, there was a broad swathe of Northern Europe "where already sixty to eighty percent of the male population could read and write," while about 40 percent of the female population could do so.[10] In some areas, as in England, literacy was nearly universal by 1800.[11] But not only that, some evidence points to the fact that Dutch levels of literacy were higher than those for Great Britain in the period between 1500 and 1800.[12] On balance, continental Europe experienced "an enormous rise in literacy" between the sixteenth and eighteenth centuries.[13]

A still more sophisticated indicator of levels of literacy is the rate of book production. Very recently, new databases have been developed that enable economists to chart rates of book production all the way from the sixth century to the end of the eighteenth[14] (more on which to follow). Book production is, however, related to the emergence of a public sphere as well as newspapers that cannot be assumed to emerge

[8] Richard C. Nelson and Edmund S. Phelps, "Investment in Humans, Technological Diffusion, and Economic Growth," *America Economic Review* 52, no. 2 (1966): 69–75.

[9] C. A. Reis, "Economic Growth, Human Capital Formation and Consumption in Western Europe before 1800," in *Living Standards in the Past*, eds. R. C. Allen, T. Bengtsson, and M. Dreibe, 195–227 (Oxford: Oxford University Press, 2005).

[10] Ibid., p. 201.

[11] For additional estimates of European literacy in this period, see Jan Luiten van Zanden, *The Long Road to the Industrial Revolution* (Leiden, Netherlands: Brill, 2009), esp. pp. 192ff.

[12] Ibid., pp. 191–92, and see his figure 23.

[13] Ibid., p. 217.

[14] Eltjo Buringh and Jan Luiten van Zanden, "Charting the 'Rise of the West': Manuscripts and Printed Books in Europe, a Long Term Perspective from the Sixth to the Eighteenth Century," *Journal of Economic History* 69 (2009): 409–45.

spontaneously around the world. Moreover, the public sphere wherein public opinion can be formed and expressed, along with newspapers, is related to the enhancement of commerce, trade, and the free flow of information. Consider then before the rise of book production the creation of a public sphere and newspapers as unique developmental aspects of the West.

The Public Sphere: Petitions and Newspapers

The emergence of a public sphere and newspapers is part of the long-standing tradition, especially in England, of submitting petitions to king and Parliament. We should not forget, however, that there was a long-standing tradition within the universities that allowed the public discussion of a surprisingly broad range of issues covering the natural world, political rulership, and the whole panoply of theological questions. This was most obviously signaled by the periodic quodlibet sessions that encouraged and allowed those present to ask "whatever you like." Public disputations were normal so that it is not surprising that Martin Luther posted his Ninety-five Theses on the cathedral door in Wittenberg. His post also contained a request "that those whoever cannot be present personally to debate the matter orally will do so in absence in writing."[15] It is difficult to imagine a similar challenge and debate regarding the Confucian legacy, the "mandate of Heaven," or the mandarinate in China posted on the door of a great temple in Beijing in any era. Heads would literally have rolled.[16]

Likewise, it is extremely difficult to imagine an Islamic scholar in the sixteenth or seventeenth century posting such a challenge on the door of the Great Mosque in Damascus (or in Istanbul), inviting scholars and others to publicly debate the fundamentals of Islam, the meaning of the Quran, and what its sources might be.

---

[15] In John Dillenberger, *Martin Luther: Selections from His Writings, Edited with an Introduction* (New York: Doubleday, 1961), p. 490.

[16] China did have rebels, such as the early-seventeenth-century Donglin group, who "urged [followers] to form convictions on the basis of truth and adhere to them uncompromisingly without regard for the consequences." But they did not bear a revolutionary message. They claimed a failure of the emperor to live up to Confucian ideals, not a new democratic ideal. For this episode and the exceedingly harsh treatment those rebels got, see John W. Dardess, *Blood and History in China: The Donglin Faction and Its Repression, 1620–1627* (Honolulu: University of Hawaii Press, 2002). The later work connects early-seventeenth-century patterns of repressing dissent with the Tiananmen Square uprising of 1989. Also see Roger V. Des Forges's review essay on that book in *China Review International* 9, no. 1 (2002): 85–91.

One could say, however, that the petitioning process considered here expanded the public sphere, bringing it down to the layman's level. Petitioning documents submitted by rich and poor alike were instruments having juridical status. In them, petitioners complained of the "miscarriage of justice or requested relief from taxes, forest laws, and other regulations."[17] A medieval ritual had been established whereby the first act of a convening Parliament was to appoint the receivers and "trier of petitions." Both individuals and collectives had the right to petition so that by the 1640s, we find 15,000 or more signatories to a single petition presented to Parliament.[18] The view commonly expressed was that petitioning is "the indisputable right of the meanest subject."[19] Nothing like this practice existed in China or the Muslim world for no domains of legal autonomy or parliaments evolved outside of Europe.

Much of this, naturally, depended on the advent of the printing press in the fifteenth century. The major transformation of pamphlets into newsbooks and then newspapers occurred in the 1640s in England. Newsbooks as modified pamphlets appeared in the 1640s following the tradition of *corantos* (currents of news) that had appeared earlier in Holland and on the Continent. Pamphlets as a source of information supplied the primary means for creating and influencing public opinion. It was a social invention of the seventeenth century. The pamphlets were cheap publications that conjoined the flow of information with commercial benefit. They became foundational in the shaping of seventeenth-century moral and political communities by constituting a public sphere of popular political opinion.[20]

In the 1640s, the pamphlet format had been transformed into the newsbook, a quarto publication consisting of sixteen pages, bound together on the left margin to make a little book. Such books reported on a variety of topics but especially domestic issues. The newspaper itself appeared in 1666 (the first being the *London Gazette*). It was larger, printed on half sheets in two columns on both sides. Once the newspaper press emerged in the middle of the seventeenth century, it has not stopped since. As

[17] David Zaret, "Petitions and the 'Invention' of Public Opinion in the English Revolution," *American Journal of Sociology* 101, no. 6 (May 1996): 1510; and Zaret, *Origins of Democratic Culture: Printing, Petitions, and the Public Sphere in Early-Modern England* (Princeton, NJ: Princeton University Press, 2000).

[18] Joad Raymond, *Pamphlets and Pamphleteering in Early Modern Britain* (New York: Cambridge University Press, 2003), p. 301.

[19] Zaret, "Petitions."

[20] Raymond, *Pamphlets*, p. 26.

the historian of English newspapers, Joad Raymond, put it, the arrival of the newsbook (in 1640) set in motion an "avalanche, which, with no more than a few weeks of interruption," continues rolling right up to this morning.[21] There is hardly a more visible symbol of the intellectual freedom assumed to go along with a public sphere than the daily press. But it cannot be taken for granted as an ordinary developmental outcome.

For, in contrast to this, no newspapers emerged either in China or the Muslim world until the early or late nineteenth century, when British and other Western exemplars were taken as models. Indeed, Gutenberg's printing press was immediately banned for use with Arabic in the Muslim world. Edicts were issued against it by Sultans Bayzid II in 1485 and Salim I in 1515.[22] This second royal edict against the printing press was issued just two years before Martin Luther would post his Ninety-five Theses on the Wittenberg Cathedral door. That document was then rapidly disseminated throughout Germany and much farther through the aid of the printing press.

The first newspapers in the Middle East appeared in 1824 (or 1828).[23] In China, the first newspaper modeled on British exemplars did not appear until the 1840s.[24] The long tradition of book printing using woodblock technology that had existed in China since the eighth century failed to give rise to the daily newspaper. Considering these alternative outcomes, it would not be surprising to discover that book production in China, despite its long legacy of block printing, lagged behind the West (on which more later).

Literacy and the Reformation

In addition to the advent of the printing press, the influence of the political petitioning process, and the emergence of the daily newspaper, we should

[21] Raymond, *The Invention of the Newspaper: English Newsbooks 1641–1649* (Oxford: Clarendon Press, 2003), p. 80.

[22] J. Pedersen, *The Arabic Book* (Princeton, NJ: Princeton University Press, 1984), p. 133.

[23] Niyazi Berkes, *The Development of Secularism in Turkey* (repr. New York: Routledge, 1998), p. 126, gives credit to Mehmet Ali for this and places the founding in 1828. Richard Bulliet puts the founding in 1824 in *The Case for Islamo-Christian Civilization* (New York: Columbia University Press, 2004), p. 79. Mahmud II is credited with founding the second newspaper in the Muslim world, in 1831.

[24] See Rosewell S. Britton, *The Chinese Periodical Press* (Taipai: Ch'eng-Wen, 1966), and Elizabeth Sinn, "Emerging Media: Hong Kong and the Early Evolution of the Chinese Press," *Modern Asian Studies* 36, no. 2 (2002): 421–65. For printing more generally in China, see Cynthia Brokaw and Kai-Wing Chow, eds., *Printing and Book Culture in*

acknowledge the extraordinary effect of the Protestant Reformation on literacy in the sixteenth century and thereafter. All the Protestant reformers had major commitments to the idea that every Christian, man and woman, should read the holy book for him- or herself and thus should be literate. From Wycliffe and Tyndale to Luther, Zwingli, and Calvin, all the reformers championed the cause of lay access to the holy scripture. And whereas commentators have sometimes spoken of harsh treatment of the reformers, in fact, their books and ideas soon triumphed across Europe. In that way, they gave an historic push to the policy of universal education that was indeed translated into public policy, especially in England, Switzerland, and Germany. Conversely, the intellectually oppressive conditions in China gave birth to no new major religious or philosophical movements.[25]

This connection between literacy and the Reformation brings to mind Max Weber's profound though always controversial thesis about the Protestant ethic and the spirit of modern capitalism. In shorthand terms, Weber believed that an ethic of hard work evolved out of the teachings of the reformers, especially Luther, Calvin, and the Puritan divines. It was natural to think that if such a work ethic had indeed formed during the spread of Reformation teachings across Europe, it would have an impact on economic activity and worldly success, just as the reformer John Wesley believed that it did. For he was aware of the paradoxical effects that Methodism could have. As he wrote in a London newspaper in 1786:

I fear, wherever riches have increased, the essence of religion has decreased, in the same proportion. Therefore, I do not see how it is possible, in the nature of things, for any true revival of [true] religion to continue long. For religion must

Late Imperial China (Berkeley: University of California Press, 2005). For the oppressive Chinese attitude toward dissent and petitions, see note 25.

[25] The radically different approach to dissent in China and its fierce repression of "heterodox views" has been extensively written about. Because of that attitude, no "reformation" or major new school of philosophy or daily press emerged. Among others, see the following: Timothy Brook, "Censorship in Eighteenth Century China: A View from the Book Trade," *Canadian Journal of History* 22 (August 1988): 177–96; Luther Carrington Goodrich, *The Literary Inquisition of Ch'ien-Lung* (Baltimore: Waverly Press, 1935); R. Kent Guy, *The Emperor's Four Treasures: Scholars and the State in the Later Ch'ien-lung Era* (Cambridge, MA: Harvard University Press, 1987); Alexander Woodside, "The Reign of Ch'ien Lung," *Cambridge History of China* 9 (1994): esp. 290–91; and the other sources listed in n16. The extraordinary restrictions on the TV broadcast of President Obama's speeches during his visit to China in November 2009 reflects the centuries-old Chinese aversion to freedom of speech and the press.

necessarily produce both industry and frugality, and these cannot but produce riches.[26]

Thus, the eternal paradox: Protestant lifeways lead to wealth, and that leads people to fall away from the straight and narrow path.

Now, however, a major new study has been published by two German economists proposing a reinterpretation of a central aspect of Weber's thesis. The suggestion is that the kernel of the Reformation's effect on economic development was its stress on *universal* education.[27] Forgotten in all the preceding research on Weber was the fact that Luther himself was actively engaged in education reform. In 1524, he published a pamphlet, "To the Councilmen of All Cities in Germany That They Establish and Maintain Christian Schools." It carried the message of alarm regarding what Luther saw as the deteriorating state of schools and education and the call to correct the situation by building and maintaining new ones. Implicit in all that Luther wrote was the assumption that it was incumbent on every believer to study and read the Bible. As early as 1520, in "An Address to the Christian Nobility of the German Nation on the Improvement of the Christian Estate," Luther suggested that "every Christian should know the whole of the Holy Gospel, wherein His name and His life are written, by the age of 9 or 10."[28] Likewise, in 1530, Luther wrote and preached a sermon titled "Sermon on Keeping Children in School." Ordinary believers were encouraged to provide an education for their children and to keep them in school, thereby avoiding idleness and wicked ways.

Luther's influence was still further amplified by the thought and career of Philipp Melanchthon, the German educational reformer. He worked closely with Luther and later served as an educator in the federal government, working to reform the German education system, especially the Protestant universities. Some of Luther's radical followers apparently wanted to abolish all education, thinking that the Bible was enough.[29]

[26] Max Weber used this passage in *The Protestant Ethic*, p. 175, but I have taken it from *The Works of John Wesley (Bicentennial Edition)*, ed. Albert C. Outler (Nashville, TN: Abingdon Press, 1984–89), pp. 9.537–30.

[27] Sascha O. Becker and Ludger Woessmann, "Was Weber Wrong? A Human Capital Theory of Protestant Economic History," *Quarterly Journal of Economics* 124, no. 2 (May 2009): 531–96.

[28] As cited in Horst Rupp, "Phillip Melanchthon," *Prospects: Quarterly Review of Comparative Education* 26, no. 3 (September 1996): 612.

[29] Robert S. Westman, "The Melanchthon Circle, Rheticus, and the Wittenberg Interpretation of the Copernican Theory," *Isis* 66, no. 232 (1975): 169.

Luther opposed this. Thanks to Luther's and Melanchthon's strong advocacy, the more balanced view that all believers should be educated prevailed. In the end, Lutheranism and other Reformation teachings not only stressed the importance of literacy but also advocated the establishment of city and town schools.

In their new study, testing the implications of just such ideas against nineteenth-century Prussian data, Sascha O. Becker and Ludger Woessmann find that Protestant economies "prospered because instruction in reading the Bible generated the human capital crucial to economic prosperity."[30] Conducting a major portion of their research using the nineteenth-century Prussian census seems particularly apt because the region includes Wittenberg. Using the 1871 census, they find a "significant, positive association" between Protestant affiliation and economic prosperity in late-nineteenth-century Prussia. They also find that "Protestantism had a strong effect on literacy" and that this literacy effect is "large enough to account for practically the entire Protestant lead in economic outcomes."[31] Using still other data, the authors show the strong relationship between Protestant background and literacy across an international sample of twenty-two countries in 1900. The data are for countries in which Protestants and Catholics constituted a majority. As the percentage of Protestants in the countries rises, so does the literacy rate. Protestant countries had near universal literacy in 1900, but no Catholic country reached full literacy, with many falling far short.[32] The strong Lutheran and broad Protestant effects on school enrollments in the Western world up to the nineteenth century stand out in estimates of primary school attendance, as shown in Table E-1.

In this way, the authors have reinterpreted Weber's thesis while aligning it with the current human capital theory of economic development. There are, for sure, multiple determinates of the rise of literacy in the sixteenth and seventeenth centuries, but here we see a factor of singular importance in that rise: the Reformation.

There is still another approach to the rise of literacy and economic development that led to the great seventeenth-century divergence between Europe and other parts of the world, especially China. That can be found in efforts to measure levels of book production.

---

[30] Becker and Woessmann, "Was Weber Wrong," p. 543.
[31] Ibid., p. 532.
[32] Ibid., p. 543.

TABLE E-1. *Primary School Enrollment: Rates per 10,000 (1830)*

Estimated Primary School Enrollment Rate ca. 1830		Dates at Which Selected Countries Equaled Enrollment in the United Kingdom ca. 1830	
United States	1,500	Indonesia	1930
United Kingdom	900	India	1939
France	700	Egypt	1960
Germany	1,700	Iran	1960
		Turkey	1960
		China	1960

Adapted from Appendix Table 1, "Estimated School Enrollment Rate by Country, 1830–1975," in Richard A. Easterlin, "Why Isn't the Whole World Developed?" *Journal of World History* 41, no. 1 (1980): 18–19.

Literacy and Book Production

Unlike measures of literacy that often rely on church parish records indicating a person's ability to sign his name or estimates of school enrollments in the sixteenth and seventeenth centuries, levels of book production indicate actual patterns of book consumption and, hence, actual reading and reading capacity. They indicate higher levels of literacy as well as the desire of literate individuals to read books. Working with these assumptions, a group of economists and social historians associated with the University of Utrecht have constructed databases of book titles extending from the sixth century to 1800 in Europe.[33]

In a sequence of studies, they have shown a dramatic and steady rise of book production in eight major European countries from just before the invention of the moveable-type Gutenberg press to the end of the eighteenth century. The rise of book production is, of course, a proxy of literacy rates and parallels more direct measures of literacy, as seen earlier regarding literacy rates in England.[34]

Moreover, the same authors have shown that there is a very high correlation between rates of economic growth and book production.[35] In

[33] Buringh and van Zanden, "Charting the 'Rise of the West'"; Jeorg Baten and Jan Luiten van Zanden, "Book Production and the Onset of Modern Economic Growth," *Journal of Economic Growth* 13 (2008): 217–35.
[34] Baten and van Zanden, "Book Production."
[35] Buringh and van Zanden, "Charting the 'Rise of the West,'" p. 410.

other words, the production of books in Europe from the sixth century to the end of the eighteenth century closely tracks estimates of gross domestic production. In the early years of the datasets, sixth to sixteenth century, manuscript production was carefully estimated. The results show that "it was exactly in the countries in which book production increased fastest... that real wages developed systematically better over the centuries before the industrial revolution."[36]

In contrast to those developments, estimates of book production in Asia reveal exceedingly low levels of production. The printing of books in Asia was close to zero in the cases of India and Indonesia because the printing press did not arrive, particularly in India, until the early nineteenth century,[37] and block printing from China had not taken off there or in Indonesia.

Looking at the data for China in the two early modern periods, the Ming (from roughly 1522 to 1644) and Qing (1644–1911) dynasties, likewise reveals low levels compared to Europe. For example, using data on book production in China assembled by Sinologists for the Ming Dynasty,[38] Buringh and van Zanden found that only forty-seven titles per year were produced in the whole of China. This is extremely low, and even multiplying this estimate by a factor of ten, Chinese production did not equal that of European countries of similar population. For in Western Europe during the period 1522–1644, "the average annual production was estimated to be about 3,750 titles," which is roughly forty times higher than the Chinese rate of the same period.[39]

For the Qing period, 1644–1911, an output of about 120,000 editions works out to 474 titles per year compared to 6,000 titles produced in Western Europe in just the year 1644.[40] Book production rates in China thus did not compare with European rates.

Even if evidence should emerge suggesting an equivalence of book production in China and Western Europe in the early modern period, we

---

[36] Baten and van Zanden, "Book Production," p. 231.
[37] Frances Robinson, "Islam and the Impact of Print in South Asia," in *The Transmission of Knowledge in South Asia*, ed. Nigel Crook (New York: Oxford University Press, 1996), p. 63.
[38] See Lucille Chia, "Mashaben: Commercial Publishing in Jiangyang from the Song to the Ming," in *The Song–Ming Transition*, eds. P. J. Smith and R. von Glahn (Cambridge, MA: Harvard University Asia Center, 2003), 284–89; Cynthia Brokaw, "On the History," in *Printing and Book Culture in Late Imperial China*, eds. Cynthia J. Brokaw and Kai-wing Chow (Berkeley: University of California Press, 2005), p. 27.
[39] Buringh and Zanden, "Charting the 'Rise of the West,'" p. 437.
[40] Ibid.

TABLE E-2. *World Literacy Rates (in Percent) in the Early Twentieth Century*

	1950	1962
Africa	15–20	16–22
America, North and South	80–81	80–82
Arab Countries	13–18	18–22
Asia and Oceania	29–33	43–47
Europe and Soviet Union	90–94	93–97

From Minedlit 5. *Statistics of Illiteracy: UNESCO Memorandum for the Teheran Conference, 1965*, as cited in Sir Charles Jeffries, *Illiteracy: A World Problem* (London: Pall Mall Press, 1967), p. 27.

have to remember that the Chinese books would be entirely devoid of the modern scientific advances (those detailed in Chapters 7–12), none of which occurred in China. Put differently, the books would have none of the modern scientific "propositional knowledge" that some economists have posited as critical for economic growth.[41]

Furthermore, there was no history in China of books equivalent to Aristotle's natural books containing the science of physics (mechanics), al-Haythm's optics, and Euclidean geometry, which had been studied in European universities since the Middle Ages. Likewise, there would be no studies of plants and animals using microscopes or accounts of the growing literature on air pumps and pneumatics or electrical studies. As noted earlier, there was in Chinese thought no equivalent overarching conception of science or Greek natural philosophy.

In sum, whether we use literacy estimates based on years of schooling or rates of book production, all areas outside Europe lagged hugely behind European standards (see Tables E-1 and E-2). The same was true with regard to the content of the books because modern Newtonian science was not to be found elsewhere. And because the evidence shows a strong positive association between literacy rates (as well as book-production rates) and economic growth, it is clear that Europe had an overwhelming literacy and human capital advantage that was directly linked to economic development. That advantage, not surprisingly, resulted in levels

[41] Joel Mokyr, *The Gifts of Athena: Historical Origins of the Knowledge Economy* (Princeton, NJ: Princeton University Press, 2002). In *The Long Road to the Industrial Revolution*, Jan Luiten van Zanden operationalized this concept, "propositional knowledge," as just that which could be found in books. See his chapter 6: "The Philosophers and the Revolution of the Printing Press."

of economic well-being far above those of Asia and the Middle East after the seventeenth century that persisted for hundreds of years.

The Scientific Legacy and the Industrial Revolution

We now have a fairly clear sketch of the path by which the scientific legacy was transformed into economic assets through education and entrepreneurial activity. During the opening decades of the eighteenth century, Newtonian principles and ideas were being disseminated in school textbooks and taught by popularizers in public lectures all over England. The contemporary historian of science, Margaret Jacob, articulated the process among Europeans, especially Britons, when she observed that people learned the new science "in schools and lecture halls; they picked up its contents from general textbooks; they read about scientists and their exploits in newspapers and journals."[42] The general populace of Europe was being increasingly exposed to the scientific worldview. In that sense, they were lifted out of the pre-Newtonian, premicrobe, and preelectric world that prevailed outside Europe. In addition, Jacob and her collaborator Larry Stewart have noted the spread of the new scientific worldview via the "marketplace for science" and the emergence of "scientific entrepreneurs" who made "a handsome living from the market for Newton's science."[43] Other lecturer-experimenters did the same, inventing new demonstration devices to be used in public lectures to reveal the new scientific principles and unmask false prophets.

Likewise, the American historian of modern technology, David Landes, admitted that millwrights in the last half of the eighteenth century, who were perhaps itinerant engineers, were more than jacks-of-all trades, for they had considerable theoretical knowledge:

They were not the unlettered tinkerers of historical mythology. Even the ordinary millwright, as [William] Fairbain notes, was usually "a fair arithmetician, knew something of geometry, leveling, and mensuration, and in some cases possessed a very competent knowledge of practical mathematics. He could calculate the velocities, strength and power of machines... "[44] as well as develop and modify mill plans.

[42] Margaret Jacob, *Scientific Culture and the Making of the Industrial West* (New York: Oxford University Press, 1997), p. 6.

[43] Margaret Jacob and Larry Stewart, *Practical Matters: Newton's Science in the Service of Industry and Empire, 1687–1851* (Cambridge, MA: Harvard University Press, 2004), pp. 64ff.

[44] David Landes, *The Unbound Prometheus: Technological Change and Industrial Development* (New York: Cambridge University Press, 1969), p. 63; the internal quote for

The Canadian economist, Richard Lipsey, and his associates put it well when they said that

at no previous time, and at no place in the West outside of Britain in the eighteenth-century, could one say such things about ordinary millwrights (or their analogues in other times and places). These millwrights had access to, and accessed, a pool of mechanical theory and applied knowledge. This knowledge underlay the great mechanical inventions of the industrial Revolution, including the steam engine and the great engineering work that preceded it in the eighteenth century.[45]

The unique scientific legacy was increasingly assimilated by being published in books, taught in schools, and applied to a broad range of technical and commercial endeavors.

Other students of the Industrial Revolution have noted the many ways in which the new science of mechanics was applied to a broad range of new products and processes such as machine tooling, gearing making, instrument making, and so on. As the British historians A. E. Musson and Eric Robinson pointed out, the new mechanics of the seventeenth century

was used not merely to calculate the movement of heavenly bodies, but also in practical arts such as navigation, cartography, ballistics, mining, and surveying, and these gave rise to the craft of instrument-making: manufacture of telescopes, microscopes, barometers, chronometers, micrometers, dividing and gear-cutting engines[46]

and so on.

In a word, the old modes of production using animal, wind, and water power were soon to be superseded, whereas the new mechanics of the Newtonian revolution were to find endless applications. At the same time, new types of commodities emerged, such as glass for windows and mirrors, lenses, eyeglasses, scientific instruments, and all sorts of newly mechanized machines. This set the stage for an extraordinary economic divergence such that at the end of the twentieth century, if not before, real output per capita in Western Europe and North America would reach

William Fairbain, *Treatise on Mills and Millworks* (London: Longmans, Green, 1865), p. 1.vi.
[45] Richard G. Lipsey, Kenneth I. Carlaw, and Clifford T. Bekar, *Economic Transformations: General Purpose Technologies and Long Term Economic Growth* (Oxford: Oxford University Press, 2005), pp. 246–47.
[46] A. E. Musson and Eric Robinson, *Science and Technology in the Industrial Revolution* (Toronto, ON: University of Toronto Press, 1969), p. 23.

"more than ten times that of many less developed countries."[47] This accumulated knowledge of the scientific revolution was far beyond anything to be found outside of Europe.

Science, education, and industry had come together in an unprecedented fashion. Schools and universities greatly expanded the literate population while adding the new spirit of inquiry to their charge. The new science and industry were located in a legal framework that was intended to support both parliamentary democracy and progressive economic development. The emergence of the broad range of new commercial products and processes created a Schumpeterian revolution that completely changed the nature of economic activity in Europe – and eventually the world. From that point of view, that devolution made all kinds of traditional products and processes obsolete. Given that new dynamic emerging in the eighteenth century, it matters little that by some estimates, China's standard of living was perhaps as high or higher than European standards prior to that time.[48] For clearly, China had not been developing the new systems of public schooling, the human capital, or the scientific capital that would be needed for the future. Similarly, its legal system was anything but progressive.

In Europe, the factor of rising literacy (joined with rising book production) was a powerful source of the new European energy. That same link between literacy and economic success was once again to power new economic trends in the twentieth and twenty-first centuries.

A recent extensive review of the literature on the effects of literacy and cognitive skills on economic growth amplifies this conclusion: "cognitive skills have powerful effects on individual earnings, on the distribution of income, and on economic growth."[49] Those results reinforce the thrust of the present study: underlying Europe's economic success from the eighteenth century onward was its singular commitment to universal literacy and its long-standing study of the sciences in universities. At the same time, the unique legal framework that Europeans worked out, starting in the medieval period, was not matched elsewhere in the world in the seventeenth and eighteenth centuries.

---

[47] Robert C. Allen, Tommy Bengtsson, and Martin Dribe, *Introduction to Living Standards in the Past* (Oxford: Oxford University Press, 2005), p. 1.

[48] This view has been argued by Wong, *China Transformed*; Goldstone, "Efflorescence and Economic Growth"; Pomeranz, *Great Divergence*. Those estimates, however, are now being reconsidered; see van Zanden, *Long Road*; Allen et al., *Living Standards in the Past*.

[49] Eric A. Hanushek and Ludger Woessmann, "The Role of Cognitive Skills in Economic Development," *Journal of Economic Literature* 46, no. 3 (2008): 657.

Comparative Literacy in the Twentieth Century

Finally, let us turn to comparative literacy rates based on estimated school enrollments that will allow us to see the gap in literacy between East and West. That gap has persisted since the seventeenth century. In Table E-1, we can see the estimated school-enrollment rates of the United States, United Kingdom, France, and Germany in 1830 in the left column. In the right-hand column are dates for six Asian and Middle Eastern countries, at which time their primary school enrollments equaled those of the United Kingdom in 1830, though its enrollments were not the highest in Europe at the time. Nevertheless, the average number of years that the six countries lagged behind the United Kingdom regarding primary-school enrollments was slightly more than 126. If the comparisons were made with the rates for Germany or the United States, the lag would be even greater.

The low levels of primary-school enrollments in Asia and the Middle East in the early nineteenth century are an inevitable outcome of the social policies of the countries in question. They did not set up systems of public schooling as Europeans did as far back as the sixteenth century. Neither China, Southeast Asia, nor the Middle East had the benefit of a Reformation movement that encouraged – indeed, demanded – universal literacy. In the absence of such a movement, or another impetus toward public schooling, illiteracy rates remained frightfully high from the early modern period all the way to the end of the twentieth century.

When we turn to the early twentieth century and more reliable data are available, the low levels of literacy outside Europe and the United States again stand out, as seen in Table E-2. We see that at midcentury, the vast majority of citizens (upward of 80%) in the Middle East, Asia, and Africa were still illiterate. In 1950, when American and European literacy rates ranged from 80 to 90 percent, the Arab-Muslim Middle East was only 13 to 18 percent literate. The literacy rate of China in 1950 stood at about 20 percent, whereas in British India (1947), it was somewhere between 12 and 20 percent.[50]

---

[50] The literacy rate for China, prepared recently by the Chinese Ministry of Education, was reported by Ted Plafkr, "China's Long – but Uneven – March to Literacy," *New York Times*, February 12, 2001. The 12.2 rate for India is reported in J. P. Naik and Syed Nurullah, *A Students' History of Education in India 1800–1973* (New Delhi: Macmillan, 1976), but J. C. Aggarwal and S. Aggarwal, *Education in India: A Comparative Study of States and Union Territories* (New Delhi: Concept, 1990), p. 107, report a 24.9% literacy rate for men and 7.9% for women in 1950.

From this, it is obvious that England, and Europe more broadly, had an institutional, intellectual, and human capital advantage that fed directly into the Industrial Revolution and would not be matched by any other population around the world for centuries.

The Legacy of the Past and Societal Development

Given the preceding review of the scientific and educational path of Europe in the seventeenth century, and especially the accompanying institutional developments, it is apparent that Western Europe was on an entirely different developmental plane than the non-West. This was probably true since the time of the Greeks. Nevertheless, as the Dutch economist Van Zanden put it, it was "a long road to the industrial revolution," one that began in the medieval legal breakthroughs centered on new sets of legal rights and procedures, property rights, the creation of a variety of zones of legal autonomy, new forms of collective action, and new conceptions of citizenship.[51] Above all, there were universities, charitable organizations, autonomous cities and towns, guilds, professional associations, and parliaments that gave Europeans many new spheres of legal autonomy on which future juridical and economic activities could be built. Those new structures gave European social actors both greater stability and greater flexibility. They did so by allowing Europeans to join together in new corporate forms of legality. They gave durability to their joint endeavors while serving both collective interests and democratic principles signified by the maxim "what affects all should be considered and approved by all."[52] The concomitant commitment to education, both at the university and local level, resulted in human capital formation from the fifteenth century onward, far ahead of the rest of the world. According to van Zanden, that lead was probably 300 to 400 years ahead of rivals such as Japan and China.

Non-Europeans were triply disadvantaged: first, whereas literacy rates rose rapidly in Europe from the sixteenth century onward (perhaps even as

[51] van Zanden, *Long Road*, pp. 200–1, and p. 294 regarding institutional efficiency from an economic point of view; also see Tine de Moor, "The Silent Revolution: A New Perspective on the Emergence of Commons, Guilds, and Other Forms of Corporate Collective Action in Western Europe," *International Review of Social History* 53 (2008): 179–212.

[52] See Gains Post, *Studies in Medieval Legal Thought: Public Law and the State, 1150–1322* (Princeton, NJ: Princeton University Press, 1964); and see Harold Berman, *Law and Revolution* (Cambridge, MA: Harvard University Press, 1983).

early as the fifteenth century), levels of literacy in non-Western countries were extremely low and remained so to the end of the twentieth century; second, there was no scientific revolution outside the West; and third, the legal and intellectual foundations for stable economic development as well as democracy and constitutional government were absent. Nothing parallel to the legally circumscribed public sphere of newspapers and public dissent appeared outside Europe before the end of the first quarter of the nineteenth century. Even then, those publications were distant approximations of the European press.

All these results run counter to those who argue that there was no cultural or institutional difference between Western Europe and China, Mughal India, or the Ottoman Empire in this period.[53]

Finally, we cannot overlook among all these assets the unique broad-based scientific curiosity that propelled modern science throughout the seventeenth century. If we take that strong inclination as a proxy for the general European disposition toward entrepreneurial activity, then we have a strong motivational asset that must have been deeply involved in all the technological innovations that came out of the eighteenth century, the classic age of the first Industrial Revolution. It suggests that the singular pursuit of modern science in Europe was intimately connected with spillover effects in applied science and technology that were to power the Industrial Revolution. The new generations of tinkers and millwrights were imbued with the modern ethos of science.

In the end, all these advantages that accumulated over time set off Western Europe as well as "Europe overseas" from other parts of the world. All of those factors go a long way toward explaining why the West succeeded as it did.

---

[53] Jack Goldstone, "Capitalist Origins, the Advent of Modernity, and Coherent Explanation: A Response to Joseph M. Bryant," *Canadian Journal of Sociology* 33, no. 1 (2008): 120–21. Goldstone writes that there were no "cultural or institutional dynamics leading to a materially superior civilization in the West" before 1850, and that "nothing like a 'course of modernizing development' can be seen anywhere before 1800, except perhaps Britain from the early 1700s...."

Selected References

Adivar, Adnan. 1939. *La Science chez les Turcs Ottomans*. Paris: G. P. Maisonneuve.

———. 1943. *Osmali Türklerinde Ilm*. Istanbul: Remzi Kitabevi.

al-Hassan, Ahmad Y., and Donald R. Hill. 1986. *Islamic Technology: An Illustrated History*. Cambridge: Cambridge University Press.

Ali, M. Athar. 1975. "The Passing of Empire: The Mughal Case." *Modern Asian Studies* 9, no. 3: 385–96.

Allen, Robert C., Tommy Bengtsson, and Martin Dribe, eds. 2005. *Living Standards in the Past: New Perspectives on Well-Being in Asia and Europe*. Oxford: Oxford University Press.

Andrewes, William. 1996. "Finding Local Time at Sea, and the Instruments Employed." In *The Quest for Longitude*, edited by William J. J. Andrewes, 393–404. Cambridge: Collection of Historical and Scientific Instruments, Harvard University.

Ansari, S. M. Razaullah. 1985. "Introduction of Modern Western Astronomy in India during the Eighteenth and Nineteenth Centuries." In *History of Astronomy in India*, edited by S. N. Sen and K. S. Shula, 353–402. New Delhi: Indian National Science Academy.

Arjomand, Said Amir. 1999. "The Law, Agency, and Policy in Medieval Islamic Society: Development of the Institutions of Learning from the Tenth to the Fifteenth Century." *Comparative Studies in Society & History* 41, no. 2: 263–93.

Ashbrook, Joseph. 1984. *The Astronomical Scrapbook: Skywatchers, Pioneers, and Seekers in Astronomy*. Cambridge, MA: Harvard University Press.

Azhar, Zahir Ahmad. 1984. "Mulla Mahmud Jawnpuri." In *Encyclopedia of Islam*, Urdu ed., vol. 20, 27–29. Lahore, Pakistan.

Bacon, Francis. 1620/1960. "The Great Instauration." In *The New Organon*, edited by Fulton H. Anderson, 3–32. Repr. Indianapolis, IN: Bobbs-Merrill.

Bagheri, M. 1997. "A Newly Found Letter of al-Kashi on Scientific Life in Samarkand." *Historia Math* 24, no. 3: 241–56.

Baten, Jeorg, and J. L. van Zanden. 2008. "Book Production and the Onset of Modern Economic Growth." *Journal of Economic Growth* 13, no. 3: 217–35.

Becker, Sascha O., and Ludger Woessman. 2009. "Was Weber Wrong? A Human Capital Theory of Protestant Economic History." *Quarterly Journal of Economics* 124, no. 2: 531–96.

Benjamin, Park. 1898/1975. *A History of Electricity (The Intellectual Rise of Electricity from Antiquity to the Days of Benjamin Franklin)*. Repr. New York: Arno Press.

Berggren, J. L. 2003. "Trigonometry in the Islamic World." In *Episodes in the Mathematics of Medieval Islam*, 127–53. New York: Springer.

Berkes, Niyazi. 1964. *The Development of Secularism in Turkey*. New York: Routledge.

Berkey, Jonathan. 1992. *The Transmission of Knowledge in Medieval Cairo*. Princeton, NJ: Princeton University Press.

Berman, Harold. 1983. *Law and Revolution: The Formation of the Western Legal Tradition*. Cambridge, MA: Harvard University Press.

Bernier, Francois. 1996. *Travels in India (1656–1668)*. Translated by Irving Brock, edited and revised by A. Constable. New Delhi: Asia Educational Service.

Biagioli, Mario. 1993. *Galileo, Courtier: The Practice of Science in the Culture of Absolutism*. Chicago: University of Chicago Press.

———. 2006. *Galileo's Instruments of Credit: Telescopes, Images and Secrecy*. Chicago: University of Chicago Press.

Birch, Thomas. 1676/1968. *History of the Royal Society of London*. 4 vols. Repr. New York: Johnson.

Blanpied, William A. 1975. "Raja Sawai Singh II: An 18th Century Medieval Astronomer." *American Journal of Physics* 43, no. 12: 1025–35.

Bodde, Derk. 1991. *Chinese Thought, Science, and Society: The Intellectual and Social Background of Science and Technology in Pre-modern China*. Honolulu: University of Hawaii Press.

Bodde, Derk, and Clarence Morris. 1967. *Law in Imperial China*. Cambridge, MA: Harvard University Press.

Bond, John David. 1921. "The Development of Trigonometric Methods Down to the Close of the XVth Century." *Isis* 4, no. 2: 295–323.

Boschiero, Luciano. 2007. *Experiment and Natural Philosophy in Seventeenth-Century Tuscany: The History of the Accademia del Cimento*. Dordrecht, Germany: Springer.

Boyle, Robert. 1675–76. "Experiments and Notes about the Mechanical Origins of Production of Electricity." In *The Works of Robert Boyle*, vol. 8, http://www.bbk.ac.uk/boyle/researchers/works/boyle_works.html.

Brentjes, Sonja. 1999. "The Interests of the Republic of Letters in the Middle East, 1550–1700." *Science in Context* 12, no. 3: 435–68.

———. 2002a. "Western European Travelers in the Ottoman Empire and Their Scholarly Endeavors (Sixteenth–Eighteenth Centuries)." In *The Turks*, vol. 3, edited by Hasan Celal Güzel, C. Cem Oguz, and Osman Karatay, 795–803. Ankara: Yeni Türkiye.

———. 2002b. "On the Location of the Ancient or 'Rational' Sciences in Muslim Educational Landscapes (AH 500–1100)." *Bulletin of the Royal Institute for Inter-Faith Studies* 4, no. 1: 47–71.

Britton, Rosewell S. 1966. *The Chinese Periodical Press*. Taipai: Ch'eng-Wen.

Brockey, Liam. 2007. *Journey East: The Jesuit Mission to China 1579–1724*. Cambridge: Belknap Press.

Brokaw, Cynthia. 1996. "Commercial Publishing in Late Imperial China: The Zou and Ma Family Businesses of Sibao, Fujian." *Late Imperial China* 17, no. 1: 49–92.

———. 2005. Introduction to *Printing and Print Culture in Late Imperial China*. Berkeley: University of California Press.

Brokaw, Cynthia, and Kai-Wing Chow, eds. 2005. *Printing and Print Culture in Late Imperial China*. Berkeley: University of California Press.

Brook, Timothy. 1988. "Censorship in Eighteenth Century China: A View from the Book Trade." *Canadian Journal of History* 22 (August): 177–96.

Bryant, Joseph. 2006. "The Rest and the West Revisited." *Canadian Journal of Sociology* 31, no. 1: 403–44.

———. 2008. "A New Sociology for a New History? Further Thoughts on the Eurasian Similarity and Great Divergence Thesis." *Canadian Journal of Sociology* 331, no. 1: 149–67.

Bulliet, Richard. 2004. *The Case for Islamo-Christian Civilization*. New York: Columbia University Press.

Buringh, Eltjo, and J. L. van Zanden. 2009. "Charting the 'Rise of the West': Manuscripts and Printed Books in Europe. A Long-Term Perspective from the Sixth through the Eighteenth Centuries." *Journal of Economic History* 69, no. 2: 409–45.

Butterfield, Herbert. 1954. *The Origins of Modern Science, 1300–1800*. Revised ed. New York: Free Press.

Cajori, Florian. 1928. *A History of Mathematical Notation*. Vol. 1. La Salle, IL: Open Court.

Calhoun, Craig, ed. 1992. *Habermas and the Public Sphere*. Cambridge, MA: MIT Press.

Cardwell, D. S. L. 1963. *Steam Power in the Eighteenth Century*. London: Sheed and Ward.

———. 1994. *The Fontana History of Technology*. London: Fontana Press.

Carlos, Edward Stafford. 1959. *The Sidereal Messenger of Galileo Galilee*. London: Dawsons of Pall Mall.

Chabás, José, and Bernard Goldstein. 2003. *The Alfonsine Tables of Toledo*. Boston: Kluwer Academic.

Chafee, John W. 1985. *The Thorny Gates of Learning in Sung China: A Social History of Examinations*. New York: Cambridge University Press.

Chamberlain, Michael. 1994. *Knowledge and Social Practice in Medieval Damascus, 1190–1350*. New York: Cambridge University Press.

Chapman, Allan. 1982. Introduction to *The Preface to John Flamsteed's Historia Coelestis Britannica or British Catalogue of the Heavens*. London: Trustees of the National Maritime Museum.

———. 1984. "Tycho Brahe in China: The Jesuit Mission to Peking and the Iconography of European Instrument-Making Processes." *Annals of Science* 41: 417–43.

———. 1990. "Jeremiah Horrock, the Transit of Venus and the 'New Astronomy' in Early Seventeenth Century England." Chapter 5 in *Astronomical Instruments*. Aldershot, UK: Ashgate, Variorum.

Chartier, Roger. 1996. "Gutenberg Revisited from the East." *Late Imperial China* 17, no. 1: 1–9.

Chia, Lucille. 2003. "Mashaben: Commercial Publishing in Jiangyang from the Song to the Ming." In *The Song–Ming Transition*, edited by P. J. Smith and R. von Glahn, 284–29. Cambridge, MA: Harvard University East Asia Center.

Chipman, R. A. 1954. "Unpublished Letter of Stephen Gray on Electrical Experiments, 1707–08." *Isis* 45, no. 1: 33–40.

Chu, Hsi. 1990. *Learning to Be a Sage: Selections from the Conversations of Master Chu*. Arranged topically, translated with a commentary by Daniel K. Gardner. Berkeley: University of California Press.

Clagett, Marshall. 1959. *The Science of Mechanics*. Madison: University of Wisconsin Press.

Clark, David H., and Stephen P. Clark. 2001. *Newton's Tyranny: The Suppressed Scientific Discoveries of Stephen Gray*. New York: W. H. Freeman.

Cohen, I. B. 1980. *The Newtonian Revolution*. Cambridge, MA: Harvard University Press.

———. 1984. "Newton's Discovery of Gravity." *Scientific American* 244, no. 3: 166–79.

———, ed. 1990. *Puritanism and the Rise of Modern Science: The Merton Thesis*. Edited with an introduction by I. Bernard Cohen, with K. E. Duffin and Stuart Strickland. New Brunswick, NJ: Rutgers University Press.

———. 1999. "A Guide to Newton's *Principia*." In Isaac Newton, *The Principia: Mathematical Principles of Natural Philosophy*, 1–368. A new translation by I. B. Cohen and Anne Whitman, assisted by Julia Budenz. Berkeley: University of California Press.

Conrad, Lawrence. 1995. "The Arab-Islamic Medical Tradition." In *The Western Medical Tradition, 800 BC to 1800*, edited by Lawrence I. Conrad, Michael Neve, Vivian Nutton, Roy Porter, and Andrew Wear, 120–21. New York: Cambridge University Press.

Dankoff, Robert. 2004. *An Ottoman Mentality. The World of Evliya Çelebi*. Boston: Brill.

Dardess, John W. 2002. *Blood and History in China: The Donglin Faction and Its Repression, 1620–1627*. Honolulu: University of Hawaii Press.

De Bary, William. 1991. *The Trouble with Confucianism*. Cambridge, MA: Harvard University Press.

D'Elia, M. Pasquale. 1960. *Galileo in China: Relations through the Roman College between Galileo and the Jesuit Scientist-Missionaries (1610–1640)*. Translated by Rufus Suter and Matthew Sciascia. Cambridge, MA: Harvard University Press.

Di Bono, Mario. 1995. "Copernicus, Amico, Fracastoro and Tusi's Device: Observations on the Use and Transmission of a Model." *Journal for the History Astronomy* 26: 133–54.

Dijksterhuis, E. J. 1961. *The Mechanization of the World Picture.* New York: Oxford.

Dobrzycki, Jerzy, and Richard L. Kramer. 1996. "Peurbach and Maragha Astronomy? The Ephemerides of Johannes Angelus and Their Implications." *Journal for the History of Astronomy* 27: 187–237.

Donahue, William. 2006. "Astronomy." In *The Cambridge History of Science, vol. 3, Early Modern Science,* edited by Katherine Park and Lorraine Daston, 562–95. New York: Cambridge University Press.

Drake, Stillman, ed. 1957. *Discoveries and Opinions of Galileo.* Translated with an introduction and notes by Stillman Drake. New York: Anchor Books.

———. 1978. *Galileo at Work: His Scientific Biography.* Chicago: University of Chicago Press.

Duncan, A. M., trans. 1976. *Copernicus on the Revolutions of the Heavenly Spheres: A New Translation from the Latin by A. M. Duncan.* North Vancouver, Canada: David and Charles.

Dunne, John H., S.J. 1962. *Generation of Giants: The Story of the Jesuits in China in the Last Decades of the Ming Dynasty.* Notre Dame, IN: University of Notre Dame Press.

Dupré, Sven. 2000. "Mathematical Instruments and the 'Theory of the Concave Spherical Mirror': Galileo's Optics beyond Art and Science." *Nunicus* 15: 531–88.

———. 2005. "Ausonio's Mirrors and Galileo's Lenses: The Telescope and Sixteenth Century Practical Optical Knowledge." *Galilaeana* 2: 145–80.

Dursteler, Eric. 2006. *Venetians in Constantinople: Nation, Identity and Coexistence.* Baltimore: Johns Hopkins University Press.

Duyvendak, J. J. L. 1948. "Comments on Pasquale D'Elia's, *Galileo in Cina* [sic]." *T'oung Pao, Second Series* 38, nos. 2/5: 321–29.

Elman, Benjamin. 1994. *A Cultural History of Civil Service Examinations in Late Imperial China.* Berkeley: University of California Press.

———. 2002. "Jesuit *Scientia* and Natural Studies in Late Imperial China, 1600–1800." *Journal of Early Modern History* 6, no. 3: 209–32.

———. 2005. *On Their Own Terms: Science in China, 1550–1900.* Berkeley: University of California Press.

Emsley, John. 2000. *The Shocking History of Phosphorus: A Biography of the Devil's Element.* London: Macmillan.

Endress, Gerhard. 2003. "Mathematics and Philosophy in Medieval Islam." In *The Enterprise of Science in Islam: New Perspectives,* edited by Jan P. Hogendik and Abdelhamid I. Sabra, 119–76. Cambridge, MA: MIT Press.

Engelfriet, Peter M. 1998. *Euclid in China: The Genesis of the First Chinese Translation of Euclid's Elements Books I–VI (Jihe yuanben: Beijing, 1607 and Its Reception to 1723).* Leiden, Netherlands: Brill.

Fairbank, J. K., Edwin O. Reischauer, and A. M. Craig, eds. 1965. *East Asia: The Modern Transformation.* Boston: Houghton-Mifflin.

Fazoglu, Ihsan. 2007. "Qushji: Abu al-Qasim 'Ali Ibn Muhammad Qushci-zade." In *The Bibliographical Encyclopedia of Astronomers*, vol. 2: 946–48. New York: Springer.

Findly, Ellison Banks. 1993. *Nur Jahan, Empress of Mughal India.* Oxford: Oxford University Press.

Forbes, Eric C. 1990. "The Comet of 1680–1681." In *Standing on the Shoulders of Giants: A Longer View of Newton and Halley*, edited by Norman Thrower, 312–23. Berkeley: University of California Press.

Forbes, Eric G. 1982. "The European Astronomical Tradition: Its Reception into India, and Its Reception by Sawai Jai Singh II." *Indian Journal of History of Science* 17, no. 2: 234–43.

Foster, William, ed. 1899. *The Embassy of Sir Thomas Roe to the Court of the Great Mogul, 1615–1619. As Narrated in His Journal and Correspondence.* London: Hakluyt Society.

———, ed. 1928. *A Supplementary Calendar of Documents in the India Office Relating to India or to the Home Affairs of the East India Company, 1600–1640.* London: East India Office.

———. 1968. *Early Travels in India, 1583–1619.* Delhi: S. Chand.

Frank, Andre Gunder. 1998. *ReOrient: Global Growth in the Asia Age.* Berkeley: University of California Press.

Franke, Wolfgang. 1963. *The Reform and Abolition of the Traditional Chinese Examination System.* East Asian Monograph Series, 10. Cambridge, MA: Harvard University.

Freedberg, David. 2002. *The Eye of the Lynx.* Chicago: University of Chicago Press.

French, Roger. 1984. "An Origin for the Bone Text of the Five-Figure Series." *Sudhoff's Archive* 68, no. 2: 143–58.

———. 1999. *Dissection and Vivisection in the European Renaissance.* Aldershot, UK: Ashgate.

Fu, Lo-shu. 1966. *Documentary Chronicle of Sino-Western Relations (1644–1820).* 2 vols. Tucson: University of Arizona Press.

Gaukroger, Stephen. 2006. *The Emergence of Modern Scientific Culture.* New York: Oxford.

Gilbert, William. 1600/1893. *William Gilbert of Colchester, on the Lodestone and Magnetic Bodies, and the Great Magnet Earth. New Physiology, Demonstrated with Many Arguments and Experiments.* Translated by P. Fleury Mottelay. New York: John Wiley.

Gingerich, Owen. 1973a. "Johannes Kepler." *Dictionary of Scientific Biography* 7: 289–312.

———. 1973b. "From Copernicus to Kepler: Heliocentrism as Model and Reality." *Proceedings of the American Philosophical Society* 117, no. 6: 513–23.

———. 1982. "The Galileo Affair." *Scientific American* 247, no. 2: 133–43.

———. 1984. "Phases of Venus in 1610." *Journal for the History of Astronomy* 15: 209–10.

———. 1985. "The Accuracy of Ephemerides." *Vistas in Astronomy* 28: 339–42.

―――. 2001. "Giovanni Antonio Magni's 'Keplerian' Tables of 1614 and Their Implications for the Reception of Keplerian Astronomy in the Seventeenth Century." *Journal for the History of Astronomy* 32: 237–62.

―――. 2004. *The Book Nobody Read: Chasing the Revolutions of Nicolaus Copernicus*. New York: Walker.

Gingerich, Owen, and Albert van Helden. 2003. "From Occhiale to Printed Page: The Making of Galileo's Sidereus Nunicus." *Journal for the History of Astronomy* 34, part 3, no. 116: 251–67.

Gode, P. K. 1969. "Some Notes on the Invention of Spectacles." In P. K. Gode, *Studies in Indian Cultural History*, vol. 3, part 2: 106–7. Poona, India: BOR Institute.

Goldstone, Jack. 2002. "Efflorescence and Economic Growth in World History: Rethinking the "Rise of the West' and the Industrial Revolution." *Journal of World History* 12, no. 2: 323–89.

―――. 2008. "Capitalist Origins, the Advent of Modernity, and Coherent Explanation: A Response to Joseph M. Bryant." *Canadian Journal of Sociology* 33, no. 1: 119–33.

Golvers, Noël, trans. 1993. *The Astronomia Europaea of Ferdinand Verbiest, S. J. (Dilligen, 1687)*. Text, translation, notes, and commentaries by Noël Golvers. Nettetal, Germany: Steyler.

―――. 1999. "Verbiest's Introduction of Aristoteles Latinus (Coimbra) in China: New Western Evidence." In *The Christian Mission in China in the Verbiest Era: Some Aspects of the Missionary Approach*, ed. N. Golvers, 33–51. Leuven, Netherlands: Leuven University Press.

Goodrich, Luther Carrington. 1935. *The Literary Inquisition of Ch'ien-Lung*. Baltimore: Waverly Press.

Govi, Gilberto. 1889. "The Compound Microscope Invented by Galileo." *Journal of the Royal Microscopical Society* 9: 574–98.

Grant, Edward, ed. 1974. *A Source Book in Medieval Science*. Cambridge, MA: Harvard University Press.

―――. 1981. *Much Ado about Nothing*. New York: Cambridge University Press.

―――. 1984. "Science in the Medieval University." In *Rebirth, Reform, Resilience: Universities in Transition, 1300–1700*, edited by James M. Kittelson and Pamela J. Transue, 68–102. Columbus: Ohio State University Press.

―――. 2007. *A History of Natural Philosophy*. New York: Cambridge University Press.

Grendler, Paul F. 2002. *The Universities of the Italian Renaissance*. Baltimore: Johns Hopkins University Press.

Gutas, Dimitri. 1998. *Greek Thought, Arabic Culture*. New York: Routledge.

Guy, R. Kent. 1987. *The Emperor's Four Treasures: Scholars and the State in the Later Ch'ien-lung Era*. Cambridge, MA: Harvard University Press.

Greif, Avner. 2006. *Institutions and the Path to the Modern Economy: Lessons from Medieval Trade*. New York: Cambridge University Press.

Guericke, Otto von. 1672/1968. *New Magdeburg Investigations [so-called] on Void Space (Neue Magdeburg Versuche über den Leeren Raum)*. Facsimile ed. Edited by Hans Schimak. Repr. Dusseldorf, Germany: VDI.

Habermas, Jürgen. 1989. *The Structural Transformation of the Public Sphere.* Translated by T. McCarthy. Boston: Beacon Press.

Hahn, Roger. 1971. *The Anatomy of a Scientific Institution: The Paris Academy of Sciences, 1666–1803.* Berkeley: University of California Press.

Hall, A. Rupert. 1992. *Isaac Newton: Adventurer in Thought.* New York: Cambridge University Press.

Hammer-Purgstall, M. de. 1841. *Histoire de L'Empire Ottoman: Depuis son Origine Jusqu'a Nos Jours.* Vol. 2. Paris: Parent-Desbarres.

Hanushek, Eric A., and Ludger Woessmann. 2008. "The Role of Cognitive Skills in Economic Development." *Journal of Economic Literature* 46, no. 3: 607–68.

Hao, Fang. 1948. *Studies in the History of the Relations between China and the West.* Peiping, China: Institutum Sancti Thomae.

Hartner, Willy. 1973. "Copernicus, the Man, the Work, and His Achievement." *Proceedings of the American Philosophical Society* 117, no. 6: 413–22.

Hartwell, Robert M. 1971. "Financial Experience, Examinations, and the Formulation of Economic Policy in Northern Sung." *Journal of Asian Studies* 30: 281–314.

Hashimoto, Keizo. 1988a. *Hsü Kuang-chi'i and Astronomical Reform: The Process of the Chinese Acceptance of Western Astronomy 1629–1635.* Kansai, Japan: Kansai University Press.

———. 1988b. "Telescope and Observation in Late Ming China." *Journal of the Division of Social Sciences (Kansai University)* 19, no. 2: 91–101.

Hauksbee, Francis. 1704–5. "Several Experiments on the Mercurial Phosphorus, Made before the Royal Society, at Gresham College." *Philosophical Transactions* 24: 2129–35.

———. 1706–7a. "An Account of an Experiment before the Royal Society at Gresham College, Touching the Extraordinary Elistricity [sic] of Glass, Produceable on a Smart Attrition of it; with a Continuation of Experiments on the Same Subject . . . ," *Philosophical Transactions* 25: 2327–35.

———. 1706–7b. "An Account of an Experiment Made before the Royal Society at Gresham College, Together, with a Repetition of the Same, Touching the Production of a Considerable Light upon a Slight Attrition of the Hands on a Glass Globe Exhausted of Its Air; with Other Remarkable Occurrences." *Philosophical Transactions* 25: 2277–82.

———. 1706–7c. "Several Experiments Shewing the Strange Effects of the Effluvia of Glass, Produceable on the Motion and Attrition of It." *Philosophical Transactions* 25: 2372–77.

Heilbron, John. 1979. *Electricity in the 17th and 18th Centuries.* Berkeley: University California Press.

Heinen, Anton M. 1982. *Islamic Cosmology: A Study of As-Suyuti's "al-Hay'a as-saniyafi' l-hay'a as-suniya."* With critical edition, translation, and commentary. Beirut: Franz Steiner.

Hellman, C. Doris, and Noel Swerdlow. 2008. "Peurbach." *Dictionary of Scientific Biography* 15: 473–79.

Henry, John. 2008. *The Scientific Revolution and the Origins of Modern Science.* 3rd ed. New York: Palgrave/Macmillan.

Hill, Christopher. 1965. *The Intellectual Origins of the English Revolution.* Oxford: Clarendon Press.

Hill, Richard L. 1989. *Power from Steam: A History of the Stationary Steam Engine.* New York: Cambridge University Press.

Ho, Ping-ti. 1967. *The Ladder of Success: Aspects of Mobility in China, 1368–1911.* Rev. ed. New York: Columbia University Press.

Hobson, John M. 2004. *The Eastern Origins of Western Civilization.* New York: Cambridge University Press.

Holton, Gerald. 1973. "Johannes Kepler's Universe: Its Physics and Metaphysics." Chapter 2 in *Thematic Origins of Scientific Thought: Kepler to Einstein.* Cambridge, MA: Harvard University Press.

Hourani, Albert. 1992. *A History of the Arab People.* New York: Warner Books.

Houston, R. A. 1988. *Literacy in Early Modern Europe: Culture and Education, 1500–1600.* London: Longman Group.

Hu, Minghui. 2002. "Provenance in Contest: Searching for the Origins of Jesuit Astronomy in Early Qing China, 1665–1705." *International History Review* 24, no. 1: 1–36.

———. 2004. Review of Zang Baichun, *The Europeanization of Astronomical Instruments in Ming and China* (in Chinese). New Century History of Science Series, 2. Shenyang, China: Liaoning Education Press, *Isis* 95, no. 4: 699–700.

Huang, Yi-Long. 1991. "Court Divination and Christianity in the K'ang-hsi Era." *Chinese Science* 10: 1–20.

Huang, Yi-Long, and Chang Chih-ch'eng. 1996. "The Evolution and Decline of the Ancient Chinese Practice of Watching for the Ethers." *Chinese Science* 13: 82–106.

Hucker, Charles O. 1985. *A Dictionary of Official Titles in Imperial China.* Stanford, CA: Stanford University Press.

Huff, Toby E. 1993/2003. *The Rise of Early Modern Science: Islam, China and the West.* 2nd ed. New York: Cambridge University Press.

———. 2000. "Science and Metaphysics in the Three Religions of the Book." *Intellectual Discourse* 8, no. 2: 173–98.

———. 2002. "Attitudes towards Dissection in the History of European and Arabic Medicine." In *Science. Locality, and Universality*, edited by Bennacer El Bouazzati, 61–88. Rabat: Publications of the Faculty of Letters and Human Sciences.

———. 2006. "The Open Society, Metaphysical Beliefs and Platonic Sources of Reason and Rationality." In *Karl Popper: A Centenary Assessment: Metaphysics and Epistemology*, vol. 2, edited by Ian Jarvie, Karl Milford, and David Miller, 19–44. Aldershot, UK: Ashgate.

———. 2006. "Some Historical Roots of the Ethos of Science." *Journal of Classical Sociology* 7, no. 2: 193–210.

Hughes, Barnabus. 1967. *Introduction to Regiomontanus on Triangles.* Translated with an introduction and notes by Barnabus Hughes. Madison: University of Wisconsin Press.

Hughes, D. W. 1990. "Edmond Halley: His Interest in Comets." In *Standing on the Shoulders of the Giants: A Longer View of Newton and Halley*, edited by Norman J. W. Thrower, 324–27. Berkeley: University of California Press.

Hunter, Michael. 1989. "The Importance of Being Institutionalized." In *Establishing the New Science*, edited by Michael Hunter, 1–43. Woodbridge, UK: Boydell Press.

ibn Anas, Malik. 1989. *Al-Muwatta: The First Formulation of Islamic Law*. Revised in whole and translated by Aisha Abdurrahman Bewley. Granada: Madinah Press of Granada.

Ihsanoglu, Ekmeleddin. 1997. "Ottoman Science." In *Encyclopaedia of the History of Science, Technology, and Medicine in Non-Western Cultures*, edited by Helaine Selin, 799–805. Boston: Kluwer.

———. 1992/2004. "The Introduction of Western Science to the Ottoman World: A Case Study of Modern Astronomy." Chapter 2 in *Science, Technology and Learning in the Ottoman Empire*. Aldershot, UK: Ashgate Variorum.

———. 2005. "Institutionalization of Science in the Medreses of Pre-Ottoman Turkey." In *Turkish Studies in the History and Philosophy of Science*, edited by G. Irzik and G. Güzeldere, 265–84. Netherlands: Springer.

Ilardi, Vincent. 1976. "Eyeglasses and Concave Lenses in Fifteenth-Century Florence and Milan: New Documents." *Renaissance Quarterly* 29: 341–60.

———. 1993. "Renaissance Florence: The Optical Capital of the World." *Journal of European Economic History* 22, no. 3: 507–42.

———. 2007. *Renaissance Vision from Spectacles to Telescopes*. Philadelphia: American Philosophical Society.

Israel, Jonathan I. 1995. *The Dutch Republic: Its Rise, Greatness, and Fall 1447–1806*. New York: Oxford.

Jacob, Margaret. 1998. *Scientific Culture and the Making of the Industrial West*. New York: Oxford University Press.

Jacob, Margaret, and Larry Stewart. 2004. *Practical Matters: Newton's Science in the Service of Industry and Empire, 1687–1851*. Cambridge, MA: Harvard University Press.

Jardine, Lisa. 1999. *Ingenious Inventions: Building the Scientific Revolution*. New York: Doubleday.

Job of Edessa. 1935. *Book of Treasures*. Syriac text edited and translated with a critical apparatus by A. Mingana. Cambridge: W. Heffer.

Johns, Adrian. 2006. "Coffeehouses and Print Shops." In *The Cambridge History of Science*, vol. 3: 320–22. New York: Cambridge University Press.

Keller, Suzanne. 1972. "William Gilbert." *Dictionary of Scientific Biography* 5: 396–401.

Kennedy, E. S. 1970. "The Arabic Heritage in the Exact Sciences." *Al-Abhath* 23: 337–44.

———. 1983. "The History of Trigonometry: An Overview." In *Studies in the Islamic Exact Sciences*, edited by E. S. Kennedy et al., 3–29. Beirut: American University Beirut Press.

Kennedy, E. S., and Victor Roberts. 1959. "The Planetary Theory of Ibn al-Shatir." *Isis* 50: 227–35.

Kepler, Johannes. 1609/1992. *New Astronomy*. Translated by William H. Donahue. Cambridge: Cambridge University Press.

———. 1618–21/1995. *Epitome of Copernican Astronomy and Harmonies of the World*. Translated by Charles Glenn Wallis. Amherst, NY: Prometheus Books.

———. 1605/2000. *Optics: Paralipomena to Witelo and Optical Part of Astronomy*. Translated by William Donahue. Santa Fe, NM: Green Lion Press.

Khan, Iqbal Ghani. 2001. "Medieval Theories of Vision and the Introduction of Spectacles in India, ca. 1200–1750." In *Disease and Medicine in India: An Overview*, edited by Deepak Kumar, 26–39. New Delhi: Tulika Books.

———. 2002. "Technology and the Question of Elite Intervention in Eighteenth-Century North India." In *Rethinking Early Modern India*, edited by Richard B. Barnett, 255–88. New Delhi: Manohar.

Kirby, William C. 1995. "China Unincorporated: Company Law and Business Enterprise in Twentieth-Century China." *Journal of Asian Studies* 54, no. 1: 43–63.

Kochhar, Rajesh. 2000. "Pre-telescopic Astronomy in India." In *History of Indian Science, Technology and Culture AD 1000–1800*, vol. 3, part 1: 171–97. New Delhi: Oxford University Press.

Koyré, Alexandre. 1952. "An Unpublished Letter of Robert Hooke to Isaac Newton." *Isis* 43: 312–37.

———. 1965. "Newton and Descartes." In *Newtonian Studies*, 53–114. Cambridge, MA: Harvard University Press.

———. 1965. "Copernicus and Kepler on Gravity." In *Newtonian Studies*, 173–75. Cambridge, MA: Harvard University Press.

Kracke, Edward, Jr. 1953. *Civil Service in Early Sung China, 960–1067*. Cambridge, MA: Harvard University Press.

Kuhn, Thomas. 1977. "A Function of Thought Experiments." In *The Essential Tension*, 240–65. Chicago: University of Chicago Press.

Kumar, Deepak, ed. 1991. *Science and Empire: Essays in Indian Context, 1700–1947*. Delhi: Anamika Prakashan.

———. 2001. *Disease and Medicine in India: A Historical Overview*. New Delhi: Tulika Books.

———. 2006. *Science and the Raj: A Study of British India*. 2nd ed. New York: Oxford University Press.

Kunitzsch, Paul. 2003. "The Transmission of Hindu-Arabic Numerals Reconsidered." In *The Enterprise of Science in Islam*, edited by Jan P. Hogendijk and A. I. Sabra, 3–21. Cambridge, MA: MIT Press.

Kuran, Timur. 2008. "Institutional Causes of Underdevelopment in the Middle East: A Historical Perspective." In *Institutional Change and Economic Behavior*, edited by János Kornai, Laszlo Matyas, and Gérard Roland, 64–76. New York: Palgrave-Macmillan.

Kurz, Otto. 1975. *European Clocks and Watches in the Near East*. Leiden, Netherlands: Brill.

Lach, Donald. 1965. *Asia in the Making of Europe*. Chicago: University of Chicago Press.

Landes, David. 2000. *The Revolution in Time*. Rev. and enlarged 2nd ed. with a new preface. Cambridge, MA: Harvard University Belknap Press.

Lattis, James. 1994. *Between Copernicus and Galileo: Christoph Clavius and the Collapse of Ptolemaic Cosmology.* Chicago: University of Chicago Press.

Lee, Thomas H. C. 1985. *Government Education and Examinations in Sung China.* Hong Kong: Chinese University Press.

Leiser, Gary. 2004. "The Madrasa and the Islamization of Anatolia before the Ottomans." In *Law and Education in Islam: Studies in Honor of Professor George Makdisi,* edited by Joseph E. Lowry, Devin J. Stewart, and Shawkat M. Toorwa, 174–91. London: E. J. W. Gibb Memorial Trust.

Link, Perry. 2009. "China's Charter 08." *New York Review* 56, no. 1: 54–56.

Lipsey, Richard G., Kenneth I. Carlaw, and Clifford T. Bekar. 2005. *Economic Transformations: General Purpose Technologies and Long Term Economic Growth.* Oxford: Oxford University Press.

Liu, Adam Yuen-chung. 1981. *The Hanlin Academy, 1644–1850.* Hamden, CT: Archon Books.

Livingston, John. 1996. "Western Science and Educational Reform in the Thought of Shaykh Rifa'a al-Tahtawi." *International Journal of Middle Eastern Studies* 28: 543–64.

———. 1997. "Shaykhs Jabarti and 'Attar: Islamic Reaction and Response to Western Science in Egypt." *Der Islam* 74, no. 1: 92–106.

Lyons, Henry. 1944. *The Royal Society, 1660–1940.* Cambridge: Cambridge University Press.

Makdisi, George. 1961. "Muslim Institutions of Learning in Eleventh-Century Baghdad." *Bulletin of the School of Oriental and African Studies* 24: 1–55.

———. 1970. "Madrasah and University in the Middle Ages." *Studia Islamica* 32: 255–64.

———. 1974. "The Scholastic Method in Medieval Education: An Inquiry into Its Origin in Law and Theology." *Speculum* 49: 640–61.

———. 1980. "On the Origin and Development of the College in Islam and the West." In *Islam and the Medieval West,* edited by Khalil I. Seeman, 26–49. Albany: State University of New York Press.

———. 1981. *The Rise of Colleges: Institutions of Learning in Islam and the West.* Edinburgh: Edinburgh University Press.

———. 1984. "The Guild of Law in Medieval Legal History: An Inquiry into the Origins of the Inns of Court." *Zeitschrift für Geschichte der Arabische-Islamischen Wissenschaften* 1: 233–52.

Malet, Antonio. 2003. "Kepler and the Telescope." *Annals of Science* 60: 107–36.

———. 2005. "Early Conceptualizations of the Telescope as an Optical Instrument." *Early Science and Medicine* 10, no. 3: 237–62.

McDermott, Joseph. 2005. "The Ascendance of the Imprint in China." In *Printing and Book Culture in Late Imperial China,* edited by Cynthia J. Brokaw and Kai-wing Chow, 55–106. Berkeley: University of California Press.

McNeil, William. 1990. "The Rise of the West after Twenty-five Years." *Journal of World History* 1, no. 1: 1–22.

———. 1998. "World History and the Rise and Fall of the West." *Journal of World History* 9, no. 2: 215–36.

Merchant, Carolyn. 2008. "'The Violence of Impediments': Francis Bacon and the Origins of Experimentation." *Isis* 99, no. 4: 731–60.

Mercier, Raymond. 1984. "The Astronomical Tables of Rajah Jai Singh Sawai." *Indian Journal of History of Science* 19, no. 2: 143–71.

Merton, Robert K. 1938/1970. *Science, Technology and Society in Seventeenth Century England.* New York: Harpers.

———. 1973. *The Sociology of Science: Theoretical and Empirical Investigations.* Chicago: University of Chicago Press.

Middleton, W. E. Knowles. 1964. *The History of the Barometer.* Baltimore: Johns Hopkins University Press.

Mitchell, A. Crichton. 1932. "Chapters in the History of Terrestrial Magnetism: [1] The Discovery of Directionality." *Journal of Terrestrial Magnetism and Atmospheric Electricity* 37, no. 2: 105–46.

———. 1937. "The Discovery of Declination." *Journal of Terrestrial Magnetism and Atmospheric Electricity* 42, no. 1: 241–80.

———. 1939. "The Discovery of Dip." *Journal of Terrestrial Magnetism and Atmospheric Electricity* 44, no. 1: 77–80.

Miyazaki, Ichisada. 1981. *China's Examination Hell.* New Haven, CT: Yale University Press.

Mokyr, Joel. 2002. *The Gifts of Athena: Historical Origins of the Knowledge Economy.* Princeton, NJ: Princeton University Press.

Monconys, Balthasar. 1665–66. *Journal des Voyages de Monsieur de Monconys.* Ca. 1647–48. Lyon, France: Chez Horace Boissat and George Remevs.

Moody, Ernest. 1951. "Laws of Motion in Medieval Physics." *Scientific Monthly* (January): 18–23.

———. 1957. "Galileo and Avempace [Ibn Bajja]: Dynamics of the Leaning Tower Experiments." In *Roots of Scientific Thought*, edited by Philip P. Wiener and Aaron Noland, 176–206. New York: Basic Books.

Moor, Tine de. 2008. "The Silent Revolution: A New Perspective on the Emergence of Commons, Guilds, and Other Forms of Corporate Collective Action in Western Europe." *International Review of Social History* 53: 179–212.

Morrison, Robert. 2003. "The Response of Ottoman Religious Scholars to European Science." *Archivum Ottomanicum* 21: 187–95.

Murdoch, John E., and Edith D. Sylla. 1978. "The Science of Motion." In *Science in the Middle Ages*, edited by David C. Lindberg, 206–64. Chicago: University of Chicago Press.

Musson, A. E., and Eric Robinson. 1969. *Science and Technology in the Industrial Revolution.* Toronto, ON: University of Toronto Press.

Nasr, Seyyed Hossein. 1968. *Science and Civilization in Islam.* New York: New American Library.

Nauenberg, Michael. 1994. "Hooke, Orbital Motion, and Newton's *Principia*." *American Journal of Physics* 62, no. 4: 331–50.

Needham, Joseph. 1959–62. *Science and Civilisation in China.* 6 vols. New York: Cambridge University Press.

———. 1970. "The Pre-natal History of the Steam-Engine." In *Clerks and Craftsmen in China and the West*, 136–201. Cambridge: Cambridge University Press.

Needham, Joseph, and Lu Gwei-Djen. 1970. "The Optick Artists of Chinagu." In *Studies in the Social History of China and South-East Asia: Essays in Memory of Victory Purcell*, edited by Jerome Ch'en and Nicolas Tarling, 197–224. Cambridge: Cambridge University Press.

Nelson, Benjamin. 1973/1981. "Civilization Complexes and Intercivilizational Encounters." Repr. *On the Road to Modernity: Conscience, and Civilizations: Selected Writings by Benjamin Nelson*, edited by Toby E. Huff, 80–106. Totowa, NJ: Rowman and Littlefield.

Neugebauer, Otto. 1968. "On the Planetary Theory of Copernicus." *Vistas in Astronomy* 10: 89–103.

Ogburn, W. F., and Dorothy Thomas. 1922. "Are Inventions Inevitable?" *Political Science Quarterly* 37: 83–98.

O'Malley, C. D. 1965. "Pre-Vesalian Anatomy." In *Andreas Vesalius of Brussels, 1514–1564*, 1–20. Berkeley: University of California Press.

O'Neill, Ynez Violé. 1977. "Innocent III and the Evolution of Anatomy." *Medical History* 20: 429–34.

Ornstein, Martha. 1928. *The Role of Scientific Societies in the Seventeenth Century*. Chicago: University of Chicago Press.

Pascal, Blaise. 1937. *The Physical Treaties of Pascal: The Equilibrium of Liquids and the Weight of the Mass of Air*. Translated by I. H. B. Spiers and A. G. H. Spiers, with an introduction and notes by Frederick Barry. New York: Columbia University Press.

Pesic, Peter. 2008. "Proteus Rebound: Reconsidering the 'Torture of Nature.'" *Isis* 99, no. 2: 304–17.

Peters, F. E. 1968. *Aristotle and the Arabs*. New York: Simon and Schuster.

Philipp, Thomas, and Moshe Perlmann, eds., trans. 1994. *Abd al-Rahman al-Jabarti's History of Egypt*. Stuttgart, Germany: Franz Steiner.

Pingree, David. 1978. "Islamic Astronomy in Sanskrit." *Journal of the History of Arabic Science* 2: 315–80.

———. 1996. "Indian Reception of Muslim Versions of Ptolemaic Astronomy." In *Tradition, Transmission and Transformation*, edited by F. J. Ragep and Sally Ragep, 471–85. New York: Brill.

———. 2003. "The Sarvasiddhantaraja of Nityananda." In *The Enterprise of Science in Islam*, edited by Jan P. Hogendijk and A. I. Sabra, 269–84. Cambridge, MA: MIT Press.

Pingyi Chu. 1997. "Scientific Dispute in the Imperial Court: The 1664 Calendar Case." *Chinese Science* 14: 7–34.

———. 2003. "Remembering Our Grand Tradition: The Historical Memory of the Scientific Exchanges between China and Europe, 1600–1800." *History of Science* 41: 193–21.

Pomeranz, Kenneth. 2000. *The Great Divergence: China, Europe and the Making of the Modern World Economy*. Princeton, NJ: Princeton University Press.

Post, Gainess. 1964. *Studies in Medieval Legal Thought: Public Law and the State, 1150–1322*. Princeton, NJ: Princeton University Press.

Poulle, Emmanuel. 1988. "The Alfonsine Tables and Alfonso X of Castile." *Journal for the History of Astronomy* 29: 97–113.

Putnam, George. 1896–97/1962. *Books and Their Makers in the Middle Ages*. 2 vols. New York: Hillary House.

Quaisar, Ahsar Jan. 1982. *The India Response to European Technology and Culture (A.D. 1498–1707)*. Delhi: Oxford University Press.

Ragep, F. Jamil. 2001. "Freeing Astronomy from Philosophy: An Aspect of Islamic Influence on Science." *Osiris* 16: 49–71.

———. 2005. "Ali Qushji and Regiomontanus: Eccentric Transformations and Copernican Revolutions." *Journal for the History of Astronomy* 36, part 4: 359–71.

Rao, Kameswara, N. A. Vagiswari, and Christina Lois. 1984. "Father J. Richaud and Early Telescope Observations in India." *Bulletin of the Astronomical Society of India* 12: 82–85.

Rawski, Evelyn. 1979. *Education and Popular Literacy in China*. Ann Arbor: University of Michigan Press.

———. 1985. "Economic and Social Foundations of Late Imperial Culture." In *Popular Culture in Late Imperial China*, edited by David Johnson, 3–33. Berkeley: University of California Press.

Reed, Donald Malcolm. 1990. *Cairo University and the Making of Modern Egypt*. New York: Cambridge University Press.

Reeves, Eileen. 2008. *Galileo's Glassworks: The Telescope and the Mirror*. Cambridge, MA: Harvard University Press.

Roberts, Victor. 1957. "The Planetary Theory of Ibn al-Shatir: A Pre-Copernican Copernican Model." *Isis* 48: 428–32.

Robinson, Francis. 1996. "Islam and the Impact of Print in South Asia." In *The Transmission of Knowledge in South Asia*, edited by N. Crook, 62–97. Oxford: Oxford University Press.

———. 2001. *The 'Ulama of Farangi Mahall*. London: Hurst.

Rogers, J. M. 1993. *Mughal Miniatures*. New York: Thames and Hudson.

Roller, Duane, and Duane H. D. Roller. 1954. *The Development of the Concept of Electric Charge*. Harvard Case Studies in Experimental Science 8. Cambridge, MA: Harvard University Press.

Rose, Mark. 1993. *Authors and Owners: The Invention of Copyright*. Cambridge, MA: Harvard University Press.

Rosen, Edward. 1947. *The Naming of the Telescope*. New York: Henry Schuman.

———. 1956. "The Invention of Eyeglasses." *Journal of the History of Medicine and Allied Sciences* 11: 13–46, 183–218.

———, trans. 1965. *Kepler's Conversation with Galileo's Sidereal Messenger*. With introduction and notes by Edward Rosen. New York: Johnson.

Rosen, S. 1989. "Public Opinion and Reform in the People's Republic of China." *Studies in Comparative Communism* 22, nos. 2–3: 153–70.

Rowe, William T. 1990. "The Public Sphere in Modern China." *Modern China* 16, no. 3: 309–29.

———. 1993. "The Problem of 'Civil Society' in Late Imperial China." *Modern China* 19, no. 2: 139–57.

Ruestow, Edward. 1996. *The Microscope in the Dutch Republic*. New York: Cambridge University Press.

Russell, Gül. 1992. "'The Owl and the Pussy Cat': The Process of Cultural Transmission in Anatomical Illustration." In *Transfer of Modern Science and*

Technology to the Muslim World, edited by Ekmeleddin Ihsanoglu, 180–212. Istanbul: Research Center for Islamic History, Art, and Culture.

Sabra, A. I. 1987. "The Appropriation and Subsequent Naturalization of Greek Science in Medieval Islam: A Preliminary Statement." *History of Science* 25: 223–43.

———. 1994. "Science and Philosophy in Medieval Theology: The Evidence of the Fourteenth Century." *Zeitschrift für Geschichte der Arabisch-Islamischen Wissenschaften* 9: 1–42.

Saliba, George. 1994. "A Sixteenth-Century Arabic Critique of Ptolemaic Astronomy: The Work of Shams al-Din al-Khafri." *Journal for the History of Astronomy* 25: 15–38.

———. 2004. "Reform of Ptolemaic Astronomy at the Court of Ulug Beg." In *Studies in the History of the Exact Science in Honor of David Pingree*, edited by Charles Burnett, Jan P. Hogendijk, Kim Plofker, and Michio Yano, 810–24. Leiden, Netherlands: Brill.

———. 2007. *Islamic Science and the Making of the European Renaissance.* Cambridge, MA: MIT Press.

Sarma, S. R. 2003. *Astronomical Instruments in the Rampur Raza Library.* Rampur: Rampur Raza Library.

Savage-Smith, Emilie. 1995. "Dissection in Medieval Islam." *Journal of the History of Medicine* 50: 67–110.

Sayili, Aydin. 1960. *The Observatory in Islam.* Ankara: Turk Tarih Kurumu Besimevi.

Shapin, Steven. 1996. *The Scientific Revolution.* Chicago: University of Chicago Press.

Shapin, Steven, and Simon Schaffer. 1985. *Leviathan and the Air-Pump.* Princeton, NJ: Princeton University Press.

Sharma, V. N. 1995. *Sawai Jai Singh and His Astronomy.* Delhi: Motilal Banarsidass.

Shaw, Stanford. 1976. *History of the Ottoman Empire and Modern Turkey.* Vol. 1. New York: Cambridge University Press.

Singer, Charles J. 1917. "Steps Leading to the Invention of the First Optical Apparatus." In *Studies in History and Method of Science*, 395–413. Oxford: Oxford University Press.

———, ed. 1954. *A History of Technology.* Vol. 2. Oxford: Clarendon Press.

Sivin, Nathan. 1984. "Why the Scientific Revolution Did Not Take Place in China – or Didn't It?" In *Transformation and Tradition in the Sciences*, edited by E. Mendelsohn, 531–54. Cambridge: Cambridge University Press.

———. 1988. "Science and Medicine in Imperial China – The State of the Field." *Journal of Asia Studies* 47, no. 1: 41–90.

———. 1990. "Science and Medicine in Chinese History." In *Heritage of China*, edited by Paul S. Ropp, 164–96. Berkeley: University of California Press.

Sluiter, Engel. 1997a. "The First Known Telescopes Carried to America, Asia, and the Arctic, 1614–1639." *Journal for the History of Astronomy* 28: 141–45.

———. 1997b. "The Telescope before Galileo." *Journal for the History of Astronomy* 28: 223–34.

Smit, P., and J. Heniger. 1975. "Antoni van Leeuwenhoek (1632–1723)." *Antonie van Leeuwenhoek* 41: 217–18.

Spence, Jonathan. 1980. *To Change China*. New York: Penguin.

Stone, Lawrence. 1964. "The Educational Revolution in England, 1560–1640." *Past and Present* 68, no. 28: 41–80.

———. 1969. "Literacy and Education in England, 1640–1900." *Past and Present* 42, no. 2: 69–139.

Stoye, John. 1993. *Marsigli's Europe, 1680–1730*. New Haven, CT: Yale University Press.

Sun, Xiaochun. 2001. "On the Star Catalogue and Atlas of Chongzhen Lishu." In *Statecraft and Intellectual Renewal in Late Ming China: The Cross-Cultural Synthesis of Xu Guangqi*, edited by Catherine Jami, Peter Engelfriet, and Gregory Blue, 311–21. Leiden, Netherlands: Brill.

Swerdlow, Noel. 1973. "The Derivation and First Draft of Copernicus's Planetary Theory." *Proceedings of the American Philosophical Society* 117: 423–512.

———. 1996. "Renaissance [Astronomy]." In *Astronomy before the Telescope*, edited by Christopher Walker, foreword by Patrick Moore, 187–230. London: British Museum.

Swerdlow, N. M., and Otto Neugebauer. 1984. *Mathematical Astronomy in Copernicus's De Revolutionibus*. New York: Springer.

Swetz, Frank J. 1987. *Capitalism and Arithmetic. The New Math of the 15th Century*. La Salle, IL: Open Court Press.

Thrower, Norman J. W., ed. 1990. *Standing on the Shoulders of the Giants: A Longer View of Newton and Halley*. Berkeley: University of California Press.

Torricelli, Evangelista. 1644. "Letter to Michelangelo Ricci Concerning the Barometer." In *A Source Book in Physics*, edited by William Francis Magie, 70–73. New York: McGraw-Hill.

Troll, Christian. 1978. *Sayyid Ahmad Khan: A Reinterpretation of Muslim Theology*. Atlantic Highlands, NJ: Humanities Press.

Turner, G. L'E. 1980. *Essays on the History of the Microscope*. Oxford: Senecio.

Usher, Abbott. 1954. *A History of Mechanical Inventions*. Rev. ed. Cambridge, MA: Harvard University Press.

Van Helden, Albert. 1968. "Christopher Wren's 'De Corpore Saturni.'" *Notes and Records of the Royal Society of London* 23, no. 2: 213–29.

———. 1974a. "The Telescope in the Seventeenth Century." *Isis* 65, no. 1: 38–58.

———. 1974b. "Saturn and His Anses." *Journal for the History of Astronomy* 5: 105–21.

———. 1974c. "'Annulo Cingitur': The Solution of the Problem of Saturn." *Journal for the History of Astronomy* 5: 155–74.

———. 1977. "The Invention of the Telescope." *Transactions of the American Philosophical Society* 67, part 4: 1–67.

———. 1976–77. "The 'Astronomical Telescope,' 1611–1650." *Annali dell'Istituto e Museo di Storia della Scienza* 1: 14–35.

———. 1977. "The Development of the Compound Eyepieces, 1640–1670." *Journal for the History of Astronomy* 8: 26–37.

————. 1983. "The Birth of the Modern Scientific Instrument, 1550–1700." In *The Uses of Science in the Age of Newton*, edited by John G. Burke, 49–82. Berkeley: University of California Press.

————. 2009. "Who Invented the Telescope?" *Sky and Telescope* 118, no. 1: 64–69.

————. 2009. "The Beginnings, from Lipperhey to Huygens and Cassini." *Experimental Astronomy* 25: 3–16.

Veselovsky, I. N. 1973. "Copernicus and Nasir al-Din al-Tusi." *Journal for the History of Astronomy* 4: 128–30.

Volwahsen, Andreas. 2001. *Cosmic Architecture in India*. New York: Prestel.

Wakeman, Frederic, Jr. 1993. "The Civil Society and Public Sphere Debate: Western Reflections on Chinese Political Culture." *Modern China* 19, no. 2: 108–38.

Walzer, Richard. 1962. *Greek into Arabic*. Columbia: University of South Carolina Press.

Webster, Charles. 1976. *The Great Instauration: Science, Medicine and Reform 1626–1660*. New York: Holmes and Meier.

Weisheipl, James. 1975. "The Curriculum of the Faculty of Arts at Oxford in the Early Fourteenth Century." *Medieval Studies* 26: 143–85.

————. 1984. "Science in the Thirteenth Century." In *The History of the University of Oxford*, vol. 1, edited by J. I. Cato, 435–69. Oxford: Oxford University Press.

Westfall, Richard. 1993. *Never at Rest: A Biography of Isaac Newton*. New York: Cambridge University Press.

Whiteside, D. T. 1970. "Before the *Principia*: The Maturing of Newton's Thoughts on Dynamical Astronomy, 1664–1684." *Journal for the History of Astronomy* 1: 5–19.

Willach, Rolf. 2001. "The Development of Telescopic Optics in the Middle of the Seventeenth Century." *Annals of Science* 58: 381–98.

————. 2008. *The Long Route to the Telescope*. Philadelphia: American Philosophical Society Proceedings, 98, pt 5.

Wilson, Catherine. 1995. *The Invisible World: Early Modern Philosophy and the Invention of the Microscope*. Princeton, NJ: Princeton University Press.

Wilson, Curtis. 1989. "The Newtonian Achievement in Astronomy." In *The General History of Astronomy*, vol. 2A, edited by R. Taton and C. Wilson, 233–74. New York: Cambridge University Press.

Wong, R. Bin. 1997. *China Transformed: Historical Change and the Limits of European Experience*. Ithaca, NY: Cornell University Press.

Woodside, Alexander, and Benjamin Elman. 1994. "Afterword: The Expansion of Education in Ch'ing China." In *Education and Society in Late Imperial China*, edited by A. Woodside and Benjamin Elman, 525–60. Berkeley: University of California Press.

Zaman, Muhammad Qasim. 2002. *The Ulama in Contemporary Islam*. Princeton, NJ: Princeton University Press.

Zanden, J. L., van. 2009. *The Long Road to the Industrial Revolution: The European Economy in a Global Perspective, 1000–1800*. Leiden, Netherlands: Brill.

————. n.d. "Economic Growth in a Period of Political Fragmentation." http://www.iisg.nl/research/jvz-economic_growth.pdf.

Zaret, David. 1989. "Religion, Science, and Printing in the Public Sphere in Seventeenth Century England." *American Sociological Review* 54: 163–79.

————. 1996. "Petitions and the 'Invention' of Public Opinion in the English Revolution." *American Journal of Sociology* 101, no. 6: 1497–1555.

Zilsel, Edgar. 1941. "The Origin of Gilbert's Scientific Method." *Journal of History of Ideas* 2, no. 1: 1–32.

Index